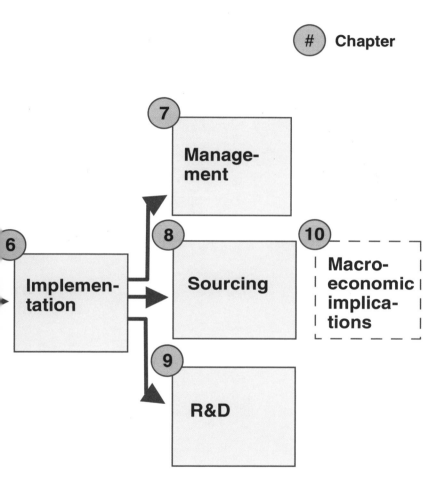

Global Production

Eberhard Abele · Tobias Meyer · Ulrich Näher
Gernot Strube · Richard Sykes

Editors

Global Production

A Handbook for Strategy and Implementation

Prof. Dr.-Ing. Eberhard Abele
Technische Universität Darmstadt
Institut für Produktionstechnik
und Spanende Werkzeugmaschinen
Petersenstr. 30
64287 Darmstadt
Germany
abele@ptw.tu-darmstadt.de

Dr. rer. nat. Ulrich Näher
McKinsey & Company
Roppongi 1st Bld.
1–9–9 Roppongi
Minato-ku
Tokyo 106-8509
Japan
Ulrich_Naeher@McKinsey.com

Richard Sykes
McKinsey & Company
21 South Clark Street
Suite 2900
Chicago, IL 60603-2900
USA
Richard_Sykes@McKinsey.com

Dr.-Ing. Tobias Meyer
McKinsey & Company
3 Tesmasek Avenue
#18–03 Centennial Tower
Singapore 039190
Singapore
Tobias_Meyer@McKinsey.com

Dr.-Ing. Gernot Strube
McKinsey & Company
Sophienstr. 26
80333 Munich
Germany
Gernot_Strube@McKinsey.com

ISBN 978-3-540-71652-5   e-ISBN 978-3-540-71653-2

DOI 10.1007/978-3-540-71653-2

Library of Congress Control Number 2007936682

Production: LE-T$_E$X Jelonek, Schmidt & Vöckler GbR, Leipzig, Germany
Typesetting: PTP-Berlin Protago-T$_E$X-Production GmbH, Berlin, Germany
Cover-Design: WMX Design GmbH, Heidelberg, Germany

Printed on acid-free paper

9 8 7 6 5 4 3 2 1
springer.com

Preface

Global Production summarizes McKinsey & Company's extensive thinking on one of management's most complex and risk-laden topics in our increasingly networked world. It is based on a large-scale survey originally conducted by McKinsey's German Office together with Darmstadt University of Technology. The "ProNet" (production network) survey included interviews with over 100 managers at 54 companies, yielding a wealth of data on best practices used by global leaders, as well as the pitfalls to avoid. The empirical data is validated by a theoretical computer model that uses all the relevant input parameters for production network relocation and setup.

The German original that appeared in 2005 received extremely positive feedback and is already being reprinted. It has been updated in this English edition with the most recent findings from our global knowledge network. We have included further international case studies and new geopolitical data.

The latitude for improvement in network optimization is astounding. We found that most companies only achieve cost reductions of 10 percent or less, despite the typical potential of 30 to 45 percent for incumbents largely based in high-cost countries before the move. The rigorously integrated approach at the heart of this book has been resoundingly successful and has been applied in numerous situations with our clients worldwide. The results are enabling companies to tap much higher rates of return, sidestepping the often crippling risks posed by the multiple factors we examine.

The macroeconomic perspective is also very revealing. Statistics show that a country's economy reaches a breakpoint when its industry attains a GDP share of 45 to 50 percent. This is the trigger for its switch to de-industrialization, in which services account for the lion's share of the "post-industrial" economy. In conjunction with increased connectivity, the globalization of production is dramatically accelerating this trend. Whereas in the past it took a century or more for a country to evolve through the three phases from industrializing and highly industrialized to post-industrial economy, this progression now often takes just half that time. This has vast economic and social implications for countries at each stage of development. Regions also need to drive their progress along all the relevant parameters much more proactively, building the prerequisites for the next stage well in advance.

Their joint effort can even fast-forward an economy's movement along the development path.

Our partner in the survey, the Institute of Production Management, Technology, and Machine Tools (PTW) at Darmstadt University of Technology, provided outstanding support throughout, and we thank all our colleagues there for their contributions. We also wholeheartedly thank the many executives who provided invaluable support in the form of insights, data, and case studies for our survey, as did our McKinsey colleagues across the globe. Above all, we gratefully acknowledge the intense dedication of the team of editors and authors, who conducted the analysis and have developed this groundbreaking methodology with such deep commitment.

We very much hope CEOs and production network managers as well as strategic planners and students in this field will find this book helpful, and we welcome any comments or queries readers may have.

Prof. Dr. Jürgen Kluge
Director

Table of Contents

Michael Stolle, Ulrich Näher, Frank Jacob, Nicolas Reinecke, James Hexter, Marina Dervisopoulos

Sebastian Simon, Ulrich Näher, Mads D. Lauritzen

FRANK JACOB, GERNOT STRUBE

1 Why Go Global? The Multinational Imperative

Summary

Globalization is not a new phenomenon. The networking of the world's economy has been evolving for centuries, with companies gradually expanding beyond their national borders. What is new is the dramatic acceleration of this process. The rapid networking of global communications is being mirrored by web-like value chains that increasingly span the world.

Global production provides an unparalleled opportunity for companies to grow into new markets while at the same time boosting their competitiveness. However, most of today's networks are legacy structures – only a fraction were strategically planned. As a result, there is huge potential to be captured from rethinking traditional structures, approaches, and supply relationships. And huge potential for getting it wrong. Our survey showed that production network redesign can cut a company's manufacturing costs by up to 45 percent – but over half the players achieved savings of only 10 percent or less.

This book focuses on the three industries covered by the ProNet survey: automotive engineering, machine tool manufacturing, and electronics. Their profiles are all very different, whether we look at the footprint and corporate history of key players, market characteristics, product and production technologies, or their cost structures. The beauty is that this breadth makes the results representative far beyond these three sectors. Their patterns and drivers can help to identify optimal global networks throughout the manufacturing industry.

This first chapter lays the groundwork by elucidating the historical background to globalization and reviewing the drivers and goals of the current race to go global. It then examines the status quo of our three focus industries, with an overview of their survey findings. The rest of the book, based on the results of that analysis, offers practical guidance for companies planning to reconfigure their global footprint.

Key questions, Chapter 1

- What different phases has the globalization of production gone through over time?

- What are the reasons for the increase in global production?

  □ What factors in the equation have changed?

  □ What are the underlying long-term trends?

  □ What influence are these factors and trends having on existing industries?

- What objectives do companies pursue with the globalization of production?

- Are these objectives realistic? What successes have been achieved so far?

- How does the status quo of manufacturers differ across the three focus industries – automotive engineering, machine tool manufacturing, and electronics?

- What implications do the developments outlined above have for these three industries?

- What is the current status of the globalization efforts of the three focus industries?

1.1 Phases of Globalization

International trade has existed since recordkeeping began. Herodotus, known as the "Father of History," wrote detailed reports about the trade in spices, silk, glass, porcelain, and incense between Asia and Europe along the Silk Road around 430 BC.[1] Highly specialized economic structures formed along the value chains of these goods in specific geographic regions. These early know-how clusters[2] led to **local production monopolies**. Large regional price differences (due to manufacturing advantages) made trade in these items attractive despite the rudimentary **transport** available. Global trade has advanced steadily ever since. Globalization only entered a new era with the dawn of the Industrial Revolution. **Three phases** can be distinguished, from cross-border trading to globalization in its current form (Figure 1.1).

1.1.1 Before 1930: Mainly Sales Offices Abroad

Sweeping technical innovations such as the railroad promoted the cross-border exchange of goods from around 1850 onwards. The simultaneous rise of **mass production** and its corresponding **economies of scale**[3] paved the way for the manufacture of large unit volumes. The introduction of stock corporations as a legal entity facilitated access to capital, loosening restrictions on freedom of movement and opening up new structural options. Stock corporations used these opportunities to expand their customer and supply markets, intensify their international trade relationships, and set up **sales outlets abroad**. However, inadequate means of communication set

---

[1] Cf. Franck (1986).

[2] Clusters are self-reinforcing networks of producers, suppliers, research institutions, service providers, and related institutions that operate along a value chain. The members are connected with one another through supply or competitive relationships and/or shared interests. One cluster with a long history is the concentration of US automotive industry players in and around Detroit.

[3] Economies of scale define the dependence of production volume on the factor inputs used. They occur when the production volume rises faster than the factor inputs and the unit costs fall with increasing unit volume, e.g., due to better utilization of machinery or labor or better purchasing terms.

limits to the expansion drive. Telecommunications was in its infancy, and information could barely move faster than goods. Foreign branches therefore mostly acted autonomously. Because corporate centers were unable to give guidance across long distances, manufacturing in foreign countries was rarely economically viable. Production networks in the current sense of the term did not exist. It was only when telecommunications became established at the beginning of the 20th century that it became possible to create a cost-effective network of production facilities in different countries. Delayed to some extent by World War I and the subsequent economic recession, production facilities abroad did not start to multiply substantially before 1930.

Siemens is a good **example**. Founded in 1847 under the name "Telegraphen-Bauanstalt von Siemens & Halske," the company found itself in a crisis in the early 1850s due to a lack of orders. Business deals with Russia and England gave it a fresh boost. In 1853, Siemens & Halske started to build the Russian telegraph network as its first ever foreign venture. In 1858, it founded a subsidiary in England. Its chief activity was laying ocean cables, produced at Siemens' first foreign plant in Woolwich from 1863 onwards.

This rapid **internationalization** had begun shortly after the company was set up and – with the exception of the Woolwich plant – consisted mainly of **sales offices**. However, the sites were still relatively independent of one another, and it was not yet possible to establish more intensive communication or supply chains. Production plants abroad did not increase significantly until 1930 onwards (Figure 1.2).

The nature of globalization has changed over time

Fig. 1.1: Development of globalization in three phases

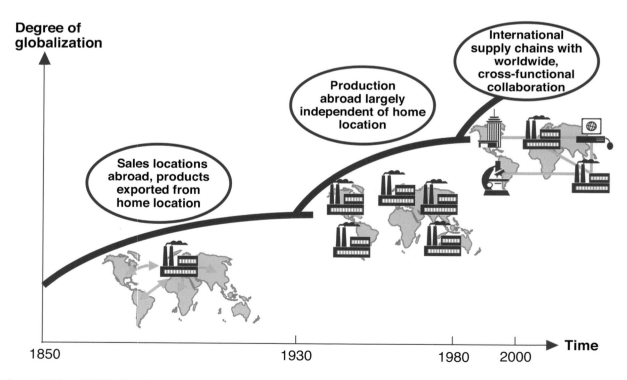

Source: McKinsey/PTW (ProNet analysis)

1.1.2 1930 to 1980: Largely Independent Production Abroad

After World War I and the world economic crisis, powerful companies arose that continued to grow fast and steadily. The triumphal march of the **brand names** began. Coca-Cola, Mercedes, and IBM became famous the world over. Increasingly low-cost, effective communication made it possible to manage companies of unprecedented scale. Organic growth and acquisitions formed industry giants that were able to tap major economies of **synergy and scale**.

Companies used their size and dominance on the home market to open up foreign markets. Production at the home factory was still not very closely integrated with production abroad. Foreign facilities mostly operated independently, aimed at developing new markets via **local production**. Their financial strength generally enabled them to implement this strategy quickly. They would often acquire competitors abroad to spare themselves the risky and time-consuming process of setting up their own sites.

General Motors (GM), for example, grew apace in its US domestic market, taking over 25 companies in the first three years of its existence. In 1931, it overtook Ford as the largest OEM in the world, and has retained this position ever since.

However, growth opportunities on the home market flattened off over time. This was barely surprising – it had a market share in the US of over 50 percent at times. The obvious course of action was to expand abroad. In 1925, GM opened its first foreign plant in Argentina, and then took over the German Adam Opel AG in 1929.

After World War II, during which GM exclusively produced military equipment, its globalization continued. It began production of Holden[4] brand automobiles in Australia in 1948, and opened Venezuela's first ever

---

Pioneers started out with sales offices abroad as early as the 19th century, but did not move much production abroad until after 1930

Fig. 1.2: Development of Siemens' foreign activities
Number of new start-ups per decade

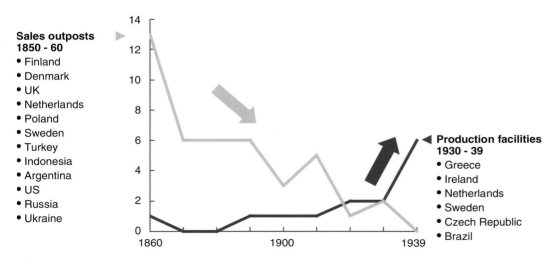

Sales outposts 1850 - 60
- Finland
- Denmark
- UK
- Netherlands
- Poland
- Sweden
- Turkey
- Indonesia
- Argentina
- US
- Russia
- Ukraine

Production facilities 1930 - 39
- Greece
- Ireland
- Netherlands
- Sweden
- Czech Republic
- Brazil

Source: Siemens

---

[4] *Holden is an Australian automobile brand founded by General Motors after World War II on the initiative of the Australian government.*

automobile factory in the same year. Its foreign plants had extensive freedom of development, production management, and product design.

1.1.3 Since 1980: Globally Networked Production and Cross-Functional Collaboration

The third era after 1980 was characterized by deregulation, a converging world economy, rapid technical progress, and declining transaction costs[5]. Trade barriers fell, GATT rounds[6] led to reductions in tariffs, and customs unions such as the EEC were founded, precursor to the EU. The economic powerhouses of the West became increasingly intertwined. It was during this period that the concept of globalization took on the significance it has today.

The internal and cross-organizational networking of companies grew in the following period much faster than markets went global (Table 1.1). **CKD and SKD assembly** were widely used.[7] Firms tapped economies of scale by manufacturing basic components centrally. Products were also tailored to customer requirements locally. Companies that grasped the opportunities of this new form of globalization quickly found themselves with a strong competitive advantage.

Global cooperation took on a new quality at the end of the 20th century. Customers no longer just exchange goods and supplies across borders. Staff at far distant locations work on the same projects on a daily basis. A business unit's functions – whether R&D, production, HR, or marketing – may well be spread throughout the world. The challenge is not just to connect individual companies and corporate units, but to set up corporate functions at the best location for each, and manage them as a network. Technologies such as the Internet and digitized communication underpin this, linking up the advantages of local know-how clusters with the factor cost benefits of distant locations. The rapid exchange of information and intangible assets is leading to a global knowledge network. And – on a historical time scale – this development has only just begun.

General Electric is the archetypal global conglomerate – not least due to the acquisition of almost one thousand different companies by long-time CEO Jack Welch. GE is regarded as a pioneer in offshoring corporate services to far-flung locations abroad. In the early 1990s, Jack Welch introduced the 70:70:70 rule. This stood for moving 70 percent of labor to low-cost locations, 70 percent of this to so-called offshore development centers, and 70 percent of that in turn to India. What this ultimately meant was that 30 percent of GE's back-office activities were relocated to India. These were primarily administrative and support functions, such as data processing, information services, operational IT consulting and support, and call centers.

As a consequence, the group's financial services company GE Capital International Services (GECIS), which originally operated from the United States, launched its globalization in 1997 with a location in India. GE put a figure of 25 to 60 percent on the savings, depending on the business segment. Further sites in Mexico, Asia, and Eastern Europe followed. In 2005 GECIS became independent and changed its name to Genpact. In 2006 it was operating with 26,000 employees in 11 countries on 3 continents.

Table 1.1: Intra-industrial trade as a share of the export trade of industrialized nations

	1954	1964	1980	1990
Germany	42%	54%	65%	79%
US	54%	71%	73%	85%
Japan	29%	34%	25%	44%
Other industrialized countries	55%	65%	71%	77%

---

[5] *Transaction costs: The costs or expenses incurred for the exchange of goods. In connection with production networks, this particularly refers to customs duties, transport costs, shipping insurance, and communication costs. The capital tied up in transportation and depreciation of the goods during transportation also count as transaction costs in this context.*

[6] *GATT: General Agreement on Tariffs and Trade.*

[7] *CKD (completely knocked down) and SKD (semi-knocked down) describe modes of manufacturing where assembly kits are produced for export. Final assembly is performed locally.*

[8] *Cf.: http://www.genpact.com*

GE does not exclude tasks requiring high qualifications from global teamwork. The concept of "Sunrise Development" has seen engineers and designers work round the clock across continents on shared projects.

Globalization is accelerating

Globalization has not just changed its face over time, it has gathered significant speed in the past few years (Figure 1.3). This is also reflected in the number of direct investments, which have risen exponentially since the mid-1980s. The foreign investment base has more than trebled within ten years. By 2003, private investors, companies, and states from across the globe had invested over USD 8 trillion in foreign companies, real estate, or finance deals. This corresponds to the combined gross national income of Japan, Germany, and France in one year. And the regional focus of investment has been shifting increasingly. In 2003, China overtook the US for the first time as the main target for direct investments.

Meanwhile, producers around the world are engaged in building up efficient global production networks. An analysis shows that the international operations of major corporations are growing faster than in their home countries (Figure 1.4). This relates primarily to revenues, but also to assets and staff as production facilities are established.

As a result of globalization, whole industries are being redefined. Within just 10 to 20 years, the focal areas of global production are shifting dramatically. Some industries – such as textiles or consumer electronics – have already completed this development.

A good example is the production of TV sets (Figure 1.5). The share of production in high-cost countries fell within two decades from 75 to 20 percent. This development was accompanied by a fundamental change in the market. New competitors from low-cost countries captured significant market shares. Brands considered established today, such as Samsung, Sharp, or Lucky Goldstar (LG), were largely unknown in the Europe of the early 1980s, and were able to gain ever more ground from domestic manufacturers

because of their attractive price-performance ratios. Former greats, particularly German consumer electronics manufacturers such as Schneider, Grundig, or Telefunken, went bankrupt. Other European manufacturers such as Thomson or Philips managed to turn themselves around only by making drastic changes in their production networks and forming alliances with attackers from low-cost countries.

A further development that will change industry structures is currently emerging, particularly for products in the electronics sector. Although traditional product suppliers often initially invested in building up their own production locations abroad, gradually toll or contract manufacturing developed into an ever more attractive option – emerging as the business model of the electronics manufacturing services (EMS) provider. EMS providers perform operational services for OEMs[9] – particularly the manufacture and assembly of products for end consumers – at very attractive terms and conditions.

The level of international integration is rising exponentially

Fig. 1.3: Direct investments* abroad
USD trillions

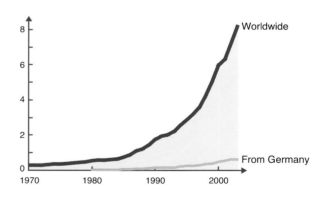

* Foreign investments worldwide
Source: World Trade Organisation (WTO)

---

[9] *The term Original Equipment Manufacturer (OEM) describes a manufacturer whose products are sold under a brand name as a single unit; an OEM normally buys components from other manufacturers, integrates them unchanged into its own products, and sells the resulting total package to end customers.*

> **Globalization has accelerated: Companies are growing abroad faster than in their home markets**

Fig. 1.4: Development of international business activities*
Percent

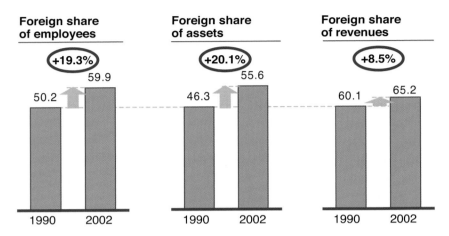

Foreign share of employees	Foreign share of assets	Foreign share of revenues
(+19.3%)	(+20.1%)	(+8.5%)
50.2 → 59.9	46.3 → 55.6	60.1 → 65.2
1990 2002	1990 2002	1990 2002

* Analysis covers BASF, Electrolux, Fiat, General Electric, IBM, Philips, Siemens, Sony, and Volkswagen

Source: UNCTAD Transnationality Company Ranking

They achieve significant cost advantages compared with OEMs via specialization, economies of scale,

> **The market entry of low-cost providers often leads to rapid relocation of an entire industry**

Fig. 1.5: Trends in global location of TV production
Percent

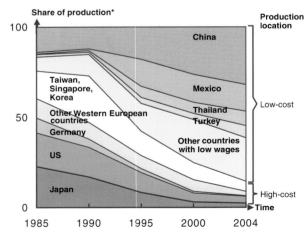

Share of production*   Production location

- China
- Taiwan, Singapore, Korea
- Mexico
- Other Western European countries
- Thailand
- Turkey
- Germany
- Other countries with low wages
- US
- Japan

Low-cost

High-cost

Time

1985   1990   1995   2000   2004

* Share of manufactured TV sets with color CR tube

Source: Reeds Electronics Yearbook, McKinsey

---

and attractive sites in low-cost countries such as Malaysia, China, Poland, Hungary, and Mexico. Flextronics, for example, the world's biggest EMS company (see Chapter 9), manufactures Sony cell phones, Hewlett-Packard printers, and Microsoft's Xbox. These providers are virtually unknown, whether Flextronics, Solectron, Elcoteq, or Hon Hai. But their customers are global brands.

The EMS sector has been acting as a **catalyst** for the radical transformation of electronics production worldwide. EMS companies are characterized by very high agility, and frequently changing network structures (Figure 1.6).

When HCC[10] incumbents award production contracts to EMS companies, this often leads to **relocation by**

---

[10] HCC refers throughout this book to high-cost countries. We define high-cost countries and high-cost locations as geographies with average gross wages for blue-collar workers at or above USD 15 per working hour. This value includes fringe and voluntary benefits. The value applies for the average working hours in the respective geography, including vacation and average absenteeism. When we convert location currencies to the US dollar, we use the long-term average inflation-adjusted exchange rate, e.g., EUR 1 = USD 1.16, to decouple the findings from the short-term impact of exchange rate fluctuations.

outsourcing. But EMS providers have been also growing in high-cost countries – particularly by taking over their customers' factories. Traditional players have to watch out that they do not lose their technological edge – and their markets. It is only a question of time before this model gains equal ground in other sectors, too.

1.2 What Are the Forces Accelerating Global Production?

In the 21st century, the globalization of production has taken on an entirely new pace, scope, and scale. The drivers just outlined are no less important, but why are companies going abroad ever faster, with ever more functions?

Diverging factor costs and growth are widening the disparity in the attractiveness of different production locations. It has become clear that the wage gap is not going to close between the new entrants to the global economy and industrialized countries anywhere soon. The political reasons are no less important: liberalizing markets and the reduction of trade barriers are shifting the centers of economic activity. Steadily tumbling transaction costs have also helped to vastly reduce the barriers to global production, with falling transportation costs and technological connectivity advancing at lightning speed.

1.2.1 Huge Factor Cost Differences

If manufacturing costs at different production locations are considered in isolation, disparities are mainly apparent in factor costs – and specifically in labor costs. The development of labor costs is clearly closely linked to prosperity: in affluent economies, wages go up; in the others, wage development is curbed.

EMS providers are the catalysts of an entire industry

Fig. 1.6: Change in the production network triggered by the three largest EMS* providers between 1992 and 2002

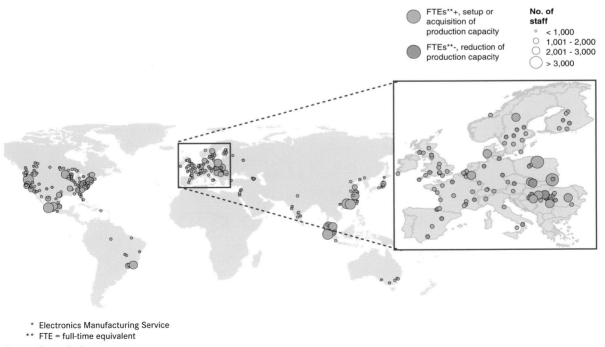

* Electronics Manufacturing Service
** FTE = full-time equivalent

Source: Press clippings

With the onset of the Industrial Revolution, growth rates in Europe and North America soared. Large parts of the rest of the world, particularly those under communist rule, experienced a very different fate. A misguided economic policy held back development of many other nations. A historically unique **prosperity gap** opened up between the industrialized countries and the rest of the world. This was accompanied by corresponding differences in local wages.

Because of the high and sustained economic growth over five or more decades, labor costs in industrialized nations are very high. Wages in other countries that have been unable to keep pace with this rapid economic development are much lower (Figure 1.7). Following initial speculation after 1990 that labor cost disparities would equalize much more rapidly, the realization has now set in that developing and newly industrialized countries will only catch up with HCCs in the very long term, if at all. In the medium to short term, the differences – in absolute terms – will in fact further escalate. Companies have no choice but to factor in these vast cost differences in their network strategy considerations – not only di-

rect but also indirect labor costs, which greatly influence the price of sourced materials.

1.2.2 High Growth in Emerging Markets

Emerging markets are experiencing very high growth in some segments – both relative and in terms of absolute market volume. Markets outside the highly industrialized world are becoming all the more attractive as a result, particularly for manufacturing companies. These enormous growth opportunities have become the key motivating force for the globalization of production. Demand for many tangible goods in major industrial players' domestic locations, on the other hand, is stagnating or growing only slowly. The main activity at home is merely the battle to carve up market shares.

1.2.3 Lower Transaction Costs

From the perspective of entire networks and value chains rather than just an individual location, a particularly important barrier for global production has been transaction costs.

Global labor cost differences are high, but the gap is slowly closing

Fig. 1.7: Development of labor cost differences
(largely proportional to GDP per capita)

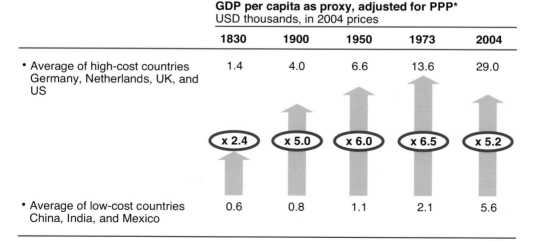

	GDP per capita as proxy, adjusted for PPP* USD thousands, in 2004 prices				
	1830	1900	1950	1973	2004
• Average of high-cost countries Germany, Netherlands, UK, and US	1.4	4.0	6.6	13.6	29.0
	x 2.4	x 5.0	x 6.0	x 6.5	x 5.2
• Average of low-cost countries China, India, and Mexico	0.6	0.8	1.1	2.1	5.6

 * PPP = purchasing power parity

Source: Maddison (2001), German Federal Office of Statistics (2005)

Transportation has historically been the main cost factor for the global exchange of goods. Up until the spread of the railroads, it was only worthwhile transporting goods with a very high value density and high margins, such as spices, silk, glass, and china. With the surge in new forms of transportation that occurred from about 1930 onwards, costs fell steadily (Figure 1.8). In 2004, the costs of ocean transport were less than 1 percent of the figure in 1830. Nowadays, even transporting goods with a low value density is cost efficient. Sending a cathode ray tube TV set with a 70-cm screen from Turkey to Germany only costs around EUR 10, or about 5 percent of its production costs. For a smaller size, higher value flat panel TV, the cost share would even be lower at around 2 percent.

Decreasing transport and communication costs are eradicating the natural barriers to globalization

The productivity gains in logistics are continuing hand in hand with falling transport costs. Ships are becoming ever larger, the crews needed for steering and loading them are shrinking due to automation, and transport risks are declining. The size of ships is leading to natural economies of scale: Less fuel is needed per unit transported. In addition, the fixed costs of supertransporters with a capacity of more than 8,000 containers – from the captain's salary through to pilotage fees – are spread out across a very large volume of goods.

The dramatic progress in communication technology has been the greatest distance killer of all. Whether orders, controlling indicators, R&D engineering blueprints – any and all intangible information can now be transmitted worldwide in an instant. The benefits are shrinking costs throughout the value chain, whether real-time datasharing, satellite-linked networking, remote maintenance or troubleshooting. It is nearly unimaginable that this technology

Transport costs have declined dramatically by historic standards, and have lost importance as a barrier to globalization

Fig. 1.8: Development of transport costs between 1830 and 2004
(mapped logarithmically)

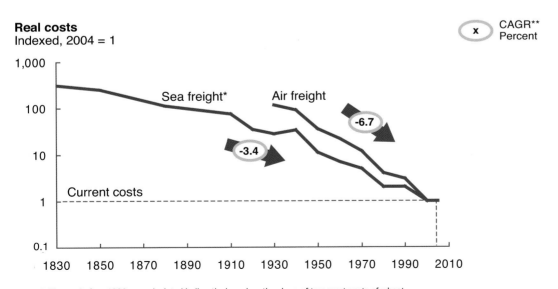

* Figures before 1930 are calculated indirectly, based on the share of transport costs of wheat
** Compound annual growth rate

Source: Baldwin (1999), World Economic Outlook (May 1997)

revolution is still in its early days. Technological connectivity has sent communication costs tumbling, to the benefit of all parties to a transaction. At the simplest level, the price of an international call has fallen to zero with Voice over Internet Protocol systems, and a tiny fraction of its previous costs even using non-VoIP telephony.

The impact of the Internet on consumer behavior is also having a knock-on effect on cost structures worldwide. Consumers increasingly have information and access to the same products and brands wherever they live. This greater demand-side transparency is putting additional cost pressure on producers worldwide, and further eradicating the significance of where an OEM is actually located.

1.2.4 Fewer Trade Barriers

Not only economically but also politically, the world has changed radically in the last two decades with the fall of the Iron Curtain and the dissolution of the East/West divide. This has been accompanied by increasing the liberalization of markets that were previously inaccessible to Western companies. Russia, Eastern Europe, and China have become attractive markets and significant importers of higher-value goods.

This development is far from over, as the example of China illustrates. China has fundamentally altered its business environment in the past 15 years, liberalizing trade, improving the protection of intellectual property, and eliminating export quotas and demands for local content.

India as an emerging economic power is also wooing companies with lucrative prospects in the competition between global locations. In 1997, India launched an initiative to reduce taxes and tariffs, improve its infrastructure, and reduce subsidies. On March 31, 2001, it lifted its last volume-based restrictions on imported goods and reduced its top tariff rate.

However, deregulation has not yet progressed very far in some arenas. In India, for example, direct in-

vestments from abroad are still regulated. Foreign investors are only permitted to have minority interests in some sectors, such as cellular telephony provision, banking, and insurance.[11] The intention behind this is to protect national companies from tough international competition.

This also applies to China, where the level of state control is heavily dependent on the specific industry (Figure 1.9). Competition is artificially restricted, preventing local manufacturers from being subjected to price pressure in many sectors. The customer pays the price. A very small number of foreign automotive manufacturers and their local joint venture partners were able to enjoy four times the margins achievable in the rest of the world there until the late 1990s. Chinese customers paid significantly more for the same automobiles than buyers in Europe and the United States.

In addition to the unilateral abolition of regulations, state trade barriers are being dismantled all over the world. Customs duties have historically been a significant source of government income. It was widely accepted, however, that they hampered the international exchange of goods. In the last few decades, the perception gradually seeped through that global trade brings more advantages to a national economy and thus the government than high customs duties. The continuous reduction of tariffs began. The basis was the GATT framework. The first was concluded in 1947. By 1994, tariffs and other trade barriers had been reduced step by step to one-fifth of their original value (Table 1.2).

The outcome of the last GATT round, the Uruguay Round, was the Marrakech Declaration that founded the World Trade Organization (WTO). The WTO commenced work in 1995. The WTO continues to apply the regulations developed under the GATT framework and further the reductions in tariffs and other trade barriers under the umbrella of the Multilateral Trade Agreement.

---

[11] See EIU (2007), p. 18.

Deregulation has not yet penetrated all sectors of industry in China

Fig. 1.9: Liberaliziation of the Chinese market by industry

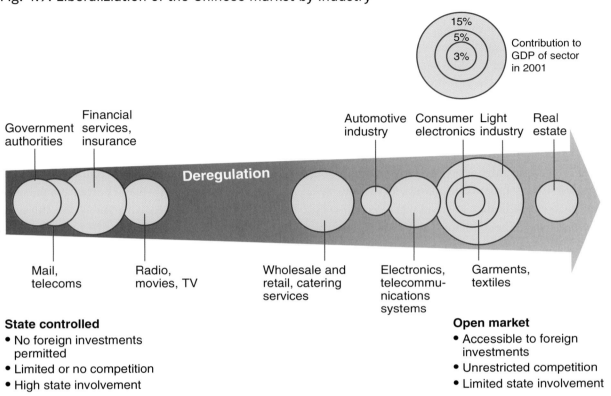

15%
5%
3%
Contribution to GDP of sector in 2001

Government authorities

Financial services, insurance

Deregulation

Automotive industry

Consumer electronics

Light industry

Real estate

Mail, telecoms

Radio, movies, TV

Wholesale and retail, catering services

Electronics, telecommunications systems

Garments, textiles

State controlled
- No foreign investments permitted
- Limited or no competition
- High state involvement

Open market
- Accessible to foreign investments
- Unrestricted competition
- Limited state involvement

Source: China today (2004)

Table 1.2: GATT rounds and the corresponding tariff reductions

	Year	Tariff reduction	Index
			100%
Geneva	1947	19%	81%
Annecy	1949	2%	79%
Tournay	1950/51	3%	77%
Geneva	1955/56	2%	75%
Dillon Round	1961/62	7%	70%
Kennedy Round	1964 – 67	35%	46%
Tokyo Round	1973 – 79	34%	30%
Uruguay Round	1986 – 94	40%	18%

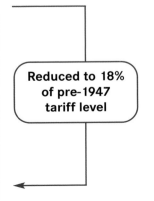

Reduced to 18% of pre-1947 tariff level

Regional economic alliances create a favorable climate for investment – a win-win for all participants

In parallel, many states came together to form economic areas during the 20th century. These alliances all aim to create a win-win situation for the member states. Companies in the member states gain better access to a larger market and are thereby able to realize economies of scale from higher production volumes. Thus free trade can lead to an improved use of resources, i.e., higher productivity and more competition. Higher productivity allows for higher wages and thus can stimulate demand whereas it also furthers the cost-efficient supply of goods.

The links forged range from pure free trade zones through customs unions (with zero tariffs on the movement of goods within the union and standard import tariffs for non-member countries) to fully in-

tegrated economic zones with a joint currency. These associations change legacy structures and have a substantial impact on the globalization choices of multinationals.

1.3 Goals of Global Production

Market development and cost reduction are generally the main motives when companies set their sights on globalization. Further reasons include the low-cost sourcing of supplied parts, high-grade knowledge and qualifications, and avoiding business risks such as exchange rate fluctuation. These secondary motives normally play a part in globalization decisions in conjunction with one of the two main aims (Figure 1.10).

Companies will choose different approaches depending on their key motivation. If they mainly wish to gain new customers in other countries, they will globalize by setting up new sales offices and strengthen-

Global production offers major opportunities, but also challenges

Fig. 1.10: The two key drivers of global production: new markets and cost reduction

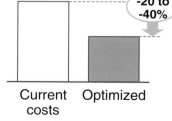

Opportunities and challenges

Market development
China: real GDP
USD billions

1.5   3.8   2.3   1.5

2003   2013 (estimated)

- Foreign countries offer major growth potential
- Companies move fast to develop markets via sales outlets
- Sustainable success often not possible with imports

Cost reduction
Production costs before/after optimization of location structure

-20 to -40%

Current costs   Optimized

- Savings potential from optimizing entire value chain is underestimated
- Companies often unable to cope with the complexity of global production networks

Source: McKinsey

Eastern Europe is developing into the key foreign location for German automotive suppliers

Fig. 1.11: Recent trends in production abroad

Automotive supply industry: imports to Germany
EUR billions

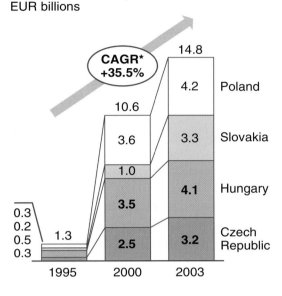

No. of staff of selected suppliers in new member countries of the EU

Poland	Bosch	500	Faurecia	3,154
	Delphi	6,039	Valeo	4,600
	Leoni	1,600	TRW	2,800
	Mahle	1,300	Autoliv	1,176
Slovakia	Beru	110	FAG Kugel-	
	Bosch	1,500	fischer	650
	ContiTech	500	Knorr-Bremse	1,200
	Delphi	1,970	Leoni	1,190
	Eissmann	390	Veritas	700
Hungary	Autoliv	840	Lear	
	Bosch	3,300	Corporation	5,068
	Continental	1,127	LUK	650
	Delphi	2,100	Michelin	2,100
	Denso	2,530	Valeo	700
	Knorr-Bremse	800	Visteon	1,200
Czech Republic	Autopal	4,415	Hella	700
	Behr	650	Johnson Control	2,900
	Bosch	6,500	Knorr-Bremse	310
	Brose	1,000	Magneton	850
	Continental	1,800	Mann+Hummel	630
	Denso	1,600	Pal	450
	Eberspächer	250	Safina	330
	Edscha	300	TRW	2,962
	Federal-Mogul	900	VDO	1,900
	Hayes Lemmerz	410		

* Compound annual growth rate

Source: McKinsey/PTW (ProNet analysis)

ing their local customer services. Occasionally they will open up production locations to support their market drive by responding promptly to customer requirements and gaining competence in manufacturing tailored products. If, however, they are primarily looking to reduce manufacturing costs in existing markets, multinationals will primarily invest in machinery and plant in LCCs,[12] or shift existing factories to the new location.

Market attractiveness has been the key reason for expansion to North America and Asia so far, while cost-cutting has been the primary attraction with Eastern Europe (Table 1.3). This is borne out by the strong growth in imports. Imports of automotive parts from Eastern Europe to Germany have risen by over 30 percent on average in the last decade (Figure 1.11, left). Other indications of the draw of

Eastern Europe are the many branches of Western suppliers with increasing numbers of staff (Figure 1.11, right).

1.3.1 The Growth Impact

It used to be that companies could grow in new markets "just" by expanding local sales and service capabilities. This is no longer true. The consensus is that production in new markets can be an important component of tapping into these markets. One reason is the transaction costs for imported products,

---

[12] *LCC refers throughout this book to low-cost countries. We define low-cost countries and low-cost locations as geographies with average gross wages for blue-collar workers at or below USD 5 per working hour. The other boundary conditions apply as for high-cost locations.*

Table 1.3: Reasons for the attractiveness of countries or groups of countries[13]

Region	Reason for attractiveness (percent)			Mentions (absolute)
	Market	Costs	Other[2]	
China/India	**52**	32	16	87
Eastern Europe (EU)	13	**59**	28	36
Other[1]	26	40	34	75

[1] *Brazil, the Philippines, Romania, Thailand (each mentioned three times), and others.*

[2] *Mentioned in questionnaire: "know-how" and "other."*

which make them too expensive. Another is that products cannot be adapted flexibly enough to local market needs. The **gain in image and trust** vis-à-vis the customer from local manufacturing is a further important argument. Another is the elimination of state regulation imposed on imported products.

Interviews during the ProNet corporate survey showed how important "soft" factors in the business context are for success in developing countries.

Local production often makes it easier to open up a new market

Even with capital goods such as machinery and plant, decision makers know local presence can become the anchor of a firm's success. These are not just hard facts like easier maintenance, and availability of spare parts. Fuzzier indicators of customer perception are also important: confidence in long-term flexibility, reliability, and the intensity and quality of customer care. Customers develop greater trust because competent contacts – including production staff – are always on site. They can count on fast reactions and short lines of communication, knowing the personnel speak their language (both literally and metaphorically).

Western companies are therefore increasingly setting up their own production facilities even for sales-oriented foreign activities. This applies particularly in Asia, because of the great distance, the high state barriers, and cultural differences. The early commitment of Volkswagen in China is a good example. By estab-

lishing a Chinese plant long before the "rush to Asia," Volkswagen managed to secure a dominant market position in the most highly populated country in the world that lasted many years (see box: "China and India – Attractive Markets if Approached Right").

For many customers, however, the connection between a brand and its nationality has intrinsic value. Porsche Director Michael Macht stated that Americans are prepared to pay EUR 1,500 more for a top car "Made in Germany."[14] There, and for that product, local production would not necessarily be the key to success. The story is different for EADS, the Airbus aircraft manufacturer. Production on site and a US image is key to success in the US aerospace market. This is also a reason why EADS focuses intensely on building activities in the US. According to an EADS spokesman, "We can only be successful if we are accepted in the US as an American company."[17] Experience shows that local presence and the link between brand and nationality often pose a conundrum.

1.3.2 The Cost Impact

Cost advantages are driving ever more companies to set up production at new locations. The decision on where to locate production operations should be based on evaluation of the parameters outlined in Chapters 2, 3, and 4. The calculation must include the total landed costs, i.e., total production and transaction costs for the entire productive value chain.

[13] *Cf. results of ProNet survey.*

[14] *http://www.staufen-akademie.de/michael_macht.html*

China and India – Attractive Markets if Approached Right

Emerging markets, particularly China, have received major attention from MNCs over the last few decades, with India moving into the spotlight more recently. Both countries share three key characteristics: GDP is soaring, their populations are very large (and thus the number of potential consumers), and factor costs – especially labor – are a fraction of those in developed countries. However, to conclude that these markets are an MNC's paradise would be overly simplistic.

If GDP is used as a measure for a country's wealth, it is true that China and India are experiencing significantly higher growth rates than developed countries. In the time frame between 2005 and 2030, expected average annual real growth rates for China and India are about 6 to 9 percent, versus 2 to 3 percent for the United States, and only 1 to 3 percent for Japan and Western European countries.[15] Looking at absolute annual GDP growth, the incremental growth in China is already higher today than in Japan and Germany, and India has just surpassed these two countries as well. Nonetheless, it will still take until about 2016 for China's GDP to outgrow Japan's, and until about 2030 for India's GDP to exceed Germany's. US GDP will still remain by far the highest of all countries. Absolute GDP growth in China will match that of the US in around 15 years (Figure 1.12).

Looking at markets rather than the size of integrated economies, the potential in emerging markets is indeed impressive. A good example is the

In terms of absolute growth of real GDP, China will overtake the US around 2024

Fig. 1.12: Real GDP and GDP growth, selected countries

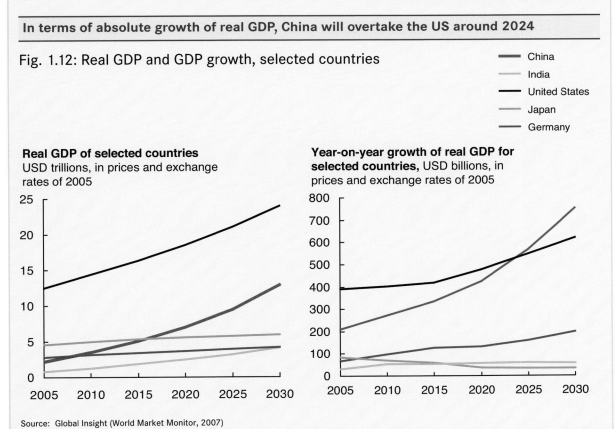

Real GDP of selected countries
USD trillions, in prices and exchange rates of 2005

Year-on-year growth of real GDP for selected countries, USD billions, in prices and exchange rates of 2005

Legend: China, India, United States, Japan, Germany

Source: Global Insight (World Market Monitor, 2007)

[15] *Source: Global Insight, World Development Indicators (World Bank)*

development of urban consumers in China. A model developed by the McKinsey Global Institute divides the key emerging middle class into a lower segment with an income of RMB 25,000 – 40,000, and an upper segment with an income of RMB 40,000 - 100,000. While the nominal currency ratio is about RMB 8 for USD 1, the different price levels create a ratio of buying power of about RMB 2 to USD 1, i.e., a Chinese household income of RMB 100,000 has similar buying power to a US household with an income of about USD 50,000. Development in China will take place in two phases. During the first wave (currently ongoing), we will see the rise of the lower-middle class reach a peak in 2009 with about 270 million consumers, about 43 percent of China's urban population. A second tran-

sition will follow in the next decade with a staggering increase in the upper-middle class. By 2025 this group will number 475 million, about 60 percent of China's projected urban population, with a disposable income of some RMB 12 trillion (Figure 1.13).

In approaching these markets, it is important to truly understand them. The tier-1 cities – Shanghai, Beijing, Guangzhou, and Shenzen – have the highest income level, at least 50 percent higher than the rest of China. However, China's rising middle class is widely dispersed, spread across some 650 cities and 10,000 towns. In addition to spending power, attitudes and behaviors also vary significantly both between the mega-cities and smaller towns as well as across the towns themselves.

The emergence of a middle class

Fig. 1.13: Share of urban households by income class in China
Percent

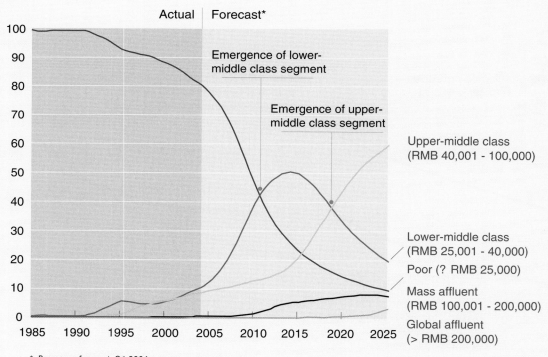

* Base case forecast, Q1 2006

Source: National Bureau of Statistics of China, McKinsey Global Institute analysis

The need for MNCs to adapt their range and pricing to local markets is always critical – and India has seen its share of success stories as well as failures recently (Figure 1.14). Hyundai has developed a clear competitive edge with a range specially tailored to the market. It offers lower power, fuel-efficient engines, tropical air conditioning, and higher vehicle clearance for road bumps. With this tailoring, it has achieved a 2 percent price premium over the local market leader, capturing significant volume as the third-largest car maker. Another major global OEM nose-dived offering a range with inferior lifetime ownership costs and a 10 to 15 percent price premium, resulting in a recent write-off of USD 100 million. McDonald's is another name in the "How to do it right" category. Offering vegetarian food as well as chicken products with a tangy, tandoori flavor, its local market

prices start at just 40 US cents. McDonald's has become India's largest fast-food chain. KFC (formerly Kentucky Fried Chicken) entered the market with its international product range and international prices, and failed to make the grade. It has since withdrawn.

In addition to the market strategy challenge, MNC operations in emerging markets often pose major challenges. Local players frequently benefit from their home advantage by producing outside the major cities, where labor costs are much lower. This puts heavy pressure on MNC prices and margins.

Another issue, particularly in China, is the country's reluctance to enforce the protection of intellectual property. In the ProNet corporate survey, a

"Indianize" the product and get the price-value equation right

Fig. 1.14: Global winners and also-rans in India
FY 2003

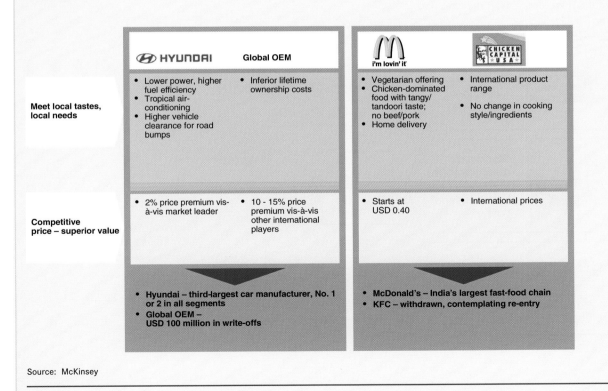

Source: McKinsey

machine tool manufacturer reported: "The Chinese started to copy our work practically during construction. They bought the same machinery and then poached our labor force six months after the start of production. We definitely won't be returning." Another manufacturer's experience of his employees' "dedication" also highlights the hardships of doing business in China: "… a short time later we found staff continuing to work at night … and selling the results to line their own pockets."

Bottom line: China and India are vast markets. They need to be on every MNC's radar. However, any approach towards the market and local operations needs to be planned and executed with painstaking care and foresight.

Many companies have managed to save costs and reduce competitive pressure by creating intelligently linked production networks. Good examples of this are the automotive supply companies' manufacturing facilities in Eastern Europe or the textile industry's relocation of production to Asia.

Companies can only survive long term by fundamentally redesigning their production networks

Particularly effective savings levers are, of course, the lower factor and materials costs, particularly wage and energy costs, but also savings in investment ex-

Cost savings of between 20 and 45 percent can normally be captured from optimizing production networks

Fig. 1.15: Production network optimization by incumbents based in HCCs*
Total landed costs, EUR millions p.a.

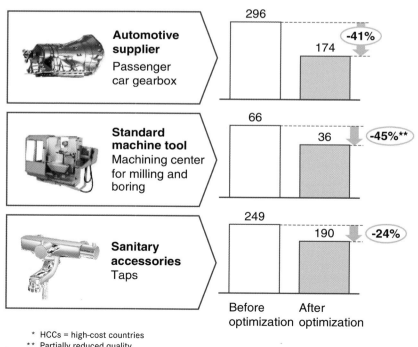

* HCCs = high-cost countries
** Partially reduced quality

Source: McKinsey/PTW (ProNet analysis)

penditure due to subsidies and tax benefits available in low-cost countries. The dominant cost lever depends to a large extent on the company's current position. Numerous projects have shown that the savings potential is generally substantial (Figure 1.15).

Just how high the savings can be is exemplified by a safety valve manufacturer that decided to set up a second plant in China to supply the local market (Figure 1.16). The cost advantages were so great that the works manager suggested even supplying the European market from China shortly after the start of production, and received approval. The transfer price set was 57 percent of the manufacturing costs in Germany.

The cost savings potential from globalization can be both opportunity and threat. Any company that wants full capacity utilization for expensive, state-of-the-art production facilities needs world-class sales volumes. Competitors who can capture market shares without expensive machinery and plant by tapping the cost advantages of globalization can threaten the economic viability of expensive production facilities for an entire segment. Companies too slow off the mark in this new constellation may find themselves without a future, as the fate of many laggards in Europe has shown.

Grundig, formerly a renowned brand in audio and video consumer electronics, failed to reshape its production network to make it more competitive for over a decade. Although the company had production facilities outside its German home base, these locations were not suited to balancing out structural disadvantages. As a result, Grundig was eventually forced into insolvency (see box: "The Grundig Example"). Rover – still one of the largest automotive

Main cost savings come from wage and materials costs

Fig. 1.16: Example: new foreign plant for safety valves in China
Percent

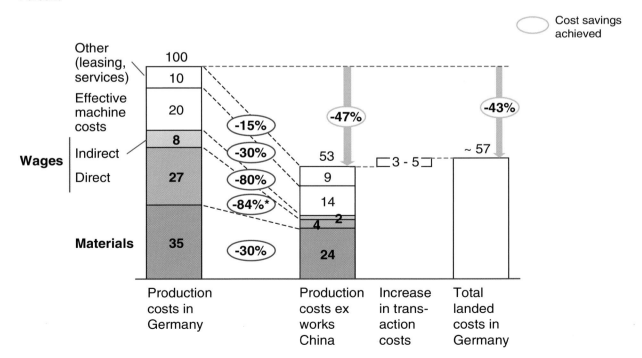

* Gross wages per worker: Germany EUR 54,000, China EUR 9,000

Source: McKinsey/PTW (ProNet analysis)

manufacturers in the world after World War II – could not keep pace with competitors for a similar reason, and went bankrupt in 2005 after a protracted decline. Sewing machine manufacturer Pfaff also failed to read the signs of the times.[15] Its production network is no longer a match for the challenges presented by new competitors from the emerging countries of Asia. The company shrank dramatically between 1981 and 2003; the number of staff fell from 9,539 to 863. Although it has meanwhile established operations in China, it remains a company with only around 1,000 employees.

The Grundig Example

With over 38,000 employees, Grundig was a renowned manufacturer of consumer electronics products at the end of the 1980s. A symbol of the German economic miracle, the company made its name selling televisions, razors, and electronic office equipment. At the start of the 1990s, the competitive landscape altered dramatically and rapidly (Figure 1.17). New brands invaded the market – impressing buyers thanks especially to their low prices. The new providers produced at low cost, mainly in Korea, Taiwan, and Turkey initially, and later in China. Grundig, on the other hand, manufactured its appliances in Germany, Austria, France, and Spain – a large proportion of them at its home factory in Nuremberg, Germany.

Grundig had a good name and enjoyed a high market share, especially in Germany and Austria. It

Grundig was unable to close the structural gap versus competitors

Fig. 1.17: Grundig's price/costs gap over time
EUR per unit*

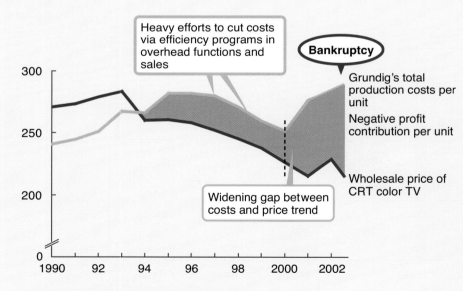

* Normed to color TV with standard cathode ray tube

Source: Annual reports, Reeds Electronics Yearbooks, McKinsey/PTW (ProNet analysis)

---

[15] Cf. Zirbik (2003).

did not see any pressing need to take action. But the situation rapidly grew more acute. The new manufacturers were gaining in experience, raising the quality of their products and increasing their cost advantage with improved processes. Between 1990 and 2004, the average price of comparable TVs fell by 2 percent each year. As other manufacturers pressed forward with the relocation of their production to low-cost countries, the price decline accelerated. This development found Grundig in a phase of increasing production costs. The company tried to keep up via additional investments in automation, but the gap between market price and Grundig's cost of goods manufactured continued to grow.

Grundig invested in cost-cutting initiatives and managed to achieve improvement rates comparable to those of other manufacturers. But the gap remained constant: costs were still higher than the prices it could charge. By now, other manufacturers had cast off the image of low-quality, cut-price providers. Grundig's share of the market was dwindling.

In response, Grundig started restructuring its own production: television assembly was discontinued in France (1992) and Spain (1993). The main factories remaining in operation were in Vienna, Austria and Nuremberg, Germany. However, this pullback did not lead to the necessary cost reductions either. In 2002, Grundig filed for bankruptcy.

An analysis of the options open to Grundig based on the annual accounts of the previous decade reveals that the company was last in a position to save itself in 1995. The funds required for restructuring and setting up new production sites were no longer available the following year – six years before the company went bankrupt (Figure 1.18). Once Grundig had fallen below the minimum liquidity limit, it could no longer be saved

The cumulative spending on relocation would have exceeded the credit line from 1996 onwards

Fig. 1.18: Credit line vs. cash-out required for the restructuring of Grundig
EUR millions

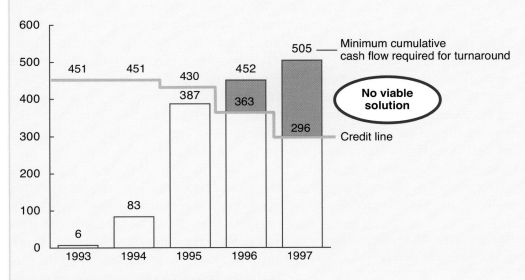

Source: Grundig annual reports, McKinsey/PTW (ProNet analysis)

without raising additional equity capital. The company's hesitation had led to a point of no return.

If this development is compared with that of other manufacturers in similar situations, one can see what might have saved Grundig: relocating production to leverage factor cost advantages.

Grundig's rival Thomson was in a similarly precarious position in 1992, but rigorously implemented a program of dramatic countermeasures. In 1996, around 80 percent of Thomson's production was in high-cost locations, but just two years later the fig-ure was only 40 percent and falling. The company has been back in the profit zone again after 1998. The toughness gained by the organization in the "manufacturing crisis" may also enable Thomson to successfully master the current difficulties.

Bottom line: Failure to take prompt action can jeopardize a company's existence. Reorganizing a production network when you are already weakened is much harder than being proactive and doing so before cash and credit line reserves are prohibitively low for a broad relocation of assets to lower-cost regions.

1.3.3 Secondary Objectives: Tapping Resources and Minimizing Risks

Access to tangible and intangible resources and the reduction of risk are examples of secondary motives that also have a major influence on the decision to go global. The term "resources" covers a wide range of factors: being close to raw materials suppliers, to the industry focus, or to technology leaders. Risk reduction includes protection against currency exposure, supply bottlenecks, and production stoppages, and also special terms offered by some states, such as direct investment subsidies and tax benefits.

1.3.3.1 Resource Access

Where tangible resources are concerned, relocating production close to the source of low-cost input products can often yield great advantages. This applies, for example, to the manufacture of metal-based products in Russia. The local availability of metal ores eliminates costs for long-distance transportation, and low labor costs are a boon both for converting ore into metal and for producing intermediate products.

> **MNCs find emerging nations' low labor costs and high growth extremely attractive**

At the same time, incumbents pursuing smart global labor strategies are finding promising talent growth in emerging nations, too. At 33 million, developing countries have more than twice the number of university-educated young professionals that developed countries do, and they can be tapped in a win-win situation for employer and employee provided multinationals install the right training and staff retention policies.

Access to intangible resources implies location close to centers of know-how in a company's industry. Companies benefit from technical and country-specific knowledge transfer and from the availability of qualified, low-cost personnel on site. When staff with specialized training are needed for low-volume production, which is often the case, companies can gain clear advantages from choosing locations where staff already have the know-how to manufacture their products efficiently. The best case is a "hat trick" or triple play: a setup that allows a company to develop products close to production, close to the market, and with a fast ramp-up.

An area with a concentration of one type of industry and a great deal of the related know-how is known as a **cluster**. Clusters act as focused pools of resources and ideas that amplify a continuous stream of innovation. Having a production site in the cluster enables companies to swiftly translate innovations into products, and is often essential if they wish to tap this know-how and play a leading industry role. These are often called lead plants or NPI (New Product Introduction) facilities. Particularly in industries where products are highly standardized for global sales, e.g., electronics, it is common to have an NPI in a facility near the R&D center, with rapid deployment to the other production sites, including those of external contract manufacturers. Well-known examples of effective industry clusters are Silicon Valley for semiconductors, "Mainboard Road" in Taiwan, or China's electrical and electronics manufacturing cluster around Shenzhen. They are also critical in emerging industries. Several centers of technology in France, Sweden, and Germany are trying to enhance their industrial growth by establishing explicit clusters around the new European Galileo satellite navigation system. Participation in the relevant clusters is critical for companies that want to play in the top league.

"More Art than Science" – Extract from an Interview on Clusters with Professor Porter

Michael E. Porter, Professor at Harvard Business School, is considered one of the world's greatest experts on competitive strategy and international competitiveness. How can regions and countries sustain and promote growth, innovativeness, and employment? Originally an aviation engineer, Professor Porter has been focusing on these central issues throughout his career.

Professor Porter, what are clusters, and why are they important for the competitiveness of an economy?
Clusters are a spatial organizational form for industry that generates greater productivity and innovation than more physically disparate structures. In a cluster, a variety of businesses and associated entities important to competition are gathered together in a relatively small area: manufacturers, suppliers, service providers, universities, and other training institutes.

What impact does a cluster have?
A cluster influences the market in three ways. First, it creates greater efficiency. Transactions can take place without high costs for logistics or transportation. Lines of communication are shorter, market participants can respond to one another faster. Clusters also produce goods that firms within the cluster can obtain relatively favorably. Anyone working outside the region has to conduct transactions and pay to access them.

Skilled staff in a specific sector are a good example. You can simply hire them, they'll move from one enterprise to another. Anywhere else, you'd have to train them first. This applies to a whole range of inputs: labor, market knowledge, technology. In a cluster they virtually become public goods to which everyone has access.

Second, opportunities drive innovation. If a large number of companies and market participants are concentrated in a small space, it is easier to detect gaps in the market. New goods or services seem to emerge all the more readily, the appropriate technical expertise is at your feet. You can also commercialize opportunities faster. All the elements of the value creation process, from the idea through to the product, can be combined in an instant. A cluster also provides better access to capital. Financial institutions that work with a cluster have sector-specific experience – from wine-growing to automotive production – and can make faster and better venture capital decisions.

Third, clusters stimulate new businesses in their field. The thresholds to market entry are lower for the reasons I've just described. It is easier to raise capital, access key suppliers, and find customers.

Source: "Mehr Kunst als Wissenschaft" by Steffan Heuer in McK Wissen 01 (2002)

1.3.3.2 Risk Reduction

A further important goal of location planning is to **minimize risks**. One way to reduce risks lies in spreading them through **diversification**. Having plants in various countries can balance out production outages in the event of political and social unrest, terrorist threats, or war, which mostly affect only one location. Diversification is also an advantage in dealing with everyday risks, such as currency fluctuations, which can threaten a company's survival.

If a company's costs are primarily incurred in the eurozone (because it has only one production location – Europe), and its sales are chiefly earned in the US dollar zone, a change in the exchange rate will have a direct impact on the company's profits. In the past five years alone, the euro/dollar exchange rate has seen swings of 40 percent. No manufacturer has that high a margin. Without countermeasures such as hedging,[16] this inevitably leads to periods of extreme losses.

Corporations can hedge on the financial markets. However, the more obvious course of action is to eliminate the imbalance via operational hedging – by aligning the currency structure of costs with the currency structure of sales. In the example above, a balance could be achieved by purchasing more parts in the dollar area, or by adding value (i.e., producing) there. Having similar currency structures eliminates the risk of exchange rate fluctuations. With global sales, global production is an obvious solution.

Companies can also use diversification to reduce sourcing risks, by using several suppliers. Depending on one supplier or even one production site only can cause severe problems and bring entire production networks to a standstill should the supplier face any number of challenges. This may happen for quality reasons or due to issues in the parts logistics. Another example of risk due to lack of operational hedging is the case of the Sony factory producing high-performance batteries for mobile phones. After a major fire in the plant in the mid-1990s, the plant ceased to supply the units for Sony and Siemens, se-

riously hampering sales in a critical phase of the exploding mobile handset market.

1.4 Current Production Networks of the Three ProNet Focus Industries

The three focal industries of this book – automotive engineering, electrical and electronics, and machine tool manufacturing – have widely differing cost structures (Figure 1.19). Almost 70 percent of the cost base of an automotive OEM are for materials and supplied parts. These items account for over 50 percent of costs in the electrical and electronics industry, but less than half in machine tool manufacturing. Since labor costs represent a relatively large cost factor in the latter, the cost pressure on in-house production is all the more intense. This explains why two-thirds of machine tool manufacturing companies – more than in the other two industries – produce abroad largely for cost reasons.

> **Most current production networks have a legacy structure, without any strategic planning**

Interestingly, both machine tool manufacturing and automotive engineering appear to be fairly successful in high-cost locations, as indicated by their high share of exports and especially their high net export surplus.[17] The numbers tell a different story in the electrical and electronics industry. Although this industry exports a significant share of its output, the share of high-tech products imported by high-cost countries is also high, sometimes making the import-export balance a zero-sum game. The position and behavior of HCC-based companies in the electrical and electronics industry are therefore quite

---

[16] *Hedging safeguards a transaction against risks such as exchange rate fluctuations or changes in raw materials prices. The person or company wishing to hedge a transaction enters into a second transaction linked with the underlying one. This normally takes the form of a forward transaction.*

[17] *The net export surplus shows how many percent more of the production value is produced than consumed in a particular country.*

different from in the automotive or machine tool sectors, as the following profiles show.

1.4.1 Automotive Industry

While automotive mass production has its origins in North America, all three US OEMs are struggling. Global markets are dominated by European and Japanese players.

The successful globalization strategies of automakers can be divided into two classes, based on the nature of their product orientation (loosely termed "premium" and "value"). Illustrating the premium product strategy, many European players have successfully leveraged their outstanding engineering skills to establish a strong position at the upper end of the market. They are realizing price premiums that allow

them to maintain an engineering and production footprint largely in high-cost countries (after making massive productivity improvements during the last industry downturn). German manufacturers are particularly strong despite the very high factor costs in their home base. Both German companies and the location of Germany itself have benefited from the strong growth of the premium segment in passenger cars.

Japanese and Korean players, on the other hand, are focusing more on the lower and middle market segments, with an emphasis on value for money. As a result, they have established global manufacturing footprints that rely increasingly on low-cost production sites.

Two highly successful companies in the automotive sector illustrate the divergent manufacturing footprint

---

Great structural differences in the industries analyzed

Fig. 1.19: Structural indicators, three selected industries

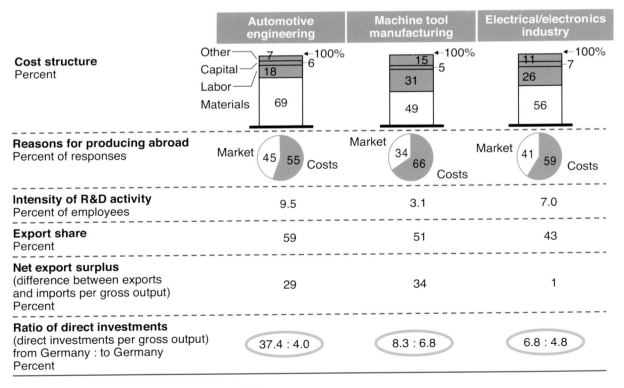

Source: Treier (2005), German Federal Office of Statistics (2005)

strategies well: BMW and Toyota. BMW pursues a strategy of producing its cars and critical large components such as engines mainly in high-cost locations with highly skilled labor forces. Most of its production is in Germany, Austria, and the UK (for the Mini), close to its engineering centers (Figure 1.20). The most recent addition was the new plant in Leipzig to manufacture some 3 Series cars and the new 1 Series. Beyond these, BMW has only two other manufacturing sites of note: a major plant in the US and a smaller one in South Africa. All of its other manufacturing operations are smaller joint ventures for SKD (semi-knocked down) in China and CKD (completely knocked down) car kits to gain eas-

ier access to markets such as Thailand, Malaysia, Russia, Egypt, and Indonesia.[18]

By contrast, Toyota pursues a much more international manufacturing footprint strategy. Strongly on track to become the largest global OEM, it is firmly established across all market segments, including the lower end. It also has tremendously high volumes – well over 9 million vehicles in 2007 – and growth. This positioning provides different imperatives for a broader production footprint geared to low costs (Figure 1.21).

Toyota still makes over 50 percent of its cars in its home base, Japan, where its plants are already the

BMW's vehicle assembly is particularly close to the market

Fig. 1.20: Production network of the BMW Group

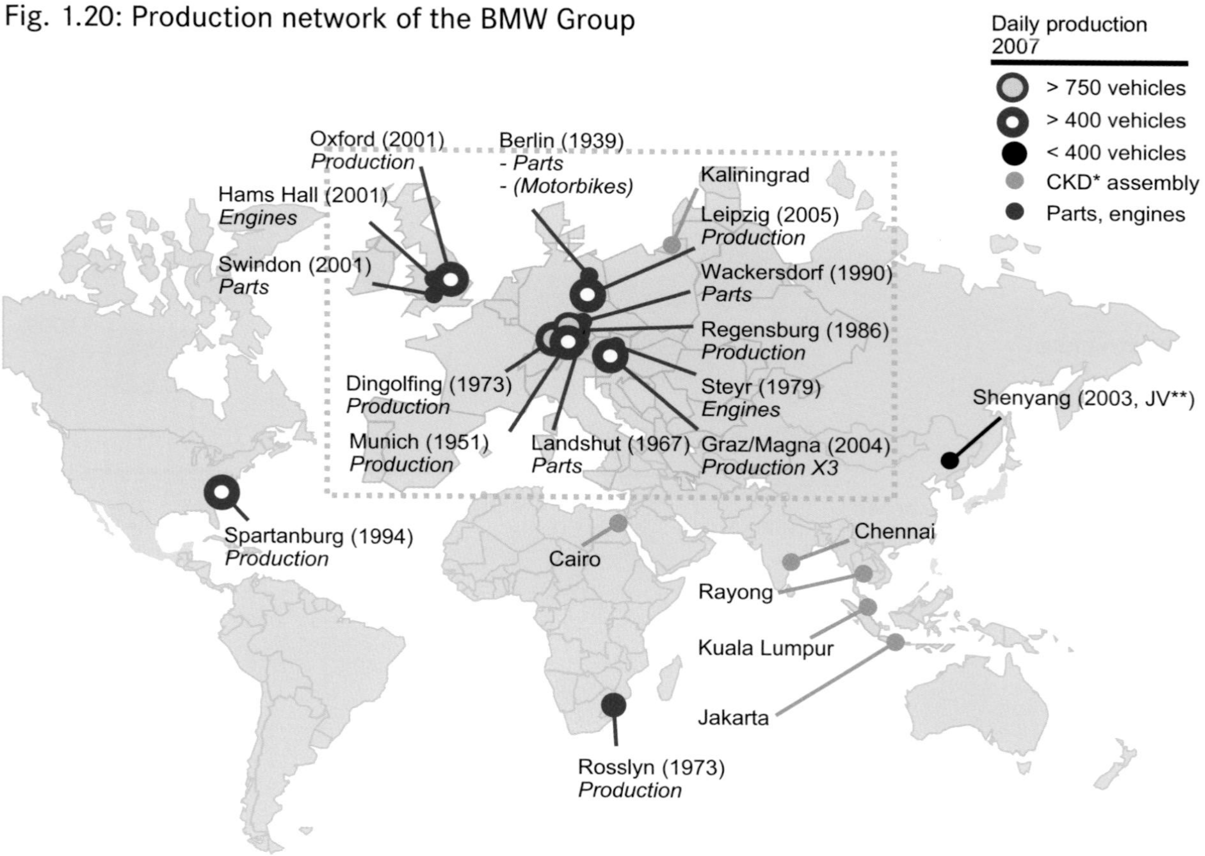

Daily production 2007

- ◯ > 750 vehicles
- ◉ > 400 vehicles
- ● < 400 vehicles
- ● CKD* assembly
- ● Parts, engines

Oxford (2001)
Production

Hams Hall (2001)
Engines

Swindon (2001)
Parts

Berlin (1939)
- Parts
- (Motorbikes)

Kaliningrad

Leipzig (2005)
Production

Wackersdorf (1990)
Parts

Regensburg (1986)
Production

Dingolfing (1973)
Production

Munich (1951)
Production

Landshut (1967)
Parts

Steyr (1979)
Engines

Graz/Magna (2004)
Production X3

Shenyang (2003, JV**)

Spartanburg (1994)
Production

Cairo

Chennai

Rayong

Kuala Lumpur

Jakarta

Rosslyn (1973)
Production

* CKD = completely knocked down
** Joint Venture with Billiance

Source: BMW

---

[18] BMW Web site.

most efficient in the world, outperforming competitor productivity by significant margins. Nonetheless, the increase in new capacity in the Toyota network to match its globally rising demand averaged around 3 percent p.a. in its Japanese plants and over 18 percent in its plants outside Japan. Its newest additions to the plant portfolio are sites in the Czech Republic, China, and Russia (planned for 2007 or 2008).[19]

To fend off the threats of lower-cost attackers, volume players based in HCCs need to rigorously improve performance along three fronts. The first and most immediate imperative is to optimize their manufacturing efficiency. Second is the ongoing drive to move additional manufacturing to low-cost locations, such as Eastern Europe. This is especially important for the growing low-cost car segment, as the success of Renault's Dacia Logan shows. Built in Romania, the Dacia Logan has plans to expand production to numerous other low-cost production sites to gain better access to new markets without compromising its low-cost position. The third imperative is to move the supply base to low-cost regions as well. Today, of the 15,000 components installed in cars made in Eastern Europe, 80 percent are imported from the West.[20] This imperative also extends to first-tier automotive suppliers, but is only truly beneficial if excess transport costs are consistently eliminated along the supply

Toyota's manufacturing footprint spans the globe

Fig. 1.21: Toyota's global presence
Thousand units p.a.

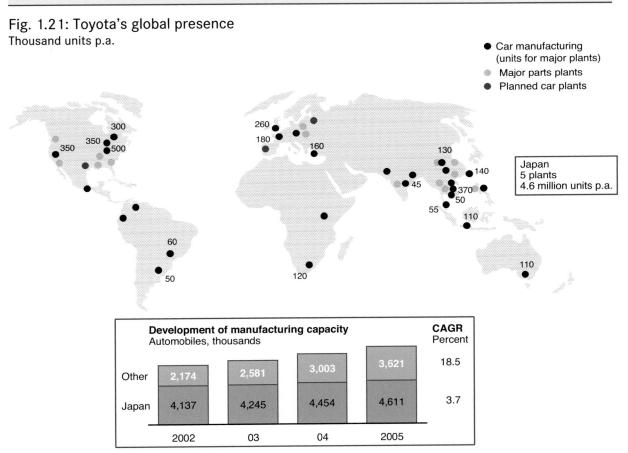

Source: Toyota Yearbook 2006

[19] *Toyota Yearbook 2006.*

[20] *Economist.com, "The big car problem," Feb 22, 2007.*

chain. In reality, this is often not the case. As the manager of an (automotive) electronics plant in Hungary reports: "The setup did not really make economic sense. We shipped 90 percent of the parts from Germany to Hungary, added about 5 percent value in manufacturing, and shipped them back to our customers in Western Europe for assembly in the vehicle."

1.4.2 Electrical and Electronics Industry

The picture in the electrical and electronics industry is very different – particularly in the growth segment of communications and consumer electronics. The share of electronics products from production in low-cost countries is growing by leaps and bounds. High-cost countries are irrevocably losing out in this field. However, to date, most of the action has concentrat-ed on the manufacture of simple components and the assembly of end products. The distribution of global value added in this industry clearly reveals that the loss of HCC market share has occurred mainly in Europe, primarily due to competition from LCCs (Figure 1.22). The US and Japan have more or less maintained their share of value added to date. However, in the future, all high-cost countries are expected to lose significant market shares in electrical and electronics production to low-cost competitors. While these trends highlight continuing country and regional differences, the absolute size of the industry has been growing significantly in all regions due to the strong growth of the global electronics market between 1980 and 2020, which is projected to see continuous growth rates of about 7 percent p.a. over this 40-year time period.

Although some regional differences in the Triad persist, electronics production is clearly moving to low-cost countries

Fig. 1.22: Value added in electronics* by region, 1980 - 2020 (estimated)
Percent

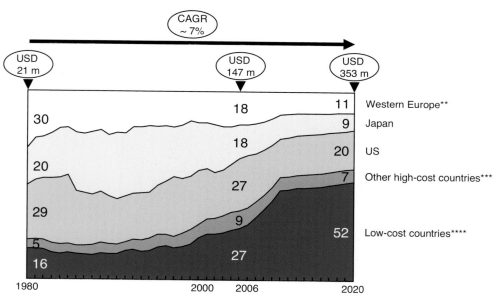

* ISIC 3832, 3833, 3839
** Austria, Belgium, Denmark, Finland, France, Germany, Greece, Ireland, Italy, Netherlands, Norway, Portugal, Spain, Sweden, Switzerland, UK
*** Australia, Canada, Israel, New Zealand, South Korea
**** Rest of world not mentioned elsewhere
Source: Global Insight (WIM, Aug 9, 2007)

The once thriving Western European electrical and electronics industry, for example, has only preserved a global presence worth mentioning in two areas: as suppliers to the automotive industry and in electricity generation and distribution. The only way to retain (and attract) the extremely capital-intensive manufacture of semiconductors in Western Europe has been high subsidies. However, European companies have been largely driven out in the greatest growth arena of the last decade, communications and consumer electronics. They are not succeeding in developing a premium segment through innovative and high-quality products, unlike the German automotive industry. Manufacturers find they cannot compensate for the comparatively poor cost structure and are losing market shares. This has already led to subcritical unit volume for mass-market products and a barely competitive cost position, and has frequently resulted in the sale or closure of factories.

Western Europe has lost significance in almost all fields of electronics, from communications and consumer electronics, office machinery, and computer segments to electrical equipment. Analysis shows that the Western European share of value added has fallen from 30 to less than 20 percent since 1980. In the field of consumer electronics, Western Europe has retained value added almost exclusively for goods with a low value density (e.g., washing machines and driers) that are very costly to ship over large distances – and even this sector has been facing increasing competition recently from locations in Eastern Europe and Turkey.

Once a manufacturing segment in this industry is gone, it is unlikely ever to return. The only opportunity for HCCs is to leverage technological breakthroughs that redefine the rules of the game of the industry for the coming one to two decades. Unfortunately, manufacturers in Germany and other HCCs missed out on the last round of such fundamental innovations in the electronics sector – whether the development of TFT and plasma television sets, DVD and hard drive recorders, or portable MP3 players – although a considerable share of the basic ingredients for these were developed in Germany. To re-

establish profitable production in high-cost locations, manufacturers in the electrical and electronics industry must find a way of minimizing the time to maturity for series production and full production ramp-up by intensifying the interaction between R&D and production.

1.4.3 Machine Tool Manufacturing

The situation in the machine tool industry is fundamentally different, though closer to that of the automotive industry. Measured against world production volume, the industry has grown nominally by an average of only 0.5 percent in the last 20 years – meaning that it has shrunk in real terms (Figure 1.23). Also, the industry is predominantly characterized by small and medium-size enterprises (SMEs). The average company employs around 160 people – compared with 863 in automotive engineering.

As the ProNet survey reveals, many companies attempted to move some of their activities to LCCs when they recognized the cost pressure and competition from emerging players. However, due to a lack of scale and limited management experience and bandwidth, these efforts were often unsuccessful. Many companies eventually retreated from their ventures abroad and refocused on their activities at home instead. In many cases, this retrenchment appears to have been successful. The market share of leading German (high-cost) manufacturers has risen in the past two decades from 17 to 25 percent[21], and their sales volume has remained about constant after adjustment for inflation. The unique value proposition of these players is their engineering expertise and mature process chain throughout the entire manufacturing process. Their operations are backed up by global service concepts, and they have succeeded in tapping attractive markets.

However, a second look reveals that this success is closely linked to that of the German automakers. In 2003, more than half of the machine tools produced

---

[21] *Excluding parts and accessories.*

in Germany went to the automotive industry and its suppliers. Machine tool manufacturers' sales figures correlate closely with automotive investment activities rather than reflecting structural strength and competitive advantage based on superior operational performance.

Consequently, a fast-growing competitor is increasingly threatening the position of high-cost manufacturers: China – now the world's fourth-largest producer of machine tools. Growth rates of over 20 percent per annum suggest that its role will continue to expand and pose a serious threat to the viability of incumbents.

Trends in the machine tool industry give indications of future development in other sectors

The reason for this rapid development, apart from manufacturing costs, is primarily the booming Chinese market. China contributes 20 percent to world demand, making it the biggest market for machine tools. This dominance of the Chinese market, which is the leader in other industries "only" in terms of growth rates, is explained by a peculiarity of the capital goods industry: Investments are always the forerunners of future production. As a result, what is happening in machine tool manufacturing previews a development that will follow in other industries. Taiwanese and Indian machine tool manufacturers are also profiting from high domestic demand, and expanding their offerings in the standard segment.

Overall, the industry situation is problematic. Lack of growth in the market as a whole makes it difficult to simply expand the network into other countries,

Global production of machine tools is stagnating; however, Germany's market share is increasing slowly but steadily

Fig. 1.23: Nominal global production of machine tools
EUR billions, excluding parts/accessories

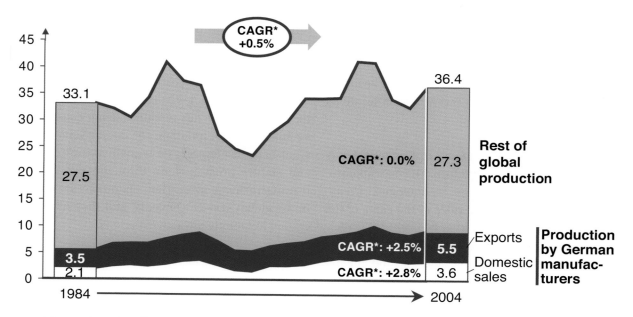

* Compound annual growth rate

Source: German Machine Tool Manufacturers' Association (2004)

since capacity utilization at existing factories would shrink as a result. If manufacturers maintain existing structures, however, they will become exposed to new competitors from emerging nations. The only path to long-term success for European manufacturers is well-planned redesign of their production networks – especially in the standard segment. Otherwise, they risk following in the footsteps of their former peers in the electronics sector.

* * *

There are many reasons for globalizing production. Most companies are aware of the potential advantages. But how familiar are they with the challenges and hurdles? Do they know how to find the right location, minimize risks, and integrate new locations into existing structures? The relationships are complex, and the answers differ widely depending on the company.

The ProNet survey showed that many companies fall down on the task (Figure 1.24). More than half achieve cost savings of no more than 10 percent with a new location. The reasons are numerous, spanning a lack of resources or experience in implementation, hesitant and incomplete implementation, and excessively low expectations about the savings potential.

Around 20 percent of the companies we surveyed, however, emerged as truly successful globalizers. They have managed to strike the right balance between high aspirations and realistic planning of available skills and resources. Analyzing the differences between what those 20 percent did and the other 80 percent provided us with invaluable insights into patterns that appear to yield success and pitfalls to avoid. In the remaining chapters, we will describe these findings through every area of the value chain, highlighting analyses and decisions that have helped companies to get it "right first time." Because companies only have one chance with a move as radical as footprint redesign.

Most companies capture less than 10 percent savings at new production sites

Fig. 1.24: Production cost savings relative to lead plant
Percent*

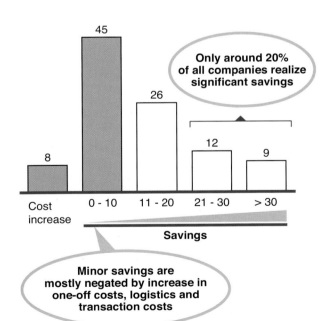

* Cost position of the 77 plants opened by survey participants in the previous five years across all countries

Source: McKinsey/PTW (ProNet analysis)

Further reading

Davis, I.: "The Biggest Contract" in *The Economist*, May 28, 2005, London, pp. 87 - 89.

Davis, I.: "Plot Your Course For the New World" in *The Financial Times*, January 13, 2006.

Drucker, P. F.: *The Essential Drucker*. Oxford: Butterworth-Heinemann, 2001.

Porter, M.: *The Competitive Advantage of Nations*. New York: Free Press, 1998.

Reich, R.: *Die neue Weltwirtschaft*. Frankfurt am Main: Fischer, 1996.

Welge, M. W. and D. Holtbrügge: *Internationales Management - Theorien, Funktionen, Fallstudien*. 3rd Edition. Stuttgart: Schäffer-Poeschel, 2003.

TOBIAS MEYER

2 Selection Criteria: Assessing Relevant Trends and Indicators

Summary

Centers of economic activity are shifting profoundly, not just globally but also within regions. Manufacturing footprints are transforming even more dramatically – and the story is not simply the transition to Asia. To make the right decisions on where to invest, it is critical for decision makers to understand what criteria matter the most, and how locations are likely to change.

The ProNet survey showed that reducing costs and tapping new markets abroad are the two main motives for globalizing production. The two characteristics that are consistently most relevant when determining the attractiveness of a production location are labor costs and market proximity. The availability of qualified workers ranks third but is growing in importance. This ranking, however, is an average that runs across multiple products and production processes. Decision makers should be aware that the relevance of the parameters we will be discussing in this chapter depends on the specific production requirements being considered. Three categories of requirements determine what matter most: the relative importance of production input factors (labor intensity, for instance), product characteristics (e.g., value density), and the geographical scope of site selection (are you choosing between continents or neighboring industrial zones?). Another perspective to remember is that while a country may not be attractive for producing parts or finished products, it may still be appealing as a market or location for corporate functions such as back-office operations.

Successful global companies manage to achieve excellent productivity and quality virtually everywhere in the world. They build on their experience, adapt their organizations to the specific circumstances they face, select appropriate manufacturing techniques, and use employment strategies tailored to the local environment. Above all, they are proactive, foreseeing the need to address these aspects early on, rather than waiting until the issues arise.

2.1 The Relevance of Selection Criteria for Global Production Locations

Multiple factors influence which location is best for producing a specific product. The decision is particularly complex when setting up facilities far from the home base in countries with very different economies and cultures.

A structured set of facts on the relevant countries is therefore vital – a road map that will help decision makers identify the key factors for or against a specific location. The art is to distinguish the information that is really crucial. Whether labor costs for a semi-skilled worker in China are currently EUR 1.0 or EUR 1.2 per hour is largely irrelevant for most industrial multinationals, despite the time spent debating these details. Much more important is whether qualified local staff can be attracted, trained, and retained to drive high-quality, productive manufacturing at the new plant.

Insight into the trends underlying the data is also crucial. How are each of the indicators developing – whether the market, factor costs, logistics costs, exchange rates, or numerous others? This chapter describes what to look for in these selection criteria and how they differ worldwide.

2.1.1 Interaction Between Location Parameters and Process Parameters

When evaluating individual production locations as well as entire networks, it is important to distinguish between location parameters and the parameters of the manufacturing process (Figure 2.1). This distinction is essential for understanding both the economics involved and the operational hurdles and requirements. Process parameters are used to weight the relevance of location parameters. If the energy intensity of a manufacturing process is low, for example, the price of energy is of little relevance as a location attribute. This means no hard and fast judgments can ever be made on how attractive a particular location is for production. It needs to be rated by

Key questions, Chapter 2

- What location characteristics and process requirements are most relevant for selecting global production locations?

- What parameters are particularly important for specific industries and regions?

- What is the current status/what trends are discernible for the following parameters?

  □ Markets: Which trends are fundamentally important, and how are they likely to change over time?

  □ Factor costs: Is the labor cost gap closing? How quickly could this happen?

  □ Productivity: Are factor cost advantages in low-cost countries negated by low productivity and poor quality?

  □ Manufacturing technology: What equipment and processes should be used? What are the implications for the optimal scale and scope of plants?

  □ Logistics: How are freight charges evolving? What is the impact of longer distances/transport times on the supply chain?

  □ External factors: What impact do taxes, subsidies, currency exchange rates, product piracy, and other risks have on the choice of locations?

  □ Migration: What expenses can be expected when setting up a factory abroad and possibly restructuring existing locations?

the requirements for a specific manufacturing step of a specific product. Oversimplifying selection entails risks. Labor costs will be the dominant criterion if the goods are simple, standard products requiring labor-intensive manufacturing. This is not the case for high-tech products with numerous variants and capital-intensive production equipment. Analyzing and selecting locations by product line and major production process (e.g., molding, processing, pre-assembly, final assembly) is therefore usually a worthwhile investment.

Location parameters reflect the characteristics of a geographic location and influence the attractiveness of a site for a specific process step for a product. Quantitative location parameters include factor costs, geographic position (determining shipping distances and therefore to a large extent transportation costs), or customs duties and taxes.

Product and production-related factors (**process parameters**) describe the manufacturing process and the characteristics of the product. Quantitative process parameters include the input factor volumes needed to manufacture a product, such as the amount of labor, energy, capital, and raw materials. The input factor volumes depend on the product characteristics and manufacturing technology. They can often be varied by, for example, altering the level of automation in production and substituting capital with labor or vice versa. Input factor volumes, prices, and other quantitative process factors have a direct impact on total production and logistics costs. Qualitative process parameters have an indirect impact on costs and reveal further location-related requirements – such as a guarantee of uninterrupted supplies or legal safeguards.

A concrete example shows how important it is to also factor in **trends**. Evaluations can change over

---

Both location and process parameters determine the optimum location

Fig. 2.1: Factors influencing site selection – location and process parameters

NOT COMPREHENSIVE

Location parameters

Factor costs	• Labor costs (by skill level)
	• Cost of capital
	• Cost of materials (parts, raw materials, energy, etc.)
Produc-tivity	• Labor productivity
	• Capital productivity
Other quantitative factors	• Distance from relevant markets
	• Potential restructuring and closure costs
	• Freight rates
Qualitative factors	• Availability of land and infra-structure, rights of ownership
	• Legal safeguards, protection of intellectual property
	• Regulations, work safety, environmental guidelines, etc.

Process parameters
(product- and production-related factors)

Input factor volumes	• Labor time (by skill level)
	• Nominal capital employed (plant and equipment)
	• Purchased parts/raw materials
	• Parts (made inhouse)
	• Space requirements (land and buildings)
Other quantitative factors	• Volume and weight
	• Delivery time requirements
	• Maintenance requirements/costs
Qualitative require-ments	• Process complexity
	• Know-how intensity and sensitivity/patents
	• Environmental requirements

Source: McKinsey

time, for instance, due to an alteration in qualitative parameters. In the 1990s, Sony shifted the production of digital cameras and camcorders from Japan to China to realize cost savings for these products, which at that time were manufactured laboriously in relatively small volumes. In 2002, production was moved back to Japan. What had changed? Product life cycles for digital cameras had shortened dramatically, and they had transformed from purely functional to high-fashion products. This made it essential to have production close to the supplier base and the main customer market. Higher volumes allowed for greater process automation. In Sony's view, China in 2002 was not mature enough, whether as a supply base or a consumer market.[1] These characteristics could of course change, improving China's attractiveness for high-tech consumer electronics OEMs.

It can be a great challenge to give adequate consideration to **qualitative criteria**, such as the protection of know-how. Comparing individual location parameters or weighting and compressing them into indices as proposed in various publications is not very meaningful. An aggregated index provides neither insights into the total production costs of a product nor the operational requirements that need to be in place to get production activities off the ground.

For some qualitative parameters, companies can and should identify quantitative relationships, such as the risk of disruptions due to political or social turmoil. An adjusted rate for the cost of capital can, for instance, take into account the expected loss of property. Security services and specialized consultancies provide country ratings and assess the likelihood of relevant events. These qualitative factors can next be quantified via mathematical correlates and then folded into an equation that includes other quantitative factors, such as the materials costs and labor intensity of a production process step.

Quantification does not, however, make sense for all qualitative factors. While quantitative factors can be weighed up against each other to a certain extent, as a higher value for one may compensate for a lower value of another, this cannot be done with all qual-

itative factors in the decision-making process. A better infrastructure cannot make up for deficiencies in legal safeguards. Both need to meet specific minimum standards, and need to be gateways for "go/no-go" decisions in their own right.

Applying minimum requirements to key location characteristics is very useful for preselecting countries. Qualitative attributes should at the very least be listed when potential sites are compared to create transparency for management. A record should be made of whether and how these parameters were included in the selection process, and whether they were incorporated into the quantitative assessment (such as total production and logistics costs of the production network configuration).

2.1.2 How Varying Perspectives Affect the Importance of Different Location Parameters

The relevance of individual location-related parameters varies for different products and production process steps, as these have diverse cost structures and levels of complexity. The significance of location-related parameters for the decision-making process may also vary by geographic region and industry.

With some location-related attributes, the company itself plays an influential role.[2] The attributes are largely but not entirely defined by the external environment. **Labor costs** – the key location-related parameter at a country level[3] – are strongly dependent on the location, but still not imposed in most

---

[1] Cf. Jiang (2003), p. 26: "[...] the latest camera design on the Chinese market is typically six months behind products on the Japanese and US markets. Therefore the manufacturer would not gain any useful information for the supply-chain-wide forecasting system by producing [...] in China."

[2] Cf. Welge (2003), p. 90.

[3] Cf. Ernst & Young (2004), p. 15, and Hardock (2000), p. 180: ("personnel costs" are rated by far the most important criterion ahead of "skills and motivation," "corporate taxation," and "labor productivity").

regions of the world. Compensation levels vary depending on the specific qualification a company requires. Similarly, a company is largely free to define its own benefits and incentives scheme. Companies can also play a role in shaping the market price for labor locally. Particularly companies with large facilities in rural, sparsely populated areas have to recognize that their demand for workers can have a dramatic impact on the local labor market.

Dealing with such **semi-external factors** as labor costs – partially determined by the company itself – can be fairly complex. To attract staff with higher qualifications, for example, the company may have to pay more. This in turn might allow the use of more complex, but also more efficient, production technology. This could increase labor productivity beyond the rise in labor costs – a trade-off worth exploring. Simplifying existing production processes and lowering qualification requirements while maintaining productivity can also work. Companies may opt for this approach especially in markets with a wide spread between the cost of skilled and unskilled labor.

> **Successful companies see global production primarily as an opportunity to reduce costs via lower labor expenses and increase sales with greater market proximity**

Likewise, productivity and quality are generally more dependent on the company than the location. When companies are gauging location factors, they should also estimate their own capability to influence them.

Survey results show that successful players with extensive experience in setting up new plants (also in developing and newly industrialized countries) are more likely to base their decision on costs. They are confident of their ability to train staff and create the local business environment they need. Labor costs are by far the most important selection criterion for these winners in the globalization arena. They also assign customer requirements very high relevance (Figure 2.2). Another sign of their stronger focus on

the opportunities and cost position of their production networks is the greater consideration they give to transportation costs, customs duties, taxes, and the possibility of subsidies.

The picture is very different for companies that are new to globalization or have suffered setbacks when starting up locations abroad. These are more likely to be guided by the risks than the opportunities. As a result, they see the availability of trained employees as a more important parameter.

> **Followers mainly see the risks: they focus more on the availability of skilled staff to ensure the feasibility of their production abroad**

How companies rate the importance of different location parameters is determined not just by the target country, but also by their country of origin. German and Japanese companies, with their home bases in countries with very high labor costs, focus more intensely on labor costs when setting up a production facility in the United States than their American counterparts do. Conversely, Americans attribute greater significance to transportation costs[4] – even when they invest in European countries, which are geographically smaller than the US. This shows that decision makers find it difficult to shed **behavioral patterns** that were successful in developing their home markets but may not be transferable to other countries.

> **The company's country of origin influences which location parameters are considered relevant**

---

[4] Cf. Tong (1980): survey on the relevance of location criteria for international companies when investing in the US (254 respondents). Transportation: 3.70 (out of max. 5); staff motivation: 3.67; room for expansion: 3.65; proximity to the market: 3.65; etc. The importance of the criteria in each case depends on the investor (e.g., share of foreign ownership, corporate center).

Another example of this gut response is that German industrial companies[5] make the availability of qualified staff their key indicator. They feel this is Germany's major advantage and automatically graft it to the top of their list of requirements abroad. Ideally, of course, sound analysis should precede that judgment. Management could also seek to make do with fewer qualified staff to manufacture simpler products more cost-effectively abroad.

The **type of product** – whether intermediate or finished – also affects how the parameters are rated. Companies planning to manufacture or source parts and intermediate products abroad will assign little importance to customer proximity, since these products will anyway go to other factories for further processing. In this case the target location for the finished product only has an indirect impact on the optimal site for manufacturing the parts. For finished products, the manufacturing location has a direct impact on the delivery lead times and flexibility the company can provide to its customers.

The parameter rating will also vary depending on the **geographical scope of the analysis** and the stage in the selection process. The importance of the criteria will differ depending on whether you are deciding between two continents (America or Asia, for instance), or looking at different countries (after having decided on the continent) – such as whether to choose Korea,

---

Leaders put more emphasis on labor costs, customer-specific requirements and taxation/subsidies

Fig. 2.2: Relevance of location criteria on a country level

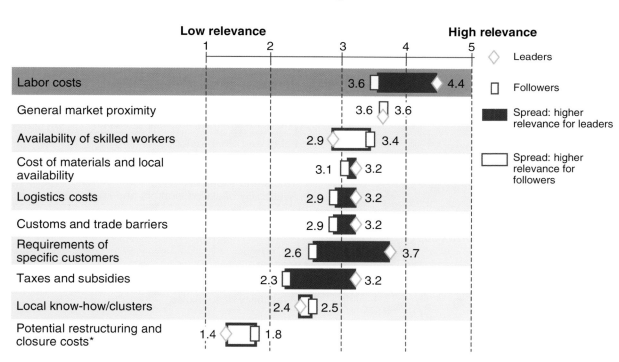

* Further criteria (classification): political and social stability (3.0/2.8), economies of scale (2.7/3.0), exchange rate effects (2.4/2.7), risks/entry barriers (2.6/2.4), land ownership/infrastructure (2.2/2.3)

Questions A3 and E7: "What are the most relevant criteria for selecting a specific country/region as a production location?"

Source: McKinsey/PTW (ProNet analysis)

---

[5] Cf. Produktion (2004): survey of 93 industrial companies.

China, or Taiwan (Figure 2.3). Taxes and customs duties are often very relevant criteria on a country level but less important at the continental or local level (with some exceptions such as the US or Switzerland, which have local tax schemes). Equally, what applies at a continental or national level is often irrelevant at a local level. In selecting a suitable suburb, industrial zone, or specific plot, differences in distance to customers and suppliers play only a minor role. Extra acreage for expansion or environmental restrictions may well be much more crucial at that stage.

* * *

A number of parameters are of general relevance on a country and regional level. Of the six categories we describe, the first five fall into a group we call the **"static perspective"** (Figure 2.4). This is the analysis of operating costs, producing an evaluation of the long-term total landed costs of a production network. These are the costs of delivering the products to the market, including materials costs, production costs, transportation, costs of carrying inventory, and customs duties. This view is useful in determining the strategic benefits of redesigning a network. The ideal position of an optimally configured network is also the best position that competitors with comparable products and similar production technology can achieve. This greenfield analysis helps assess the competitive threat that the incumbent might face, particularly from new entrants with a home base in low-cost countries.

A **dynamic perspective** includes the costs that apply to the transition phase, i.e., the setup and closure of facilities involved in the relocation. The last of the parameter categories contains these "transition financials," made up of the three parameters: investments (capex), ramp-up, and restructuring.

Relevance of the different parameters changes depending on level of analysis

---

Fig. 2.3: Scope of analysis and relevance of location parameters   ■ Focus of Chapters 2 to 4

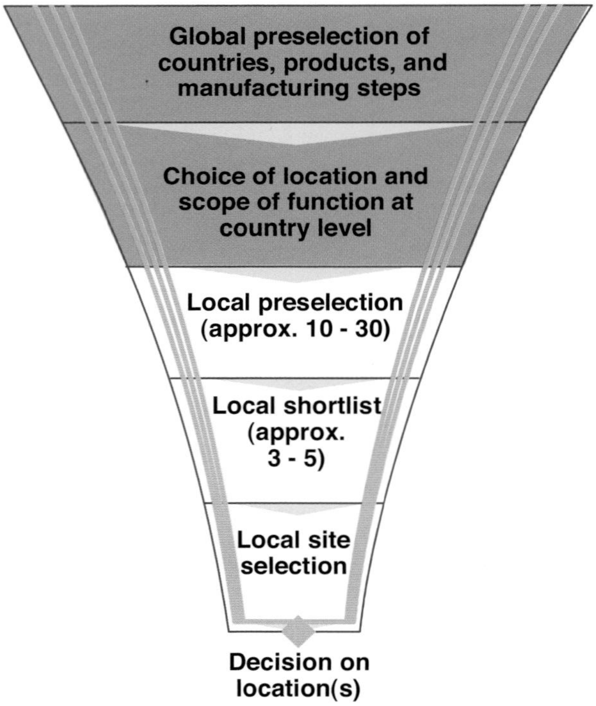

Most relevant parameters

Funnel stage	Most relevant parameters
Global preselection of countries, products, and manufacturing steps	• Political stability/market access • Geographic position/transportation costs and times • Minimum requirements concerning the market (size and maturity), infrastructure, or costs
Choice of location and scope of function at country level	• Labor and other factor costs • Size and growth of market, customer requirements • Logistics costs (incl. customs duties) • Taxes and subsidies • Availability of skilled workers and know-how
Local preselection (approx. 10 - 30)	• Local labor costs, staff availability, and qualifications • Geographic position and transport links
Local shortlist (approx. 3 - 5)	• Local labor costs, staff availability, and qualifications • Prices of land and buildings • Availability of subsidies
Local site selection	• (Detailed comparative analysis based on all relevant factors)

Decision on location(s)

Source: McKinsey

The following sections describe what to look out for in these parameters – both the underlying structures and long-term trends. We consciously distinguish between low-cost countries and developing/newly industrialized countries. The economic development of countries in Eastern Europe has been held back particularly due to the influence of Communism. As a result, labor costs in most of these countries are still low, but their economic development is otherwise advanced. They do not present the same hurdles to multinational companies as developing and newly industrialized economies. This is particularly the case with developing countries in Asia, where most unskilled workers have little or no experience with Western lifestyles and work methods. We believe that insufficient acknowledgment of the fundamental historical and cultural differences between countries is a major source of the frustration that Western companies have experienced when setting up operations in Asia, particularly China. The reverse side of this coin would also explain Western companies' relatively positive view of locations in Eastern Europe.

2.2 Markets and Market Development

The development of demand in geographical markets is a crucial driver in the globalization of production. Companies want to directly benefit from market prospects abroad, open up opportunities to increase sales, and increase their margins. They are therefore increasingly opting for a global structure, seeing the entire world (or at least large tracts of it) as their playing field. It is important, however, to assess underlying market fundamentals in order to successfully plan foreign investments and market entry strategies.

> **The production footprint strategy supports one goal: to have the best possible access to the relevant sources of supply and demand at the right time**

Input parameters to access the economic attractiveness and feasibility of production network design

Fig. 2.4: Relevant parameters for the optimization model

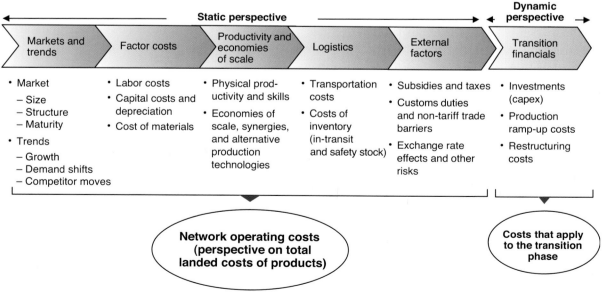

Source: McKinsey/PTW (ProNet analysis)

The development of entire industry sectors and individual product segments can be estimated by analyzing interdependencies and comparing countries. Such comparisons can produce forecasts of astonishingly high accuracy. Companies should use structural analyses of this kind when planning to enter new markets at the right time and with the right products, expand their involvement, or withdraw from a market.

2.2.1 Market Growth – Industrialization and Transition to a Service-Based Economy

The industrialization of North America, Western Europe, and Japan encompassed an economic area with approximately 500 million inhabitants, and a time frame of roughly 140 years (from around 1830 to 1970). The 25 years after 1970 saw the development of relatively stable (and partially oligopolistic) markets. Since the mid 1990s, a **new dynamic** has been emerging due to further technological innovation and surging growth in China, India, and parts of Southeast Asia. Their huge population – around 3 billion inhabitants in total – makes these countries appear particularly attractive as markets for industrial products, especially goods that long ago reached saturation in more developed economies. Paradoxically, the markets are so attractive in some areas that the intensive pressure on MNCs to enter and develop the market is leading to highly competitive market structures. As a result, virtually none of the companies involved earn adequate profits or generate a positive net cash flow in the segments that are open for investment and presumed the most attractive, at least in the short and mid-term. Particularly in capital-intensive industries that require ongoing investment to support growth, from airlines to semiconductors, local companies and subsidiaries struggle to become financially viable, independent of constant support and cash injections.

The rapid growth of developing and newly industrialized countries needs to be put into perspective. Growth rates are high, but their baseline is often very low. Absolute figures tell a different story. Highly developed industrialized nations are still seeing considerable growth in absolute terms, despite moderate growth rates – due to much higher baselines. The US economy, for example, was still nearly six times as large as China's in 2006, and absolute GDP growth will remain greater in the US than in China for another 15 years or so. So the US is still a very attractive market – though less so for manufacturing than service companies.

The reason why Asia – and particularly China – rightfully attracts so much interest from industrial companies can be seen by examining global economic trends of the last 10 years and extrapolating growth trends through to 2015 (Figure 2.5). Increasing industrial production is a major contributor to the high growth rates of developing economies in Asia. For China, strong exports have been stimulating industrial growth for quite some time. However, this export-driven economy is also demonstrating ever heavier demand for consumer durables such as washing machines. Sales of these goods in the US, Western Europe, and Japan have long since reached saturation point. Chinese companies have replaced local manufacturers in those countries. Often exports kick-started industrial production, but are now increasingly being substituted by local consumption as a key driver of growth in this segment. Only few economies, especially India with its historical focus on import substitution (rather than exports), have experienced a more consumption-driven development right from the start.

> **Developing and newly industrialized countries are experiencing strong growth in the industrial sector, while highly developed economies are expanding almost exclusively in the service sector**

In contrast to this, growth in highly developed economies is mainly in the service sector. This means manufacturers have less to gain from economic growth in high-cost countries, and have to achieve their growth objectives via diversification, predatory competition, or expansion into foreign markets where demand for industrial goods is still growing fast.

The stronger focus on the service sector in highly developed industrial countries is accompanied by much lower investment rates. Gross investment as a share of GDP in China is above 40 percent, at a par with private consumption. In contrast, the investment rate in Germany, the US, and other industrial countries is only around 18 percent; private consumption dominates, particularly for services.

GDP composition can provide numerous insights for assessing the market potential of economies for specific product categories, while comparison with other countries often reveals fundamental trends and inflection points.

2.2.2 Market Growth in Developing and Newly Industrialized Countries – General Trends and Regional Specifics

On a highly aggregated level, all countries pass through a similar development pattern while manufacturing from mainly agricultural economies to highly developed industrialized ones. The development pattern of overall market volumes for product categories is often very similar. It is therefore comparatively easy to draw up estimates. The demand curve for goods, particularly consumer durables, is fairly similar for economies all over the world. As a certain GDP per capita level and inflection point is reached, demand grows very fast until saturation is attained. This is often defined by a particular number of units per capita. Depending on whether the products are subject to substitution by an item of higher value that has the same basic functionality (for example, a car for a motorbike), demand either remains stable or deteriorates after the saturation phase.

Countries' appetite for specific goods always develops along a very similar curve

Depending on the price and relative utility of the goods, the inflection point for a rapid increase in

---

China has the largest market growth in industrial products

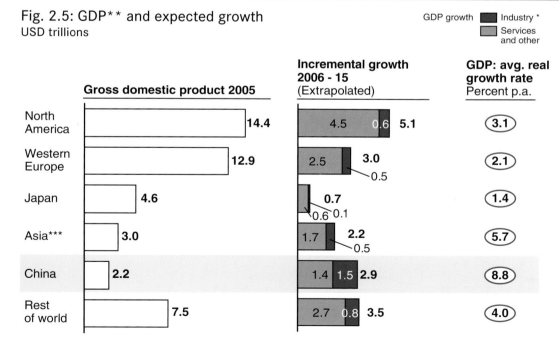

Fig. 2.5: GDP** and expected growth
USD trillions

GDP growth ■ Industry *
□ Services and other

	Gross domestic product 2005	Incremental growth 2006 - 15 (Extrapolated)	GDP: avg. real growth rate Percent p.a.
North America	14.4	4.5 / 0.6 / 5.1	3.1
Western Europe	12.9	2.5 / 3.0 / 0.5	2.1
Japan	4.6	0.6 / 0.1 / 0.7	1.4
Asia***	3.0	1.7 / 2.2 / 0.5	5.7
China	2.2	1.4 / 1.5 / 2.9	8.8
Rest of world	7.5	2.7 / 0.8 / 3.5	4.0

* Manufacturing industry, excl. construction industry
** In 2005 prices and exchange rates
*** Asia, excl. China, Japan, (Russia), and Middle East

Source: Global Insight (WIM 2006)

their demand can be very early (e.g., TV sets and bicycles) or later on (e.g., automobiles) in the development of the economy.

When GDP per capita reaches around USD 150, the demand for TV sets will pick up. Saturation for traditional, tube-based sets is reached at a GDP level of around USD 1,000 per capita. From around USD 400 GDP per capita[6] upwards, motorcycles replace non-motorized transport (bicycles, for example). From USD 800 to USD 1,500 of GDP per capita, automobile purchases increase, and these only exhibit saturation effects from around USD 20,000 per capita GDP and upwards – in other words, at a fairly late stage of development.

Besides the absolute level of GDP per capita, other factors, particularly the distribution of income, influence demand trends. The following pattern ap-

plies for many consumer durables: first, demand for the goods from the next stage in development is triggered – at a relatively low level – among segments of the upper and upper-middle classes. Demand then soars as the product becomes established in the population at large as a standard household appliance, for individual mobility or for entertainment. The strong growth phase ends when general saturation sets in, at which point few new sales are made. Ultimately, even the need for replacement drops, due to substitution by more highly developed products. Market decline is only avoided by products from the highest level of development, such as the automobile, which represents the most highly developed product for individual mobility.

Managers have tended to neglect this **maturity curve**, despite the fact that it is relatively straightforward. Take the motorcycle industry: The majority of com-

China is 4 to 5 years ahead of India in mobile phone communications

Fig. 2.6: Mobile teledensity in China and India, 1996 - 2006*

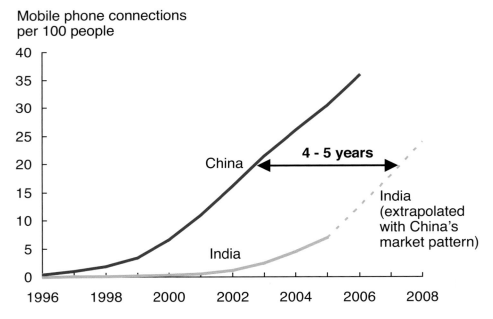

* 2006 figures estimated based on June 30 subscription figures

Source: ITU World Telecom Indicators (1996 - 2004), TRAI (India 2005), Yankee Group (China 2005), Asia Times, Aug 8 (China 2006)

[6] In 2002 prices.

panies that dominated the Western European and the US market in the 1940s and 1950s have gone out of business – instead of diversifying into automobiles and expanding into less developed markets abroad (such as Honda did).

An example that illustrates the concept well and can also be applied to other countries in a similar situation is the Chinese two-wheeler market. Sales of bicycles in China have barely grown since the end of the 1980s. Production reached saturation point at around 40 million units p.a. (40 percent of global production). On average, each household owns around 1.8 bicycles. Domestic demand has exhibited a slight downward trend over the past years. In contrast, motorcycle production since 1990 rose from fewer than 1 million units per year to an estimated 15 million units in 2003. While production and also demand have still been growing since then (to some 19 million units in 2006), growth rates have been gradually declining. In contrast, the number of automobiles per household – still very low in 2004, averaging approximately 0.04 units (the comparable US figure is around 1.9 units) – has been growing ever faster. China has only just reached the inflection point from which automobile sales could take off.

Two other examples illustrate the insights that can be gained by comparing different countries and economies.

First, the mobile phone markets in India and China have strong parallels. In India, conditions are similar to those in China four to five years ago, such as the level of mobile phone subscription fees and other costs in relation to average income. Because the country's fundamentals are similar, the Indian mobile phone market can be expected to grow in a pattern similar to that seen in the Chinese market (Figure 2.6).

The second example is based on the relationship between a country's steel consumption and its GDP. During the period of "classic industrialization" in Europe, the US, and South Korea, steel consumption rose sharply as GDP increased. Consumption reached its peak at a GDP of some USD 15,000 per capita. With further GDP growth, per capita steel consumption would decline and reach stability at a much lower level. This pattern has significant implications for producers. After reaching peak consumption, much of the demand can be covered by so-called "minimills," which mostly process scrap and thus return recycled steel to the materials cycle. Developing countries today – particularly China – are currently experiencing a strong rise in steel consumption because of their growing prosperity (GDP per capita). Other countries will follow. This includes India, which has a per capita consumption that is only one-eighth of China's. Yet with the potential substitution of steel by other materials (e.g., plastics and aluminum) and the earlier rise of the service sector's contribution to total GDP, it is doubtful that they will reach the temporary peak consumption of "classic industrialized" countries. Some basic fundamentals have changed over the 30-plus year period that lies between the development of countries such as Singapore, South Korea, and Taiwan (previously), and where India is today (Figure 2.7).

The question of the **right timing for market entry** cannot be answered in general terms. For manufacturing companies, this question always has two aspects. It requires a commercial perspective on the location as a market for the companies' products, as well as an operations perspective on the country as a potential location for the production and sourcing of parts and services. The commercial perspective needs to take into account the size of the market and its potential. It also has to consider the expected market structure and market conduct, which will ultimately drive price levels in the respective marketplace. The factors that matter most from an operations perspective were discussed in section 2.1. How strongly these two aspects are interlinked depends heavily on the characteristics of the product, particularly on economies of scale in production, value density (monetary value by weight), and order specificity of the products (make-to-order vs. make-to-stock). The size and capabilities of the company as a whole also play a role.

Generally speaking, very early market entry and local production in emerging markets appear to be steps associated with a high level of risk but also high average returns. Building businesses up from a small scale in emerging markets can generate significant value – high even for large MNCs. After the inflection point of an industry is reached, the valuation of local companies often attains extreme highs in expectation of continued high growth rates. As competition gets stiffer, early entrants have already established strong brands, and may have built up a capable, loyal workforce. Market entry at that point through an acquisition becomes costly, and late entry via organic growth requires a set of capabilities and resources that only few companies have.

The lesson: If the business climate and regulatory environment are conducive to entry and the company can sustain setbacks, it should enter a market ear-

ly, rather than waiting. Most manufacturers should aspire to globalize ahead of their competitors, making sure they are among the first to enter a country and set up production. In Asia, for instance, they should not only still have China on their radar screen, but also at least one large ASEAN country (Indonesia, Thailand, or Vietnam) and India. These countries are currently creating the conditions for growth and making themselves attractive as production locations by eliminating trade barriers, fighting corruption, expanding their infrastructures, and introducing less restrictive capital market and investment conditions.

However, **early market presence** does not automatically ensure long-term success. Volkswagen's recent loss of market share in China is one example. The situation in the Chinese automobile market illustrates how difficult it is to find the right strategic positioning, particularly in apparently attractive but still par-

The relationship between the trend in steel consumption and GDP is similar in many countries

Fig. 2.7: Intensity of consumption of finished steel per country, 1980 – 2005

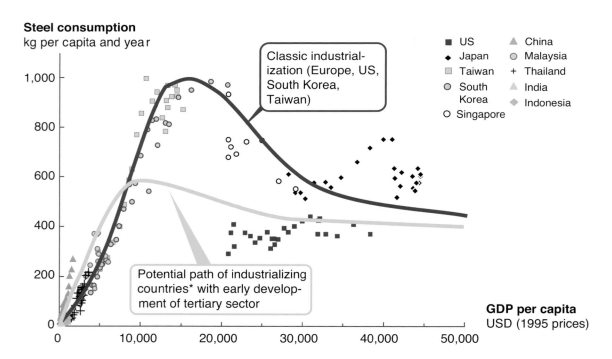

* Reasons: substitution of steel by alternative materials; earlier rise in the share of services in GDP

Source: IISI yearbooks, McKinsey analysis

tially regulated markets, and industries with major economies of scale. Excessive investments in markets that are small but open for foreign investment and poised to grow create temporary overcapacity. This in turn leads to phases in which none of the players achieves a reasonable return. Today, even markets with fast-growing demand can drown in the capacity added at a rapid pace by ever more players. Development of the Indian domestic airline industry since 2003 is an excellent example. This situation is often fueled by large investors eager to put money into these markets. The expected moves of competitors as well as the market structure and conduct that will result should therefore be analyzed before investing. Recognizing the strategic dilemma clearly does not resolve it. But transparency can help companies improve their chances by limiting their exposure, remaining flexible in markets that are overcrowded, and investing where conditions are most favorable.

In the past, some companies have managed to be in the right place at the right time. Nokia, for example, expanded its market share in China when the total market was experiencing a strong growth phase. It is now starting to focus on India, rapidly expanding its marketing, sales, and production activities. The bicycle manufacturer Hero anticipated the substitution of bicycles by motorcycles in India more than two decades ago, and transformed itself into the motorcycle manufacturer Hero-Honda.

In determining the right timing for entry, comparative trend analysis across different countries can be very helpful, particularly for established consumer durables with long life cycles and capital goods. It can provide useful insights even for some newer products, as in the comparison of the Indian and Chinese mobile phone markets described above. The demand trends for entire segments have clear parallels between countries, revealing what market trends are likely in less developed countries.

Comparing demand trends for different countries in relation to macroeconomic performance provides clear pointers to the development of market volumes for entire merchandise segments, but this does not apply to specific product types. The similarity in demand curves by merchandise group during the transformation of agriculturally based developing countries into industrialized and post-industrial economies should not conceal structural differences that exist in **customer preferences** for certain products. Globally similar customer preferences have only been found in very few areas to date. Preferences for certain product types and characteristics remain a very local matter, so it is a dangerous fallacy to try to simply transfer results from one country to another. Companies have to understand specific customer preferences before they enter markets, and be able to offer regional variants that meet the requirements of those buyers.

Successful products have to meet local requirements – comparisons across countries do not provide much insight

Although 10 years have passed since mobile phone manufacturing became a global mass market dominated by a few players, the share of clamshell cellular phones in North America is much higher than in Europe and Asia. And the sober, compact cell phones preferred by European customers are shelf-warmers in Southern and Southeast Asia, where shriller colors and ring tones are preferred and designs change constantly. Major differences in customer preferences are also evident in the demand for automobiles. In different geographies, the various product segments represent significantly different market shares. In North America, for example, 45 percent of vehicle purchases are SUVs or pickup trucks, while compact cars account for only 1 percent. In Japan, the ratio is more or less the opposite, despite comparable prosperity levels (Figure 2.8).

Figuring out customer preferences is something that large MNCs spend huge research budgets on, very often with good returns. Sometimes, this is possible even with limited means. Today, parts and figures for Hindu family altars (used in the great majority of Indian households) are largely manufactured by small and mid-sized manufacturers in China. This demonstrates that even very localized preferences may not be too complicated for foreign companies to decode.

Adjusting to local tastes and requirements has obvious implications. The number of products or product variants developed specifically for selected countries and regions is growing, and manufacturers are working hard to both provide customers with ample choice but at the same time standardize manufacturing, e.g., by using a platform concept. Worldwide, the number of automobile models, for example, has risen by around 60 percent since 1999 (depending on how "models" are classified).

Observing market trends and evaluating the attractiveness of markets also includes studying segment development over time. In the German automotive market, the medium-price segment in the compact class (-segment, e.g., VW Golf, Ford Focus, for example) has declined from 93 to 61 percent. This market share has been captured in part by premium

manufacturers (such as BMW with the 1 Series) and in part by manufacturers of basic models, such as Kia, Hyundai, or Skoda. The transition of the German market from a bulging midriff to an hourglass figure with high shares of premium and basic models is being repeated at a similar pace in other markets.

Understanding the needs of customers abroad and accurately assessing market size, growth potential, structure, and competitive conduct are only some of the capabilities that globalizing companies need to develop. Beyond this, decision makers have to realize that globalization can have further implications that are less obvious. The ability to produce more product types and variants could be one. This may require significantly altering R&D and manufacturing processes as part of the firm's globalization strategy.

2.2.3 The Elephant and the Dragon – Asia's Impact on the World Market

The high growth dynamics in China, India, and Southeast Asia have had an enormous impact on global manufacturing. The key player among these is China, its huge trade surplus with countries such as the US providing clear evidence of its success. Fluctuations in China's and – to a lesser extent – India's supply/demand balance for mass-produced goods will continue to have a vast influence on the world market due to the size and rapid pace of market development.

China's steel industry is a good example. The strong growth in steel consumption per capita has propelled China up the ranks of steel consumers (and then producers) worldwide. From 1992 to 2005, steel consumption in China grew fivefold. In 2005, it was more than double the demand in all of Western Europe – an economic zone with five times the GDP of China (Figure 2.9). Over the same period, demand in Western Europe grew by a total of only around 20 percent.

For some years prior to 2005, China's domestic production lagged behind its soaring demand, and China was the world's largest importer of steel. In 2005,

---

Regional product preferences differ markedly

Fig. 2.8: Market shares – vehicles under 6 t, 2003
Percent

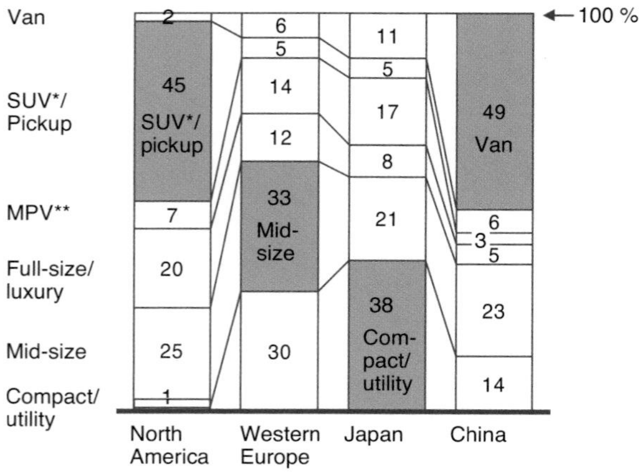

* SUV: Sports and utility vehicle (e.g., Mercedes M class, BMW X5)
** MPV: Multi-purpose vehicle (e.g., VW Touran, Opel Zafira, Renault Espace)

Source: McKinsey analysis

the situation changed fundamentally: domestic steel production in China rocketed. The surge in supply led, for the first time ever, to a significant net export of steel made in China. In 2006, China's production represented around one third of world production or four times the output of producers in the US. Given its massive output and the volatile nature of demand, China's surplus of simpler steel products could well grow to a considerable percentage of world demand for such products. As in other capital-intensive industries, fluctuations in supply and demand for steel have had a dramatic impact on prices. In the 1990s, the average price of hot rolled coils had been oscillating within the bandwidth of USD 240 to USD 400. From mid-2003 to late 2004, the average price of steel rose by nearly 100% to a level of around USD 700 per ton. Prices started to come down again at the same time as production in China expanded rapid-

ly and faster than consumption. During 2006 and 2007, worldwide prices for steel have risen again driven by the higher cost of raw materials. Prices for basic steel products in China, however, were still lower than almost anywhere else in the world. China has become a major net exporter of relatively unsophisticated steel products. Even though they export only a fraction of their production, Chinese steelmakers have become major players on international markets.

Low-cost countries are breeding grounds for aggressive competitors

The example of the steel industry demonstrates: fast market growth and high market potential are only one side of the economic dynamic of developing and newly industrialized countries. Its long-term impact on the structure and competitive conduct of indus-

China's emergence as a global steel giant

Fig. 2.9: Trends in crude steel consumption/production in China and Western Europe
Million tons p. a.

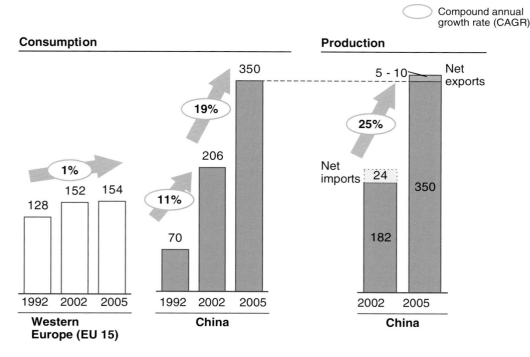

Source: IISI (Steel Statistical Yearbook, 2006)

tries worldwide is the other. The structural cost position of LCC competitors is different to those of incumbents who have most of their corporate centers and production in high-cost countries. Large, low-cost countries will become home to ever more global champions. Some Chinese companies have initiated their international expansion strategies based on strong positions in their home market, and will increasingly compete with incumbents on the world market. This trend has already affected industries such as communication electronics, computer software and hardware, domestic appliances, and steel. Going forward, the automotive and aerospace industries will face increasing competition from producers based in LCCs, though this will probably only have a significant impact on traditional markets after 2010.

These young, "greenfield" companies benefit from the fact that they can build competitive value chains without taking into account legacy structures. They can focus their manufacturing activities on areas where they are especially competent and efficient. Attackers can become more agile than their incumbent rivals by using relatively new business concepts such as the extensive use of manufacturing service providers. HCC players can often defend their higher prices for complex industrial products via their know-how advantage and a broader product portfolio. However, manufacturers that rely too much on specialty products and fail to achieve sufficient production volumes in the mass segment will find their competitiveness deteriorating. Without a strong foothold in the mass segment, R&D payback, for example, has to come from higher-value products alone.

Established players with a production footprint in high-cost countries have few defenses against low-cost imports. Even their marginal costs of production are often uncompetitive because of higher labor costs. The front line of this trend, which applied initially only to simple products, will move further towards higher-value, more complex products over time. Established players need to be proactive and stay ahead of these developments.

In some industries, it may be necessary to secure a presence in specific markets on the grounds of strategic competitiveness. Competing with new entrants in emerging markets can provide insights and surface opportunities that could help prevent the emergence of a strong, low-cost champion in a company's domestic market later down the line. Competing in the mass segment – from simple steel to small utility vehicles – will remain important for established manufacturers, though it may become even more challenging. As soon as demand in China and India sees a temporary decline, producers from these countries will have an even greater incentive to export and compete with established manufacturers in their home markets.

> **In some industries, it is vital for companies to gain an edge in specific markets to prevent future competitors from growing unchecked**

Companies should monitor the impact globalization is having on their competitive environments and regularly evaluate strategic location options (also see section 4.1). Incumbents that are expanding internationally should analyze not just market growth and market potential, but also long-term changes in the competitive environment resulting from the rapid growth of LCC competitors. This change is confronting HCC incumbents with major challenges that will continue to increase over time. How can HCC players develop and manufacture products cost efficiently that are also marketable outside Western Europe and the US? Developing or acquiring this capability will be a major driver of success.

2.3 Factor Costs – Labor, Capital, and Materials

Local factor prices have a crucial influence on the cost of goods manufactured.[7] This particularly applies to production processes with a high share of value added. Labor costs are the key location pa-

---

[7] Cf. Gutenberg (1965).

rameter in most manufacturing sectors. The share of labor is usually substantial across most of the production chain, and the differences in labor costs between locations are high. Labor costs also have an important indirect impact that is often overlooked (or underestimated): their influence on the prices of sourced materials.

Where capital costs are concerned, the key aspects to consider are refinancing costs, as well as country-specific risks to the value and operational utility of investments. This includes the market value of property and equipment, but also other items such as the default risk on receivables. With the third main element of factor costs – materials – it is useful to distinguish between processed intermediate products and raw materials when determining their impact on a location decision. If a manufacturer plans to source processed products locally, the main challenge is selecting and developing local suppliers. Other factors count for raw materials, such as natural availability, taxation, and regulation, as well as the local competitive structure.

2.3.1 Labor Costs

There is no denying that locations in developing and newly industrialized economies have very significant labor cost advantages. This will remain the case in the long term, despite the rising salaries in some of these countries due to their booming economies.

However, when the average labor cost level is considered, it is often forgotten that special qualifications and skills carry a relatively high price tag in many low-cost countries too – because they are scarce (often this is the case worldwide). The wage spread in developing and newly industrialized countries is far larger than in highly developed industrialized nations, particularly compared with countries in Western and Northern Europe and Japan. With India's and China's looming talent gap unfolding, the wage spread will only widen and be ever more important in location decisions. The costs and availability of skilled personnel is going to become a more crucial factor in the economic viability of a produc-

tion location than the differences between wages for ordinary workers. Particularly in the case of production in low-cost countries like China, India, Indonesia, Vietnam, Romania, or the Ukraine, the absolute labor costs for unskilled workers are very low compared to highly developed countries. The differences among these countries are relatively insignificant to MNCs. Much more important criteria are whether manufacturing and logistics processes will be reliable, despite the specific difficulties each country may pose, and what costs will be incurred for skilled and managerial personnel, including expatriates and temporary staff from factories at home.

The structural labor cost differences among developing and newly industrialized countries are also significant, even within a particular country. Differences between highly populated areas and rural regions are found in highly industrialized countries, too, but these are usually more marked in less developed countries.

> **A 30 percent variation in labor costs around the USD 1 per hour mark is irrelevant for almost all multinationals**

To make correct location decisions, consideration of the labor factor needs to be less about painstakingly optimizing individual components, and more about recognizing the key differences and trends of the truly relevant drivers. What matters is that company requirements and practices are adjusted appropriately to local conditions (and vice versa). It is vital that the complexity of production is in line with the local labor cost structure to generate a cost advantage and ensure reliability and quality.

2.3.1.1 Labor Cost Levels in Industrial, Developing, and Newly Industrialized Countries

Published figures[8] vary because the personnel categories or cost elements being focused on differ, but the labor cost difference from the viewpoint of West

---

[8] *Cf., for example, ifo (2005), p. 20, BCG (2004), p. 19, and the German Statistical Office (2005), ILO (2004), EIU (2004).*

Table 2.1: Average labor costs per actual working hour (estimates for 2006, at long-term average exchange rates)

Countries	Semi-skilled worker	Experienced skilled worker
Very high-cost countries, e.g., West Germany	Approx. USD 27	Approx. USD 39
High-cost countries, e.g., the UK	Approx. USD 20	Approx. USD 27
New EU members from Eastern Europe, e.g., Poland	Approx. USD 5	Approx. USD 10
Other Eastern European countries, e.g., Romania	Approx. USD 2	Approx. USD 5
Asian low-cost countries, e.g., China	Approx. USD 1	Approx. USD 4

Germany or Japan is roughly a factor of 5 to 10 for Eastern Europe and a factor of 10 to 20 for Asian low-cost countries (Table 2.1).[9] These differences vis-à-vis Eastern Europe and Asian countries would be somewhat lower for the US and UK, but of the same order of magnitude.

To determine the labor cost rate, i.e., the costs per effective hour worked, it is essential to consider the number of hours worked per year in the relevant countries. This factor is as important as the wages and ancillary wage costs per year, month, or week. The effective working hours per annum in Western European countries are only around 1,500 hours. This figure can be up to 2,300 in Eastern Europe and Asia. The labor cost difference per hour for these countries is therefore even higher than the differences in gross income per annum indicate. The main reasons for the higher number of working hours per year are longer hours per week, fewer vacation days, and lower absenteeism. Some companies in HCCs are already experiencing the positive impact of greater flexibility where working hours are concerned, and wage structures geared to the needs of production facilities.[10]

High labor cost differences will continue over the next 20 to 30 years. This period will be more like 50 years for China and India

Labor costs in developing and newly industrialized countries will only catch up with those in HCCs in the very long term, if at all. In the mid- to short term,

the difference – at least in absolute terms – will further escalate. A wage increase of 3 percent for a worker in Germany, for example, equals around USD 1 per hour. This corresponds to an increase of almost 100 percent in average Chinese labor costs. In view of the high baseline of current industrial countries, nations like China and India will need around half a century to draw close, even with rapid economic growth (Figure 2.10).

Even in the more highly developed **Eastern European countries** that have joined the EU, there are no signs that labor costs will rapidly equalize. With an average annual growth rate in labor costs of 6 percent, and an annual growth rate in Germany of 2 percent, it will take more than 20 years for the labor costs of an ordinary worker in Romania and Poland to rise to half the German level. This means significant approximation is unlikely for two to three decades.

Accession to the EU only led to a short-term rise in relative labor costs in Spain and Portugal

The accession of Spain and Portugal to the EU in 1986 was, to a certain extent, comparable to the more recent accession of the Eastern European countries. In the short to mid-term, it led to an increase in la-

---

[9] Cf., for example, UBS 2003.

[10] Cf. work-time models in the automotive industry, e.g., at Volkswagen.

In absolute terms, the labor cost differential between HCCs and LCCs will continue to grow in the near future

Fig. 2.10: Trends of successful developing countries

Standardized trend curves

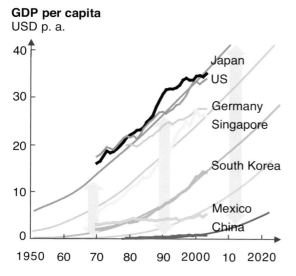

GDP per capita
USD p. a.

Japan
US
Germany
Singapore
South Korea
Mexico
China

GDP per capita
Relative to richest G8 nation

Japan
US
Germany
Singapore
South Korea
Mexico
China

Source: Global Insight, McKinsey

bor costs relative to those of other member states. However, compensatory effects came into play in the longer term. In the case of Spain and Portugal, higher inflation and the corresponding adjustment of exchange rates in the early 1990s largely compensated for the rise in relative labor costs. Over the long term, relative labor cost levels in Portugal and Spain, at around 20 percent and 50 percent of German labor costs respectively, remained largely stable (Figure 2.11). In line with this, the new Eastern European entrants to the EU have seen quite an increase in labor costs in the years 2005 to 2007 (this can also be expected for Romania and Bulgaria in 2007 to 2009). But predictions are that these will scarcely accelerate the long-term trend on a permanent basis. Whether the situation in the Iberian and Eastern European states will actually be comparable remains to be seen. Some indicators seem to point towards a similar development. Wages in the new Eastern European entrants rose relatively fast from 2003 to 2006, by an average of around 20 percent in total (in local currency) – more in Hungary than in Poland. Interestingly, the value of their currencies has actually increased

Accession of Portugal and Spain to the EU did not greatly change their relative wealth and labor costs vs. other EU countries

Fig. 2.11: GDP per capita*
Relative to Germany, index = 100

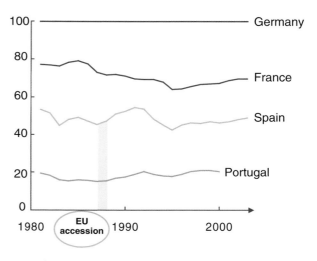

Germany
France
Spain
Portugal

EU accession

* Correlates with the average labor cost level
Source: Global Insight, McKinsey/PTW (ProNet analysis)

slightly against the euro, making the euro-based increase a little higher.

In countries with slower economic development, the gap versus prosperous industrialized nations may remain the same or even increase in the long term, due to significant devaluation of the local currency. We have seen this happen in Mexico and Brazil.

Ancillary wage costs are rising at a disproportionately high rate in industrialized nations

Other factors are also contributing to the slow pace of equalization. The rise in labor costs in HCCs is often not as moderate as a glance at net wages would indicate. Ancillary wage costs are rising overproportionally due to their aging populations and the obligation of firms to continue making high contributions towards staff social entitlements, especially in traditional industries. Surprisingly, this factor plays an even greater role in the US than in other developed countries. The costs of health insurance for employees at Boeing went up in four years by 30 percent to USD 1.7 billion, or over USD 8,000 per employee in 2004 and 2005. This corresponds to around 3.3 percent of sales – almost as high as the company's total profits. The situation at large automotive OEMs in the US is similar. Some are now spending over USD 1,000 per vehicle on the health of present and former employees.

2.3.1.2 Labor Cost Trends – Employment Structure as the Primary Driver

A realistic evaluation of labor cost development has to include **economic structure by sector**. In contrast to highly developed industrialized nations or countries such as the Czech Republic or Hungary, the share of employees in agriculture in countries like China, Romania, and the Ukraine is relatively high (Figure 2.12). In China, for example, around 50 per-

Employment in agriculture is still very high in many countries

Fig. 2.12: Comparison of the employment structure by sector, 2003
Percent

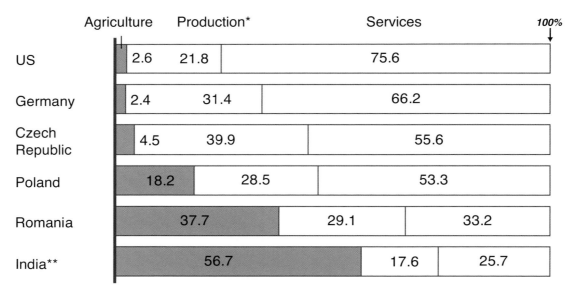

* Incl. construction industry
** For 2000 (database: Indian Ministry of Labour)

Source: German Statistical Office (statistical yearbook 2004 for foreign countries)

cent of the workforce still works in agriculture – a sector that only generates 15 percent of the gross domestic product. Similar figures apply to India, Indonesia, and other developing economies in Asia. Agriculture accounts for 19 percent of GDP in the Ukraine and 13 percent in Romania. In these countries, too, the share of staff in the agricultural sector is significantly higher than the sector's proportion of GDP.

The employment structures of low-cost countries differ hugely

Productivity increases in agriculture, e.g., via simple mechanization, will greatly reduce the need for labor. In Malaysia, which has developed faster than average in the last 30 years, the share of agricultural staff has dropped from around 54 percent (share of value added: 29 percent) to 15 percent (value added: 8 percent). The manufacturing sector has absorbed many of these former farmers.

For many developing countries, this structural shift still lies ahead. The supply-side pressure on the employment market exerted by low-skilled workers will increase further as a result, continuing to depress labor cost levels.

The fall in agricultural employment will lead to many "hidden unemployed" who will not appear in any official statistics. In India, for example, the official unemployment rate for 2003 was 10.4 percent. Another 15 percent of the employable population (aged between 16 and 60) were estimated to be seeking work. They are not included in the official statistics, however, because they are not entitled to support, or are regarded as belonging to the "hidden reserve."[11]

The expected decline in employment in sectors with low productivity will continue to depress labor costs for low-skilled staff

China has a similar discrepancy between published and actual figures. The official unemployment rate (which only applies to the cities) was cited as just 3.9 percent in 2001. Authorities claim full employment (by definition) in China's rural regions. If de facto unemployment in the rural regions is included, the unemployment rate could well be around 7 percent. Adding unemployed migrant workers would jack up the total still further. In China alone, there are an estimated 100 million rural migrant workers seeking temporary work in urban regions. As a result, the high number of poorly qualified job seekers severely depresses wage development for unskilled labor.

In countries like the Czech Republic, both the number of employees in sectors with low productivity (particularly agriculture) and the number of people seeking employment but not included in official figures are much lower. Increased demand for labor from MNCs setting up export plants will therefore have a more significant impact on wage levels for unskilled/semi-skilled staff.

2.3.1.3 Labor Cost Structure – Differences by Skills, Industry, and Region

When choosing a location, analyzing the average national labor costs may not be sufficient, since there could be major differences depending on the region, industry, or skill levels. This is nothing entirely new for companies with corporate centers in industrial nations, since these countries also have similar differences (Figure 2.13). Average labor costs in West Germany are 45 percent higher than in East Germany. In some industries, such as textiles, the difference can be as high as 60 percent. In the US, labor costs in the Midwest are around 30 percent higher than in the Mississippi Delta.

Labor costs also vary within countries in terms of specific skills, regions, and industries

---

[11] This is the share of the population willing to work, but not actively seeking a job on the labor market – whether due to the bleak prospects of actually finding an attractive occupation, or other reasons.

In developing countries, however, the gap is often much greater. A manufacturing foreman in Germany earns almost twice as much as a regular worker in the same industry, and a supervisor or shift leader receives almost three times as much. In China, the difference can be up to a factor of 10. While in Germany and other industrialized countries wages for ordinary workers are kept artificially high, in China the lack of qualified staff and an ample supply of unqualified labor lead to a much greater wage spread.

In LCCs, the rapid rise in labor costs for staff with special qualifications will intensify in the next few years. Education and migration will take a long time to boost the supply of highly qualified personnel and halt further widening of the wage gap. Over the coming years, companies that are building large production facilities in rural areas of Eastern Europe or seeking qualified sales staff in China, for example, should expect a tight labor market and higher labor costs. They need to take these factors into account when calculating the economic viability of relevant investment projects. A talent gap is looming particularly in China: the expected demand for highly-qualified, English-speaking graduates will far outpace supply.

One factor that should not be underestimated is the cost of managing production sites in LCCs. A large number of **expatriates** may be needed because suitable staff for **skilled and managerial positions** are often in very short supply. However, using expatriates is much more expensive – even more than at the home base. The employee will expect to be compensated for being transferred, the additional costs for supporting a family in a foreign country, and

Labor costs in different countries vary structurally

Fig. 2.13: Structural differences in labor costs – examples
Gross labor costs in USD per hour

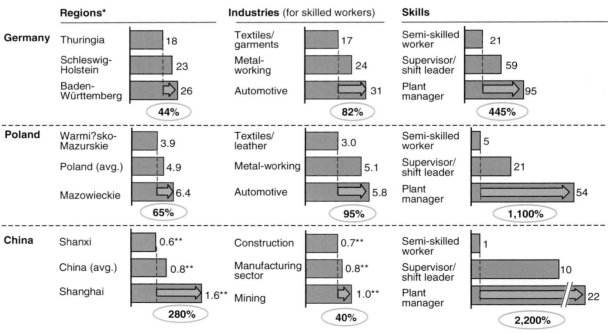

* Excl. the company's obligatory and voluntary social insurance contributions
** Manufacturing sector

Source: German statistical yearbook 2004, China Labor Statistical Yearbook 2004, Polish Central Statistical Office; converted using long-term average exchange rates (e.g., EUR = USD 1.16); McKinsey analysis

more difficult working conditions. The travel and change can also affect their productivity.

The labor cost gap is much greater in low-cost countries

Many companies have been prevented from setting up foreign sites by the high costs involved, and the fact that experienced staff are needed at home. This particularly applies to mid-sized companies, which find it hard to meet their need for managers in foreign assignments for two main reasons. In their domestic markets, these companies are often not perceived as international businesses. They tend to attract staff who are not particularly keen on international assignments. Second, these companies do not have a strong employer brand in emerging markets and often do not know how to build one, which is critical to attracting and retaining local talent.

MNCs, in contrast, often have an employer brand that attracts young, mobile staff who are interested in foreign assignments to further their careers. Access to

this type of employee pool holds down the costs of expatriates, which until recently often amounted to two to three times their home market costs. Relocation, private schooling for children, and quiet, spacious accommodation can add to the costs substantially. High expatriate expenses have also induced companies to employ local staff wherever possible, or lower-cost managers from other countries (also see section 6.3).

The average labor cost level in low-cost countries is often of little significance to MNCs

Considerations specific to LCCs also have to be kept in mind. Labor costs for multinationals in LCCs are often higher than country averages. As a rule, the lower the average labor costs, the higher the relative premium paid by MNCs (Figure 2.14). Why are their labor costs higher than those of local businesses?

- **Region:** MNC operations are mostly in densely populated areas with higher labor costs.

International companies pay a high premium in LCCs

Fig. 2.14: Labor cost premium at international companies
Percent of average labor costs per country

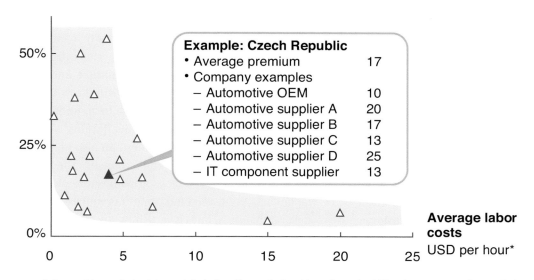

Example: Czech Republic
- Average premium — 17
- Company examples
  - Automotive OEM — 10
  - Automotive supplier A — 20
  - Automotive supplier B — 17
  - Automotive supplier C — 13
  - Automotive supplier D — 25
  - IT component supplier — 13

Average labor costs
USD per hour*

* Average labor costs, i.e., labor costs incl. all ancillary costs, for a blue-collar worker (at long-term average exchange rates)

Source: Corporate data, McKinsey analysis

■ **Industry sector:** MNCs often play a pioneering role in LCCs. They are more likely to place high demands on staff, and qualified locals are scarce.

■ **Qualifications:** MNCs generally focus on complex products with sophisticated manufacturing processes, outsourcing the simpler operations. As a result, the skills required of their ordinary workers are higher – as are the wages required.

■ **Image and employment policy:** MNCs are seen as demanding and willing to pay more. Wage demands compared with local levels are correspondingly high. MNCs also have to pay higher wages to retain employees, who become more attractive on the labor market due to their training and experience.

A long-term employment strategy may be essential for complex production processes, despite higher wages

Companies should be aware of the pros and cons of their multinational image in the local labor market. Would alternatives such as local JVs be preferable to building their own facilities? The labor cost gap is quite significant, as are the differences in companies' ability to retain high performers. In China, the branch offices of MNCs pay their general administration staff (in Controlling, Accounting, etc.) around 30 percent more than companies that have JVs with Chinese partners, and some 50 percent more than Chinese companies. In India, leading local companies manage to retain their staff much longer than MNCs. Often, an adequate human resources road map is the missing link in an otherwise promising globalization strategy.

2.3.1.4 Employment Strategies – Options

A company makes a conscious choice when selecting a location: it decides which labor pool to access. It can select from two very different strategies, depending on its size and experience. The decision is important, as it offers the trade-off between a long-term cost advantage and an easier, more rapid ramp-up. This is of particular importance to companies that need more highly qualified staff and would there-

fore be affected by a fast increase in labor costs for skilled personnel.

A "do-it-yourself" strategy for selecting a location requires critical mass and experience in low-cost countries, but has the greatest potential

"Do-it-yourself strategy": MNCs with critical mass can establish new locations in underdeveloped regions that have good basic prerequisites (e.g., quality school education). These companies can fully tap the local talent pool at minimal costs and provide training to staff to meet their company's standards. Targeted training and the systematic development of staff for leading positions will further strengthen the company's image in the labor market. Should competition for talent pick up as more companies select the region as a production location, the strong brand will help attract and retain employees and limit the need for wage increases.

This strategy requires experience with HR management in LCCs, and a critical mass of staff to be able to conduct suitable training and development programs efficiently. General Electric is one such example. They set their sights on India very early and already had over 20,000 staff there by 2003.

"Ready-made nest": Companies with little experience abroad (and particularly in developing countries) should consider adopting a "ready-made nest" strategy. Settling in areas with a better infrastructure, a more mature employment market, and more efficient administrative structures will throw up lower hurdles than opening a location in an underdeveloped hinterland. The greater availability of local suppliers and service companies can also help minimize initial investments and expenses. The potential of the site may not be as great because of higher labor costs, rent, and land prices (as well as possibly lower subsidies), but risks will be lower, too. Ramp-up costs could also be significantly lower due to the reduced need for expatriates.

Companies can also pursue differing strategies towards expected **tenure:**

A **long-term employment** strategy means more complex manufacturing processes can be used, as employees can be trained over time. To justify the training costs required and achieve high productivity, companies should offer appropriate incentives to enhance staff loyalty.

The contrast is a strategy of **minimal labor costs**, taking a high churn rate for granted and merely minimizing its negative impact. Manufacturing with a rigorous division of labor can reduce training to a few hours. This strategy is used in consumer electronics assembly, for example. Young women with limited qualifications in China and Southeast Asia are typical candidates. Retaining these employees often turns out to be difficult, as many of the women only wish to work a few years before devoting themselves to their families.

Employment strategy should be discussed explicitly during the planning process, and one of these options selected. The decision will have significant implications, both for the choice of location and design of the operating systems, as well as the equipment required (also see section 6.3.2).

2.3.2 Capital Costs and Depreciation

Unlike most input parameters, there is no de facto market price that allows direct evaluation of the cost of capital for a given investment location. Investments in production plants will always represent a certain financial risk. Operations may be disrupted by political or social conditions, jeopardizing payback.

In contrast to the ongoing expenses for labor and material, investments are largely sunk (and thus irretrievable). The interest on loans or targeted return on equity alone would not cover such risks. The interest rate, for instance, is determined by general market conditions and the company's perceived ability to repay. The rate reflects the risks the company is exposed to overall, not in one specific location. It therefore seems appropriate to use specific cost-of-capital rates that cover a potential loss of value by es-

timating the default risk for each potential location. An estimate of the default risk can be made, for example, by using indicators of social and political stability[12] or the country's credit status (Figure 2.15).[13]

As with **location-specific capital costs**, differences between the expected depreciation periods at individual locations should be factored in. If expectations on the useful economic life of machinery and plant vary between locations, the depreciation rates should also be taken into account as a location parameter. If their operating life is comparable, different accounting regulations for depreciation are a relevant location parameter only if optimizing tax advantages of the production network (cf. section 2.6.1).

2.3.3 Cost of Materials

Materials generally account for between 50 and 80 percent of the cost of goods manufactured. A distinction has to be made between product-specific processed materials from suppliers and standardized intermediate products and raw materials. The former normally dominate from a stand-alone perspective. However, if you consider entire production networks or supply chains, the costs of the latter typically account for a share of 10 to 20 percent. The remainder is supplier value added, and should be explicitly considered when optimizing the production network.

Both categories are relevant to selecting suitable locations for production sites. In selecting supplier locations, a company also has to decide early in the planning process whether to develop suppliers locally or opt for global sourcing (cf. Chapter 8). Depending on the industry, other factors and constraints beyond costs will have to be considered, such as know-how availability.

The prices of standardized intermediate products, raw materials, and energy can vary significantly

---

[12] *Cf. WEF (2005) and IMD (2003).*

[13] *Cf. OECD (2005) as well as rating agencies (e.g., Moody's) for loan loss rates.*

between locations. The market prices of raw materials and energy differ for two main reasons:

- **Costs** vary according to natural availability (example: ores are much cheaper in countries where they occur naturally due to the shorter distances that need covering).

- **State regulation** and taxation lead to cost variations, especially for utilities and electricity.

Natural availability has a particularly high impact on the local market prices of raw materials of relatively low value density. For intermediate products like steel, low value density also leads to substantial global price differences. Water and electricity prices vary widely because of natural availability but even more so due to taxation and regulations governing the quality of supply and disposal (Figure 2.16). The prices of these input factors are an important location parameter for companies whose production depends heavily on raw materials, water, and energy.

Prices of raw materials and energy differ globally, sometimes by more than a factor of 10

Fig. 2.16: Comparison of the costs of raw materials and energy

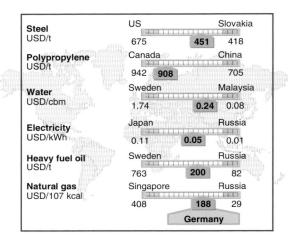

Source: Various databases 2005, McKinsey analysis

Multiple investment risks can be factored in indirectly using country-specific cost-of-capital rates

Fig. 2.15: Risk indices and country-specific cost of capital/risk premiums

| Country | Indices (by category) | | | | | Rating Index points | Risk premium Percentage points |
	Corruption	Legal system	Economic policy	Accounting/ management	Market regulation		
Finland	3	11	23	17	9	13	**-1.83**
Great Britain	20	3	25	33	13	19	**-0.44**
Hong Kong	26	12	14	33	15	20	**-0.21**
US	28	19	27	20	10	21	**0.00**
Germany	28	14	33	17	32	25	**0.86**
Japan	38	24	31	22	22	28	**1.51**
South Africa	55	34	28	33	18	34	**2.85**
Thailand	72	33	29	20	21	35	**3.11**
France	39	47	33	33	32	37	**3.53**
Poland	63	35	47	40	19	41	**4.43**
Italy	52	32	45	63	24	43	**4.94**
Argentina	65	64	33	30	27	44	**5.06**
Russia	78	44	39	40	31	46	**5.64**
China	74	39	39	56	43	50	**6.49**
Lebanon	83	60	65	44	42	59	**8.47**

Source: Based on Kurtzman (2004)

Different **competitive structures** in countries also influence the prices paid for intermediate materials. In industrialized nations, oligopolies have formed for some simple input products and basic services. In many developing countries, in contrast, the manufacture of these products is subject to intense competition. This is driven by a large number of small suppliers using labor-intensive manufacturing processes. Their low wage levels allow them to compensate for scale disadvantages and still produce small batches at competitive costs. As a result, products such as aluminum and gray cast iron parts, plastic injection-molded parts and plastic film, and packaging materials, as well as handling and logistics services, are typically 10 to 50 percent cheaper in developing countries than in industrialized nations. There, high wages necessitate the use of large-scale plants and capital-intensive equipment. The high fixed costs create entry barriers, which can encourage oligopolies.

A company's **market access** also impacts materials prices. Companies that lack a local image and knowledge of supply markets, with a subcritical pool of staff, pay higher prices than local companies. Western companies are estimated to still pay some 3 - 8 percent more for standardized electronics components in Southeast Asia and China than companies based locally.

2.4 Productivity and Economies of Scale in Manufacturing

Low factor costs can only be leveraged if sufficient **productivity** and **quality levels** can be achieved at manufacturing locations. Corporate experience demonstrates that world-class companies achieve high productivity and quality virtually anywhere – but it can take many years for them to acquire the necessary skills.

Particularly with production in LCCs, harmonizing the choice of location and manufacturing technology is crucial. The task goes beyond choosing the right level of automation; it also means folding in the level of training and experience of local staff.

In many industries, economies of scale can make gradual growth at new locations difficult. If the productivity and capacity of manufacturing systems at existing factories are high, these systems may be more cost-efficient – despite the higher factor costs – than foreign plants producing in limited volumes. In circumstances like these, companies should examine alternative manufacturing technologies that could operate at lower fixed costs, achieving cost efficiency at smaller unit volumes.

2.4.1 Physical Productivity[14] and Skills

There are different definitions of productivity, but not all of them are relevant or helpful for selecting production locations.

From a macroeconomic perspective, productivity in developing countries is so low that manufacturing there hardly appears to make sense. In India, for example, value added per capita (in terms of GDP), at approximately EUR 450 p.a., is around 80 times lower than in the US. If the figure is adjusted for differences in purchasing power, it provides a more realistic basis for evaluating manufacturing productivity (related to simple goods and services). However, the difference is still a factor of 12. Different rules come into play if MNCs leverage their experience in LCCs and their technical know-how to optimum effect. With professional management, productivity levels can be achieved in LCCs that correspond to those in HCCs. Lower labor productivity combined with higher capital productivity can be very cost-efficient and make low-wage locations desirable.

Even if **low labor and capital productivity** in LCCs represents a considerable risk to realizing cost advantages, this risk is controllable. Companies such as Hero Honda, Flextronics, Bosch, and Hyundai evidence this day in day out. The physical output per employee depends much more on the company than on the location. This is demonstrated by a comparison

---

[14] *Physical labor and capital productivity: the physical output of products per input in the form of working hours, capital invested, kilowatt hours of electrical energy, etc.*

of productivity in automotive factories in India (Figure 2.17).[15] While local companies in old factories only achieve around 7 percent of the level companies in the United States attain, the productivity level of new plants built in collaboration with a foreign partner is almost comparable to that of the US.

Productivity depends first and foremost on the company – location is only secondary

Some Japanese OEMs actually expand the scope of their operations and use their technical and management know-how beyond their own production locations, particularly in LCCs. They use a special program to help **suppliers** transfer the OEMs' optimized, mature production systems to their own production processes. This support, which can range from plant configuration, production planning, and production control to product development, improves supplier skills. This of course has knock-on benefits for the OEMs, too, who benefit from the high quality and low costs of the parts supplied. In one model case, a supplier's labor productivity improved by 45 percent within three years. The error rate fell

Productivity depends more on the company than the location

Fig. 2.17: Labor productivity* (indexed)
Example: Automotive manufacturing in India**

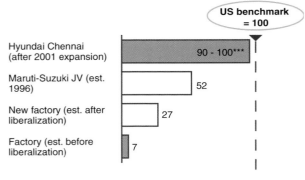

US benchmark = 100

Hyundai Chennai (after 2001 expansion)	90 - 100***
Maruti-Suzuki JV (est. 1996)	52
New factory (est. after liberalization)	27
Factory (est. before liberalization)	7

* Adjusted for the influence of different degrees of automation
** Vehicles per employee, indexed and adjusted according to the differing vertical integration of OEM plants
*** Estimate

Source: McKinsey Global Institute (MGI) FDI Report 2003: Automotive

from 1,000 ppm (parts per million) to 50 ppm over the same period.

World-class companies can achieve high productivity and quality virtually anywhere

Excellent productivity and quality can also be achieved in LCCs – but it is a long struggle. Companies have to put in more effort up front than in HCCs. This applies both to building their own production facilities and developing local suppliers. Typically, local suppliers need more supervision when developing parts and support in financing equipment and tools. Companies that operate world-class production facilities in LCCs today have accumulated the necessary know-how over decades.

Productivity is determined by numerous factors – only the key aspects are discussed below:

- **Labor and capital productivity** are directly dependent on one another. These input factors can partially be substituted for one another, through automation for example, and have to be considered together. Comparing labor productivity in production facilities in different countries is not sufficient. To generate meaningful analyses, it is crucial to adjust the figures for different capital intensities.

- The **availability and quality of skilled staff and managers** has a direct impact on manufacturing productivity. However, the alignment of supply and demand is even more important than the availability of qualified personnel. Successful companies use manufacturing processes geared to the new location and tailor their personnel policies accordingly. As a result, they have suitable staff available from the start.

- The **macroeconomic development** of LCCs influences the manufacturing environment in multiple ways. It influences the general level of education, for example, which in turn affects productivity. There are indications of self-reinforcing feedback,

---

[15] Cf. MGI (2003), particularly the "Auto" section, pp. 4, 6, and 105.

i.e., that conditions improve rapidly in fast-growing developing countries.

■ A location's **infrastructure** has considerable impact. Infrastructural weaknesses can call for contingency arrangements such as emergency power systems. These can help prevent a substantial decline in productivity but come with additional costs and investments.

All relevant input factors have to be taken into account when comparing the physical **labor and capital productivity levels** of specific locations. It is misleading to compare only one factor if input factors can be substituted for one another, such as capital (in the form of increased automation) for labor. Labor productivity in Mexican automotive plants, for example, is 30 to 35 percent lower than in the US.[16]

Much of this gap is due to the deliberate use of more labor-intensive manufacturing processes. As a result, production in Mexico is less capital-intensive and allows the use of smaller plants. Capital productivity is higher than in comparable locations in the US due to lower capital expenditure.

A comparison of the productivity of electronics plants also illustrates the strong influence that different capital intensities have on labor productivity. LCC factories only have around half the labor productivity of HCC sites on average before factoring in the differing capital intensities. The picture changes significantly after this is done. If the cost of capital is deducted from the value added, an HCC's labor productivity advantage is only around 10 to 15 percent – while its labor costs are around 500 percent higher (Figure 2.18). The comparisons show how attractive

Adjusted for capital intensity, the labor productivity of multinationals is similar in HCCs and LCCs

Fig. 2.18: Comparison of electronics manufacturing locations

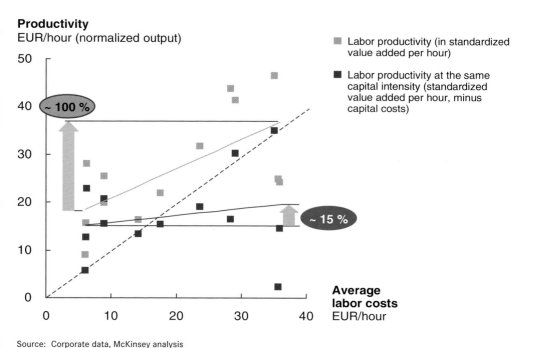

Source: Corporate data, McKinsey analysis

---

[16] *General Motors, DaimlerChrysler, and Ford plants form the basis of comparison; cf. also MGI (2003), p. 60.*

LCCs are, particularly for producing simple products with low logistical requirements and labor-intensive manufacturing processes.

If this is the case, which country is particularly suited to manufacturing a specific product, and when is the best entry window? An analysis of the trends in educational standards in individual countries and of GDP per capita (as a first approximation of labor costs) over a protracted period shows a typical **development path** (Figure 2.19). At the beginning of economic development, the share of the population with basic school education rises steeply. Then there is a significant time lag between this stage and the start of a phase of vigorous growth in prosperity and labor costs. Ultimately, education in the schools and universities reaches the level – at least quantitatively – that has largely been achieved in countries like the US and Norway.

Non-complex manufacturing will expand globally and be based where education and infrastructure are sufficient, together with the lowest labor costs

The rapid increase in basic education prior to a significant rise in prosperity and labor cost levels offers manufacturing companies very attractive opportunities. This explains the gradual shift of relatively simple manufacturing processes. Textiles and consumer electronics products, for example, have long been produced mainly in LCCs, even though regulations, customs duties, and non-tariff trade barriers have put the brake on development. For example, much of the textile production that was traditionally based in the south of the US was first moved to Mexico. Once the economic development of Asia's LCCs had made enough progress to provide suffi-

Countries reach a clear window of opportunity for manufacturing: relatively high educational standards and low labor costs

Fig. 2.19: Development of education standards and income levels
EDU index*, relative to the US in 2000

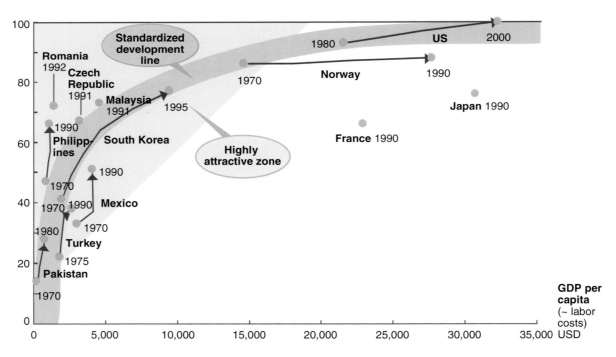

* Integrative indicator for the weighted share of the population having completed first-, second- and postsecondary-grade education

Source: UNESCO, Global Insight, McKinsey/PTW (ProNet analysis)

cient accessibility, an adequate infrastructure, and a supply base, a large share of the textile industry migrated there. Other industries, such as consumer electronics manufacturing (cf. Chapter 1 for the example of TV set manufacturing), have experienced similar continuous shifts in the dominant production locations.

The market for highly qualified employees is tight: only one in nine graduates in low-cost countries is both suitable and available for employment at an MNC

Access to talent will also become increasingly important. Location strategy needs to go hand in hand with a **global HR strategy** that aims to gain privileged access to qualified personnel at low costs. With a growth rate of approximately 5.5 percent p.a., the supply of graduates in LCCs is rising much faster than in HCCs, where figures are stagnant or even declining (such as in Western Europe). However, the supply of staff appropriate for skilled and managerial positions in MNCs still tends to be limited. Training in many developing and newly industrialized countries does not meet MNC skill requirements. On average, less than a fifth of graduates in LCCs appear suitable. In addition to poor English language skills (particularly in Brazil and China), the quality and type of education and training provided is a handicap.

The country that receives the most criticism for its overall **quality of education** – apart from top universities – is India. In other countries such as China and Russia, education is too theoretical. These graduates often cannot be assigned to factories right away – they have no grasp of the practical skills required for machine operation and work processes.

A further factor is that around 40 percent of potentially suitable graduates in LCCs are not available because they do not live in the regions where MNCs generally settle and are not prepared to move. MNCs have to compete with local companies and state institutions for the candidates that remain. However, this competition has little impact on MNC econom-

ics in developing and newly industrialized countries, because these companies can easily pay wages that, while well above the local level, are still very low by international comparison (Figure 2.20).

In addition to the shortage of appropriate university graduates, the scarcity and inadequate training of **skilled workers** in many LCCs is a key problem. The lack of experienced skilled workers can severely jeopardize the crucial ramp-up phase of a new factory. HCC incumbents normally use complex, highly automated production technology at home. If they plan to continue using this technology abroad, it is absolutely vital they find the skilled staff they need before going ahead.

The **generally** low level of education and scarcity of experienced skilled workers could be compensated for by in-house initiatives (e.g., training at a company's

Highly skilled staff are also in short supply in LCCs

Fig. 2.20: Number of young, highly skilled staff*, 2003
Millions (estimated)

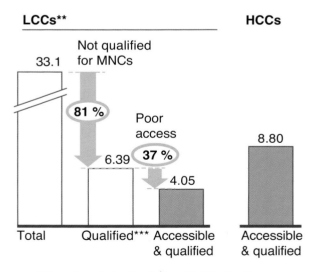

* Graduates in Engineering, Business Administration/Economics, Biology, Statistics, etc. with ≤ 7 years of professional experience
** Incl. BR, BU, CL, CN, CO, CR, CZ, ET, HU, IN, ID, LV, LT, MY, MX, PH, PL, RO, RU, SK, SL, SA, TH, TK, UR, VE
*** Incl. graduates who want to work outside their major

Source: McKinsey Global Institute (MGI)

home-based factories). However, this only makes sense if the churn rate is kept low enough. This is difficult to accomplish and becomes the core challenge in many developing countries.

An alternative approach is to develop a product design geared to the location and less complex production technology (cf. Chapter 5). This typically leads to a different, generally more incremental division of labor and reduces the skill levels and orientation periods required. While this approach means redesigning the product and costs time and money up front, it is often more cost-efficient in the long run. Another reason to consider it is because the lack of qualified skilled workers may well be a long-term issue in many developing countries. Estimates suggest that the increasing availability of these staff will be outweighed by a sharp growth in demand for the foreseeable future.

If all goes well in the Indian automotive supply industry, for example, an estimated 560,000 additional skilled workers will be needed by 2015.[17] The **training capacity** currently available, however, only ensures around 50 percent of the expected personnel requirements. This is also a great issue in China, which is attempting to meet this challenge by promoting university courses in technical disciplines. In 2001, the number of Mechanical Engineering graduates in China was already around 100,000, approximately 11 times higher than in Germany, and the figure has continued to grow since then. However, most technical education in China fails to fulfill all company requirements, due to its lack of practical application.

It is not only practical know-how that is in short supply. Knowledge of technological improvements and social trends – the bedrock of successful innovation – is not widely available. Access to relevant know-how outside the company depends heavily on exchange with other companies and research institutions in the region.[18] The structural knowledge base in industrialized countries is much stronger, especially in traditional industries such as automotive engineering, and a key reason why these sectors continue to grow.

An increase in labor and capital productivity is also anticipated as developing countries undergo **social transformation** into consumer societies. This goes beyond the obvious aspects such as infrastructure improvement and legal safeguards. There is a striking correlation between the drop in lost working hours due to labor disputes, e.g., strikes and lockouts, and the indebtedness of workers due to consumer borrowing.[19] The need for a regular income to pay off credit installments and the insight that productivity increases are the only way to a higher income in a highly competitive environment[20] evidently lead to higher motivation and productivity.

The relatively poor **infrastructure** at many low-wage locations can have substantial effects on productivity or require major additional expenditure. The effects are longterm and often not restricted to one resource. Bottlenecks shift with rising demand, e.g., from electricity to water to transportation. This makes it hard for companies to plan accordingly.

The greater Delhi area, for example, has an electricity shortfall amounting to around 10 percent of peak demand, while the water supply is around 16 percent below demand. Capital to finance expansion is lacking (price regulation being a major reason),[21] so power and water cuts will continue for the foreseeable future. Companies can only achieve an uninterrupted supply with their own independent infrastructures (entirely possible – at a cost). In India as a whole, the situation is deteriorating: power cuts are on the rise in vast metropolitan areas such as Mumbai. These areas are dependent on power supplies from the hinterland: in addition to scheduled power cuts, they often experience irregular shutdowns due to faults in the distribution

---

[17] Cf. ACMA (2005), p. 28.

[18] Cf. Porter (1990).

[19] Cf. Economic Times (2005), p. 1 and p. 12.

[20] Cf. also Lewis (2004).

[21] There was a shortfall in the supply of finance of Rs 2,440 crore in 2005/06 (approx. USD 590 million; estimated annual sales of approx. USD 1 billion).

grid, which is also underdeveloped. Similar conditions apply in large parts of China. As a result, the economic viability of investment in these regions has to be evaluated based either on lower machine availability and higher scrap rates or on a higher budget that includes backup power systems. This inevitably has a negative impact on capital productivity. The business circumstances – from training to infrastructure – that impact a location's productivity also influence the quality levels that can be achieved.

Overall, however, companies have made huge strides in production quality in LCCs over the last decade. This used to apply only to simple standard parts, but nowadays the same increasingly goes for complex systems, too. If tackled right, there is often little difference between quality at legacy locations and new plants and suppliers (Figure 2.21).

Automotive suppliers: high quality even in LCCs

Fig. 2.21: Production quality by country of origin

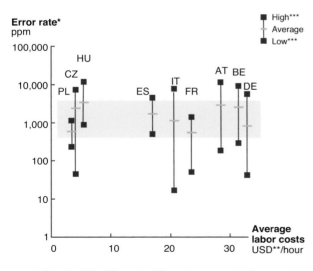

* Based on 100 million automobile parts; every country is represented by at least 10 suppliers
** With a long-term, inflation-adjusted average of USD 1.16 to EUR 1
*** 10:90 percentile; without adjustment for different parts/production complexity

Source: Company data, McKinsey analysis

2.4.2 Economies of Scale, Synergies and Production Technology

It is often possible to simplify the production technology used to suit a particular location and the skills available there, as previously explained. Such simplification frequently lowers capital intensity and the fixed costs of production. However, automated systems with high capital intensity and high fixed costs offer considerable economies of scale that favor centralized production, making a successive shift to foreign locations difficult.

The main factors discussed in this section are summarized in Table 2.2.

2.4.2.1 Economies of Scale and Synergies

Part of the costs of maintaining administrative and support processes – from plant management to the canteen or company physician – are largely independent of production volume. These fixed costs can make low-volume facilities uneconomical. The allocation of **fixed costs** can increase the costs of manufacturing products in small unit volumes so much that they outweigh the advantages that may be offered by the variable costs. If local demand is too low and export opportunities too limited, setting up full-blown production locations in developing or newly industrialized countries will prove uneconomical.

The average utilization of a plant with production machinery and tools of dedicated capacity increases with the production volume and number of machines per production stage. Figure 2.22 shows an analysis of economies of scale in the manufacture of an automotive component. The effects of the dedicated machinery and tool capacity (Figure 2.22, left-hand graph) blend into a much smoother curve in a summarized analysis of all production processes (right-hand graph).

Economies of scale exist in many production processes. In steel production, increasing the size of the blast furnace leads to a better volume-to-surface ratio and thus to improved energy efficiency. A similar

Table 2.2: Overview of the impact of different production technologies, economies of scale, and synergies with the choice of location

Factor		Impact (examples)
Economies of scale of various types	Scale advantages for fixed costs	Distribution of fixed costs across higher unit volumes
	Dedicated plant and equipment capacity	Greater utilization of dedicated production capacity due to higher unit volumes
	Marginal increases in productivity with scale	Higher marginal productivity or efficiency (i.e., higher output per input) with higher production volumes, e.g., greater energy efficiency in steel production due to larger furnaces
	Economies of scale in purchasing	Leverage of suppliers' scale advantages related to both their fixed costs and unit costs
	Economies of scale in transportation	Scale advantages from use of own assets and lower prices (e.g., due to volume discounts) for forwarding and shipping services
	Dynamic economies of scale	Learning curve effects, technical progress, and streamlining, which all increase productivity
Synergies		Synergies between corporate functions (e.g., R&D and production) and companies
Alternative production technology		Substitution of labor by capital, e.g., via automation

Fixed costs for relatively sophisticated production processes are high

Fig. 2.22: Production and materials costs, taking into account discrete capacity of machinery
EUR/unit

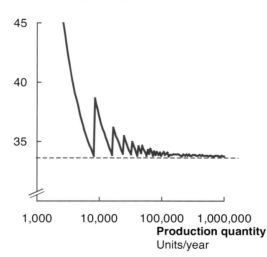

Metal cutting, casing

Cumulative throughout the process chain (production of parts and gear box assembly)

Source: McKinsey/PTW (ProNet analysis)

effect applies to aircraft and ships, where the bigger the plane/vessel, the less fuel it burns (proportionally) per trip. **Specialization**, too, can lead to savings due to larger production runs. Manufacturing only one product on a machine eliminates tool changes, reducing unproductive time.

Similar economies of scale are available to a company's **suppliers**. This is why it is generally more economical to procure large volumes from a single supplier than to divide up the volume among several suppliers. However, there is one caveat. Concentrating on one or a small number of suppliers can change the structure of the market, weaken your own negotiating position in the long term, and increase the risk of supply difficulties later.

In general, the opportunity to realize economies of scale in purchasing depends heavily on the supplier's cost structure. In China, for example, the opportunities to capture economies of scale in sourcing are more limited than in Western Europe and North America. This is due to the lower fixed costs in manufacturing and less automation. Even if volume-based economies of scale are relatively hard to tap at

suppliers, though, companies should not disregard their internal fixed costs when examining whether to take on a new supplier. These are costs incurred from supplier audits and administration, as well as the effort that goes into contract negotiations and inspecting the components supplied. Such expenses can be high. It can be worth concentrating on a small number of suppliers and interfaces.

As with production, **economies of scale exist for all means of transport.**[22] Less-than-container loads (LCLs) carried by sea are around 40 to 50 percent more expensive per kilogram of payable weight than full container loads (FCLs). The price difference between a 50 kg and 500 kg airfreight consignment is similar (Figure 2.23).

Major economies of scale can be captured in intra-company transportation and in networks for sourcing or distribution. With increasing volume in the network, the average pick-up/drop volume per customer increases (pick-up/drop factor), the need for consolidation via sorting centers (handling events) declines, transportation detours are avoided (detour factor), and the utilization of vehicles (load factor) and other machinery increases. From the consignor's perspective, high freight rates per consignment can be compensated for by accumulating volumes over a certain period. The downside is that this leads to lower delivery frequencies and higher inventory.

Learning curves, technical progress, and streamlining represent **dynamic economies of scale**. Production becomes more efficient with increasing, cumulative manufacturing volumes.[23] Production costs per unit fall. While technical progress and streamlining are not strictly linked to how long a plant has been in operation, learning curve effects can make a large difference. In aircraft production, for example, manufacturing costs and throughput

---

Larger airfreight shipments incur lower costs on a per kg basis

Fig. 2.23: Airfreight rates (net, airport-to-airport)
Indexed (overall average = 100)

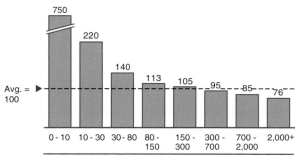

Weight category (kgs of payable weight per consignment)

Source: IATA Air Cargo Annual (CASS data), McKinsey analysis

---

[22] *Base: corporate data and IATA (2003).*

[23] *Cf. Coenenberg (2003), Chapter 7: "Erfahrungskurve als Instrument der Kostenkalkulation" (Using experience curves as an instrument of cost calculation) (p. 185 ff.).*

times are reduced by approximately 15 percent every time unit volumes double. Workers become more familiar with the activities involved and can perform these tasks faster. Learning curve effects cannot usually be transferred to a new location with a different workforce.

Consolidating different corporate functions at one location can lead to **synergies**. The physical proximity of production to product and process development, for example, often engenders a more lively exchange of ideas and experience. This in turn leads to more production-friendly product design and easier-to-handle production technology. Shorter and more direct lines of communication make the company more responsive and robust. The causes of errors can be identified and eliminated faster.

Evaluating economies of scale and synergies quantitatively for entire factories and factoring them in when optimizing production networks is not an easy matter. To get an estimate, economies of scale and synergies can be folded into location planning by considering fixed costs and alternative production technologies. Economies of scale in purchasing can only partially be mapped using these methods. The influence of the market structure, e.g., pricing in oligopolistic markets, is typically beyond the reach of simple analysis and parameterization.

While it may not be practical or necessary to consider economies of scale and synergies in great detail, companies should use their experience in determining which aspects will impact production costs the most. These effects should be approximated and included in calculations when selecting production and supplier locations.

2.4.2.2 Alternative Production Technology

Using a **tailored production process and level of automation** is advisable to reduce production costs when the factor costs at each location vary. However, this means the input parameters – especially capital and labor – must be exchangeable. Input parameters should also have a constant or decreasing marginal

utility, which is the case with many – but not all – production technologies.

Smaller blast furnaces and power stations are less efficient than large ones, and only allow limited reduction of capital intensity. The case is different for many assembly processes. Here, relatively expensive factors, e.g., capital, can often be substituted quite easily by cheap ones, e.g., labor, although this will reduce the economies of scale. This clearly evidences the potential for a more cost-effective alternative to a large, capital-intensive plant if labor costs are lower.

Production sites in countries like the Czech Republic, Poland, Hungary, and Malaysia can only benefit from alternative production technology. Staff are available locally who can run automated and complex operating systems, provided these make economic sense or are essential for quality reasons.

However, real-life examples show that companies also use comparable production processes at locations that have very different factor cost structures. The cost efficiency of developing and using alternative production processes cannot be assessed by considering the variable cost of production alone. Key features that can make it cost-efficient for companies to use identical production processes at different locations, despite wide variations in factor costs, include: fixed costs for the development, approval, and quality management of alternative production processes, and efficiency losses due to low standardization of processes and machinery. Semiconductor manufacturing is an example of a process where redesign and tailoring to a location is typically uneconomical.

Alternative production technology with lower fixed costs can provide the opportunity for smaller factories at attractive locations

---

Companies should also think about **expanding their technology portfolio**. This is particularly important if they have never opened an LCC factory before.

Many multinationals view the development of alternative manufacturing processes as controversial. Can the same quality really be achieved with a largely manual process? This doubt is certainly justified for certain manufacturing processes.[24] But reducing automation and capital requirements is usually at least partially feasible, for instance, by performing ancillary and monitoring processes manually.

* * *

The track record of global leaders shows companies can achieve high **productivity** and **quality** virtually anywhere in the world. However, in developing countries this takes a specific skill set and often more effort than in industrialized economies. Companies sometimes have to take on tasks – from providing education to building an infrastructure – that would be performed by the state or other institutions in their home country. Using alternative technology can make production abroad more attractive and easier to implement operationally. But MNCs should adjust themselves to a long haul. Developing and institutionalizing the skills needed to ramp up new sites and achieve high productivity and quality in developing countries generally takes around five years. With highly complex manufacturing processes, building the know-how and anchoring the skills may take over ten years.

2.5 Logistics – Direct and Indirect Costs

Logistics costs and lead times can be critical factors in site selection. A key factor is the distance between the site and the markets to be supplied. However, lead time restrictions and the number of product variants are also of high and increasing importance. These factors determine the mode of transport required to transport the products, which then largely defines the transportation rates (e.g., cost per ton-kilometer). Product characteristics such as value density and the ability to forecast demand (as precisely as possible) determine the importance of logistics costs in choosing a location for a specific product. Make-to-order production with short lead-time re-

quirements typically imposes severe restrictions on the distance between a production site and a market. When calculating logistics costs, we include the following cost items:

■ **Direct transportation costs.** These include the freight rates for air and sea transport, costs for handling and distribution over land, and expenses for scheduling and organizing transportation and interim warehousing.

■ **Inventory costs.** Inventory costs consist of the cost of tied capital, depreciation, and the market-side opportunity costs of extended delivery times. Higher cost of tied capital is incurred by longer transportation, transshipment and loading times, and by the higher safety inventories required to maintain service levels. The costs of value depreciation during transportation can be significant, particularly for products with a short product life cycle and sharply falling prices. Longer delivery times, especially for make-to-order products, can also lead to flagging competitiveness and necessitate price reductions. Obsolescence costs and lost sales (due to stockouts) also have to be considered in this context. These costs are not typically seen as part of inventory costs, but are definitely related.

We will first discuss how to determine direct transportation costs and times as exactly as possible. The next section examines methods for calculating and influencing stockkeeping costs. We conclude by looking at opportunities to boost performance of the production network and/or reduce costs via targeted use of various modes of transport.

2.5.1 Direct Transportation Costs

The importance of transportation costs in choosing production locations depends heavily on the **value density** of the products to be transported. Other determinants of transportation costs are the number of variants and delivery time requirements. These

---

[24] *Hero-Honda, for example, also uses automatic welding machines in India for quality reasons, e.g., to weld fuel tanks (see Chapter 5).*

factors are forcing companies in many industries to use airfreight for intercontinental transportation so that they can operate with relatively short delivery times and low inventories, even from production sites overseas.

Calculating transportation costs accurately is often a difficult undertaking when analyzing production locations. If the number of potential locations is not heavily restricted a priori, the routes between all factories and markets for which transportation costs and times have to be determined run into the hundreds. It may also be necessary to consider different modes of transport, e.g., airfreight and seafreight. Even if the current freight rates for all potentially relevant routes can be obtained from logistics service providers or shipping companies, the results will not necessarily be precise.

> **The relevant transportation costs to consider when making decisions on locations and suppliers are the expected long-term costs, not current rates**

Figure 2.24 gives an example based on seafreight, showing all the main determinants of the short-, medium-, and long-term freight rate development. This chart highlights the considerable volatility the rates are subject to. Although short-term fluctuations are irrelevant to long-term investment decisions, they make it difficult to accurately estimate the long-term transportation costs based on current data.

The use of **consistent transportation cost assumptions** is advisable for reliably analyzing the economic viability of production networks. There are two ways to obtain these:

Only the long-term drivers of transportation costs are relevant to selecting locations

Fig. 2.24: Seafreight rates – input factors

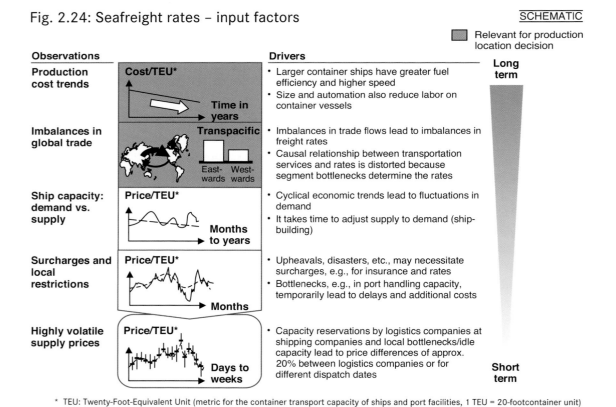

* TEU: Twenty-Foot-Equivalent Unit (metric for the container transport capacity of ships and port facilities, 1 TEU = 20-footcontainer unit)

Source: McKinsey

- Bottom-up costing: This estimates shipping rates **based on the costs** for providing the transportation service. This is, for instance, most often suitable for transportation by truck. The trucking market is highly competitive because of its polypolistic[25] market structure, and most providers use "cost plus" pricing logic.

- Statistical analysis: This technique uses **statistical data** if possible for a large number of contracts or shipments. This approach is suitable for transport markets with a more oligopolistic structure, e.g., airfreight and ocean freight routes, which relatively few carriers serve.

Factors with significant long-term influence within the payback time have to be included to determine transportation costs and times by either method. Indian ports, for example, are still much less efficient than the ports in Singapore, Hong Kong, or Shanghai, despite partial privatization over the past few years. The average turnaround time of two to four days[26] for container ships in Indian ports considerably lengthens transport time, and it is essential to factor in longer times for customs clearance and overland transport. These disadvantages will not be eliminated completely within coming years, despite efforts by Indian operators. They should therefore be included in any model calculations performed to support location decisions.

If using the statistical approach, data should be taken from one or several consistent sources. The use of irregularly sampled data from different unaligned sources is not advisable because of the high short-term volatility of freight rates. This could lead to distortions that give some locations an unfair advantage or disadvantage compared to others. This can be prevented by systematically calculating the long-term transportation rates for all relevant destinations based on one set of data – even if this means including estimates and approximations. Although the actual level for a specific route cannot be predicted any better with this calculation, it avoids relative deviations between the individual destinations.

Transportation from LCCs is not necessarily more expensive than from HCCs

When looking at a global market scenario, transportation costs are not necessarily to the detriment of LCCs. Obviously, the cost effectiveness of transportation depends on where your markets are. If considering one location for the entire world market (i.e., weighting all countries, states or provinces with their respective GDP), the results for sea and overland transportation correspond roughly to Figure 2.25. Central America, Venezuela, and Columbia are attractive in terms of transportation costs because US coastal cities are easy to reach by ship. The Czech Republic is positioned very centrally for Europe. Its truck drivers charge less than their peers in Western Europe, so it can supply the EU at lower costs.

Intercontinental freight rates have been declining long term

In the six to eight years before 2002, nominal freight rates fell some 2 to 3 percentage points on average for both air and ocean freight (Figure 2.26).[27] This long-term trend did not continue between 2002 and 2005. Airfreight rates stabilized or increased mainly due to higher fuel costs that are being passed on to customers via fuel surcharges.[28] High fuel prices will likely remain a concern for shippers. The availability of low-cost cargo capacity in passenger aircraft (belly space) and the continued addition of dedicated airfreight capacity, however, will keep up the pressure on prices, particularly in deregulated markets. In India, for example, three additional airlines announced they were entering the cargo business in 2006 alone and planning to double domestic and international uplift capacity in the coming years.

---

[25] *Polypolistic markets are characterized by a large number of market players (in contrast to oligopolies and monopolies).*

[26] *Cf. CII (2004), p. 71. Sometimes shorter times in privatized terminals (when not bottlenecked).*

[27] *Analysis based on Drewry (2004) and IATA (2003).*

[28] *Cf. announcements to that effect by DHL, FedEx, and Lufthansa Cargo at the end of 2004.*

Transportation from LCC locations need not be more costly

Fig. 2.25: Average transportation costs to the global market* by sea/land

Transportation costs for container unit (TEU)
USD (door to door), estimated for 2004

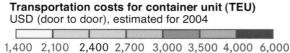

1,400 2,100 2,400 2,700 3,000 3,500 4,000 6,000

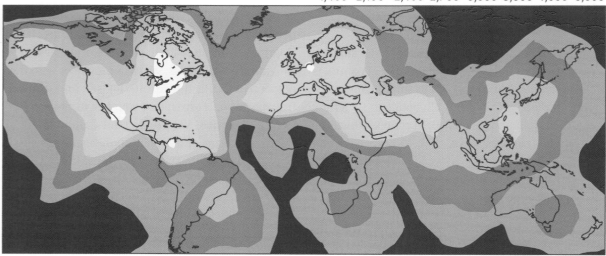

* Each country is weighted with its respective GDP

Source: Drewry (2003), shipping companies, McKinsey/PTW (Logistics Model v1.3) (ProNet analysis)

High demand for container ships enabled ocean freight rates to rise sharply from an all-time low in 2002. The sudden surge in rates led to record profits for ocean carriers, which started ordering large additional vessels. More capacity already started to affect rates in 2006, and is continuing to exert downward pressure in 2007. Going forward, the cyclic pattern the industry follows will likely keep rates volatile. But there is little indication that rates will rise long term, except possibly fuel prices.

Nominal freight rates have been declining long term

Fig. 2.26: Transportation costs – long-term trends

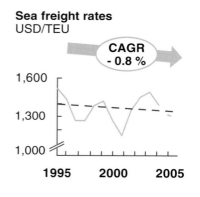

Sea freight rates
USD/TEU

CAGR
- 0.8 %

1,600

1,300

1,000

1995   2000   2005

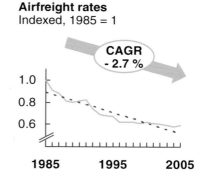

Airfreight rates
Indexed, 1985 = 1

CAGR
- 2.7 %

1.0

0.8

0.6

1985   1995   2005

* 2006 estimate; 2007/2008 forecast based on ship capacity available and expected demand

Source: Drewry (2005), Boeing (2006)

Higher fuel costs are also affecting overland transportation costs. Trucking costs in Europe are additionally being influenced by new state regulation and tolls, which are likely to keep costs rising.

Uncertainty about the future: Transportation will be more efficient but fuel prices higher

As a result, long-term **transportation cost trends** remain hard to predict, chiefly due to the uncertainty around energy prices. Some 15 liters of diesel fuel are needed to truck a 20-foot, standard sea container 100 km; sea transport only uses around a third of this. Medium-size container ships with a capacity of approximately 3,500 TEUs consume around 70 to 100 tons of heavy fuel oil a day. Some 800 liters of fuel are required to transport a 20-foot container from Germany to the coastal regions of China; an increase in the price of crude oil by USD 10 per barrel increases the direct cost of sea transportation by around 2 - 3 percent. The effect on air transportation is even greater because it consumes significantly more fuel. Around 20,000 liters of fuel are required to transport 10 tons of goods from Western Europe to China[29]. A USD 10 per barrel hike in the price of crude increases transportation costs by around USD 0.15 per kilogram, or around 5 percent. For goods with a high share of transportation costs, it is therefore advisable to perform a sensitivity analysis using various scenarios to gauge their effects on the production network. Generally, however, a network structure remains stable even if transportation costs rise by 20 to 30 percent. Higher cost for fuels (driven by the need to pay for carbon dioxide emissions, for example) will therefore likely have relatively little impact on the structure of global supply chains except for goods with very low value density such as raw materials, construction materials, and basic chemicals. Companies will, however, find it increasingly attractive to use surface transport rather than airfreight.

The ability of the supply chain to ensure timely delivery should also be scrutinized. This can be a vital criterion, especially where on-time delivery is make or break for a production chain, and the goods cannot be substituted.

Before 2001, intercontinental logistics had reached a high point in terms of reliability. Since then, structural changes have become sources of concern and could lead to disruption going forward. Terrorist attacks and related countermeasures by governments could lead to significant import/export delays. The low inventory level in modern supply chains and lack of buffer capacity in many parts of the transport chain, such as at ports, makes such distortions a real threat for global supply chains.

Increasing trade in intermediate and finished goods – mostly shipped by container – has greatly increased the demand for container handling and transportation services. Goods handled in Chinese ports (excluding Hong Kong) grew by an estimated 30 percent in 2004.[30] Shenzhen and Shanghai are now the biggest container ports in the world after Hong Kong and Singapore (as major transit port), each handling a volume corresponding to Rotterdam and Hamburg put together. This rapid growth and the increasing dependence of domestic production and sales on deliveries from abroad make logistics all the more susceptible to force majeure. Delays caused by strikes and congestion at West Coast ports in the US in 2004 considerably hampered onward operations. Some ships had to wait up to two weeks to unload in the ports of Long Beach and Oakland. The accident of a single ship in the busy Suez Canal led to huge delays in international container traffic.

The high degree of efficiency and reliability in international transportation achieved over the past few decades cannot be taken for granted going forward. Companies will have to design more robust supply chains, and consider the impact of distortions. This may, for instance, require more production sites at different locations and production closer to the market. The trade-offs between, say, resilience versus higher costs, have to be analyzed carefully (using techniques such as scenario planning).

---

[29] On old generation freighter aircraft, e.g., MD 11.

[30] Cf. Drewry (2004), p. 22.

2.5.2 Costs of Inventory – Tied Capital and Depreciation

Companies also need to accurately capture indirect costs in the supply chain. Added together, these can far exceed direct transportation costs. The costs of inventory tied up in transit and safety stock can be approximated by applying a rate to the value of the inventory. This rate should include the costs of handling and storage in warehouses, packing, and inventory management. These additional indirect costs can be substantial, particularly if companies have to switch from make-to-order to make-to-stock production because the goods are being produced too far away from the markets.

> **Packing or producing product variants with a slow turnaround in local markets can increase effectiveness and reduce inventory**

The safety stock required to maintain delivery capability is considerable, especially with a high number of variants and highly volatile, unpredictable demand. If this is the case, it is crucial to coordinate site selection with the variants of the product manufactured there. Just continuing to boost inventory levels will become costly given the rising number of variants and volatility of demand in many market segments. The faster devaluation of goods due to shorter product life cycles makes this issue even more pressing.

Companies should therefore seek new avenues when designing their global supply chains. Delaying the generation of variants is one possible option. One baseball cap manufacturer sources its undyed preliminary products from China. At its production location in California, it then makes tens of thousands of caps to order within 24 hours by dying them and using an automated process to stitch on lettering. Production to order is unavoidable because sales figures for baseball caps correlate closely with the success of the teams in the play-offs – which are, of course, impossible to forecast with decent accuracy. One capital goods manufacturer uses close-to-market

production locations to assemble to order products with low demand per variant. It concentrates its manufacture of high-turnover product variants in a few locations worldwide, producing them on a make-to-stock basis.

Leading companies build production networks that are cost-efficient, robust, and allow short lead times. Automotive OEMs in Western Europe often have their first-tier suppliers located within a two-hour drive from their assembly plants. The plants of their second- and third-tier suppliers are located further east in regions with lower labor costs. Particularly when it comes to supplying the Western European market, a graduated structure of plants and suppliers (home market, EU accession countries, Eastern Europe/Turkey, Far East) ensures high logistical performance while keeping costs down.

2.5.3 Modes of Transport – Untapped Optimization Potential

Evaluating transportation costs and times is an essential step in developing a production network strategy. MNCs can use this effort to also take a critical look at how they can improve operational procedures in the short and mid-term. Significant cost savings can be realized from the use of alternative modes of transport at almost any company. Transport costs by air, truck, rail, inland shipping, ocean freight, or special service vessel are very different. Similarly, different safety stock levels are required to ensure delivery availability for lead times of six weeks compared to one day. The main factors for selecting the optimum mode of transport are the value density of a product, the number of variants, volatility of demand, error rates, and the consequences of non-availability.

Companies can utilize two levers to use or combine different modes of transport more effectively:

- **Parallel multi-modal transportation:** Some of the transportation volume for each route, i.e., for each transport link from A to B, is shipped using one mode of transport, while the rest is sent by another. The parallel use of airfreight and seafreight, for example, can significantly reduce the need to keep safety inventories. Just in case, a shipment is also sent by airfreight or air express.

■ **Serial multi-modal transportation:** One mode of transport is chosen for one stretch of the journey, while another method takes the goods the rest of the way. In Europe, for example, combined transportation by rail and road are commonplace for moving goods across the Alps. Carriage by sea and air can be used serially for intercontinental relationships. For instance, products can be shipped from the Far East to the Persian Gulf using seafreight and from there to Europe via airfreight.

> **Parallel multi-modal transportation can reduce logistics costs by up to 50 percent for goods with a value density between EUR 15 and EUR 80 per kilogram**

If goods with a value density of more than around EUR 80 per kilogram (gross) have to be transported intercontinentally, it makes sense to opt for airfreight or – at even higher value densities – air express. The savings achieved from lower tied capital more than compensate for the higher transportation costs. The value density threshold for using airfreight can be lower for products with high demand volatility, rapid depreciation due to short product life cycles, and high non-availability costs. Products with a low value density (less than EUR 15 per kilogram) can only be transported cost-efficiently between continents by sea. For products that fall within the value density range of roughly EUR 15 to 80 per kilogram, it is usually a good idea to transport the demand baseload by sea and peak demand by air. This use of **parallel multi-modal transportation** can help companies dramatically reduce their safety inventory and lost orders.

Transportation of a small share of products by airfreight can also be used to reduce quality risks. If a portion of a production batch is injected into the succeeding production step early on, errors can be detected sooner. This can help avoid the type of

nightmare experienced by one supply chain manager in the automotive industry. Defective parts were produced and shipped over a period of seven weeks. The faults were not discovered until the first parts from these batches arrived after their long journey at another plant and were installed in the end product – and defective parts continued to arrive at the factory weeks later.

Serial multi-modal transportation takes advantage of the specific benefits that particular modes of transport have on various sectors of a route. In transalpine traffic, for example, trucks are subject to restrictions such as bans on night driving. Shifting to rail-based transport would allow the cargo to keep moving. When containers are unloaded from the vessel at a seaport, there are typically several options for onward transportation, including inland waterways, rail, and road. The mode of transport can be varied depending on costs and needs. Some companies even decide the onward transport on a consignment-specific basis in line with relevant requirements. This makes the supply chain more robust and efficient. One automotive supplier uses inland shipping to transport containers from Rotterdam to the south of Germany. If, however, delivery of the parts is urgent, the total transportation time can be shortened by two to three days by using trucks (though at somewhat higher costs). Similarly, a mail-order company for apparel at times unloads ocean containers in Singapore and the Persian Gulf region, then transfers some of the goods to air transportation, reducing transportation time by one to three weeks. The company profits from favorable airfreight rates because substantially more goods are flown into the Gulf region (overall) than the volume available for the return flights. This means rates on the Middle East to Europe leg are relatively low.

* * *

While transportation and communication have become more cost-effective and reliable in recent decades, the demands on the logistical performance of production networks have also risen sharply. Vehicle manufacturers now allow their customers to change a vehicle's features up to just a few days before delivery. And higher numbers of variants and shorter product life cycles make it even more difficult to produce remote from the market. Companies therefore have to properly assess the importance of logistics in their choice of production locations, integrating location and logistics to form an efficient whole (also see Chapter 7). Logistics have to be aligned to the product architecture and design, enabling later product customization and, as a result, later allocation of products to specific orders.

2.6 External Factors – Boundary Conditions and Risks

Leveraging globalization opportunities inevitably entails additional risks. New knowledge and skills are required to assess and manage these risks adequately.

Subsidies are one possible way to reduce the outlays on new locations. Like **taxes**, they are factors relevant for decision making, particularly at the regional and local levels. As with tax and customs arrangements, the amount of subsidies that can be obtained is subject to discussion with the authorities, at least in part, in almost every country. It is therefore advisable for companies to negotiate the terms of inward investments with several states and regions in parallel.

While **customs duties** and other barriers to trade are not as significant as they used to be, they are still crucial in some industries and should therefore be taken into account when selecting locations. Companies are still much too inclined to perceive **exchange rate effects** as a risk – even though they can definitely represent an opportunity if the right action is taken. The same cannot be said of risks resulting from the transfer of knowledge or the **violation of intellectual property rights**. The primary concern here is to limit the potential damage.

2.6.1 Subsidies and Taxes

When it comes to designing production networks, the relevance of subsidies, taxes, customs duties, and

non-tariff trade barriers is largely determined by two factors: the industry or product category and the locations involved, both as potential production sites and as primary customer markets. Generally, customs duties and subsidies should be included as approximate figures in the quantitative evaluation of production locations. Striving for a precise assessment would typically increase the effort dramatically and require consulting experts and government officials.

The inclusion of tax effects is a similar matter: the choice of location can have substantial tax implications for some process steps,[31] e.g., packaging and shipping, and these should at least be estimated. Companies should explore the legal options for minimizing the tax burden when they choose a location, even if only approximately. However, attempting to optimize tax situations and fully integrate them into the design of the production network only makes sense in exceptional cases. The endeavor hugely increases complexity, not only in terms of decision making but also in terms of coordination within the company, and there is little value to be gained from using comprehensive tax data as opposed to an approximation.

In virtually all regions (with the possible exception of the EU), negotiations on **subsidies** should be conducted with federal and local governments. Subsidies are mainly awarded in the form of tax rebates, infrastructure measures, training grants, research funding, and preferential loans. Cash subsidies, e.g., to finance the initial investment, are rare. Exceptions to this can be found in European countries, especially East Germany, where investment grants are allowed as an instrument for economic development, as well as in the high-tech and the automotive sectors, which are regarded by many governments as having national strategic importance.

AMD, for example, received grants of around EUR 500 million for establishing its new production

---

[31] Cf. Murphy (1998).

facility Fab 36 in Dresden. One automotive producer conducted parallel negotiations with several governments and succeeded in obtaining direct subsidies of around EUR 100 million each for two investment projects in Eastern Europe. Even if it is difficult initially (if not impossible) to estimate subsidies when deciding on a location, companies should explore potential subsidies and maximize the economic attractiveness of their investments through targeted negotiations with several potential locations.

Nominal **corporate tax rates** in most highly developed industrial countries are about 40 percent. The level of global taxation that is realistically achievable is well below this. Lowering the effective rate of taxes on earnings improves the company's free cash flow situation and enhances its freedom of action. Besides its impact on the valuation of a company, tax optimization is also of interest to managers whose performance objectives are based on after-tax profits. However, tax optimization instruments only have high relevance for a company if profits are retained. If profits are distributed, it is usually the shareholder's tax domicile that matters most. Trading off operational efficiencies against tax advantages might therefore not be as favorable as it appears.

Tax aspects should primarily be taken into consideration when choosing locations with distribution centers

The influence of taxes on the choice of international locations is complex. Depending on the place chosen for a specific process step, implications may arise for the production process itself, as well as for the tax base and the tax rate for other processes in the value chain. The use of a principal trading company in a low-tax country such as Switzerland, the Netherlands, or Singapore can clearly reduce a company's **taxes on earnings**. The flexibility (albeit limited) in setting transfer prices means that profits can be accumulated in the part of a company where the tax rate is beneficial.[32] Some countries (such as Belgium) provide an investment incentive for selected industries by allowing companies to move some locally generated profits to a tax-beneficial country abroad.

It is also possible to minimize **sales taxes** in some cases by the targeted geographic positioning of corporate functions such as order acceptance, packing, and shipping. This criterion is mainly relevant when choosing between countries. One exception to this rule is the US, where sales taxes and real estate tax rates are fixed, at least in part, at a local level by the individual states, counties, and cities.[33] Alaska and New Hampshire, for example, do not levy any sales taxes, while Fort Collins in Larimer County, Colorado, imposes a sales tax of 6.7 percent, set partly by the state, partly by the county, and partly by the city. One factor that led computer manufacturer Dell to decide on its US location was this type of tax. The tax has a direct impact on the gross sales price of its computers.

Optimizing the tax burden should also factor in uncertainty. The tax systems of developing and newly industrialized countries are sometimes particularly prone to confusing changes. These can greatly impact balance sheet valuations and P&L statements. In India, for example, lowering the rates for diminishing balance depreciation will markedly increase tax-relevant profits in the years to come. This will reduce free cash flows due to greater tax burdens[34], and have a direct impact on a company's liquidity. In states where the legislative majorities and government are liable to (frequent) change, investors cannot be confident of a continuous tax and finance policy.[35] Such risks should be reflected in a company's long-term optimization strategy (cf. section 2.3.2).

[32] Cf. Perridon (1999), p. 93 – 95.

[33] Cf. Karakaya (1998): survey (focused on the US) examining the relevance of 27 location factors. Availability of skilled workers: 1.94; transportation: 1.84; regulations and tax rates of states, and real estate tax: 1.80; etc.

[34] Cf. Raghunatha (2005).

[35] The Indian Government's 2005 budget bill made a compromise by including both socialist (e.g., employment guarantees, state loans, etc.) and capitalist measures (e.g., reduction in customs duties and corporate taxes, acceptance of foreign investors, etc.).

2.6.2 Customs Duties and Non-Tariff Trade Barriers

Customs duties should be explicitly considered when choosing locations and optimizing production networks. A few exceptions apply, such as when selecting locations within one trade zone (like the EU), or in an industry with only limited or no customs duties. Rates are nominally fixed and generally non-negotiable (with a few exceptions, such as for special economic zones).

However, the allocation of goods to a particular **customs category** and the value on which duty has to be paid are sometimes determined by subjective assessment. This can give them some optimization potential. Importing components instead of finished products, for example, and targeted management of the country of origin can significantly lower customs expenses. If, for instance, parts are manufactured in Italy and assembled with low value added in the Ukraine or Russia, the country of origin does not necessarily change. Consequently, when the assembled product is reimported into the EU, the company does not have to pay customs duty on the total value of the product. If, on the other hand, a product is assembled in Romania from components sourced in Asia, the country of origin should change if sufficient value is added in Romania. This is an advantage if the end product is to be sold within the EU. Imports to the EU from Romania, formerly an associate member of the customs union and a full member of the European Union since January 2007, are free of duty.

While it is true that the **significance of customs duties** has clearly lessened globally, they still remain a dominant location criterion for some regions, countries, and industries. In the 1980s, the unweighted averages[36] of the duty rates applied in Latin America and the Far East still amounted to roughly 30 percent and an astonishing 65 percent in South Asia. Even in highly developed industrialized nations, customs duties are still so high that they are often relevant to location decisions. This becomes particularly evident if duty rates are compared to the share of

value added by the OEM, which usually amounts to only 15 - 40 percent. In 2004 in the EU, the unweighted average duty rate was 5.6 percent of the value of the goods imported. For the US, the figure was 4.8 percent. Negotiations among WTO members ("Doha Round") have made little progress since then. No major revisions to the global tariff scheme were made during 2005 and 2006.

With the exception of a few industries with globally negligible duty rates (e.g., structural components in the aviation industry), companies should consider at least the main implications of local customs clearance requirements, i.e., the impact on costs and lead time, when evaluating locations and designing production networks. This can be a very laborious undertaking, because customs data is not always easy to establish. The rates for customs duties are geared towards trade relationships (country to country) and product groups, and are correspondingly numerous. Extensive databases are available from commercial providers, but the task of classifying products and identifying possible optimization levers is difficult and cannot be handled by databases.

It is often necessary to consult with customs authorities to be absolutely sure of the classification of a product and the duty rate that applies. If customs duties account for a significant share of total landed costs, then it may make sense to analyze the effect of complicated duty refunds and their impact on the location configuration.

> **Customs duties can be minimized by coordinating the selection of production locations and product design**

Optimizing product design, assembly sequence, and the choice of location to minimize customs duties and taxes can pay off. Managers should focus on

---

[36] *I.e., the average of all customs duty categories, without considering the value of imports/exports per category. A weighted average would factor in the value and thus determine the actual average rate paid. The weighted average rate is typically somewhat lower than the unweighted average rate.*

bringing finance staff, internal auditors, product designers, and production planners to the table to jointly evaluate potential and develop a coordinated approach.

When designing production networks, production planners and engineers often think in terms of equipment, operational sequences, and component features, while business graduates, lawyers, and internal auditors are more concerned with legal entities, legal forms, financial statements, and contractual agreements. The greatest challenge is to align and harmonize the views (and possibly conflicting objectives) of those involved to benefit the entire company. Adjusting the product design and production sequence to minimize custom duties is clearly one such cross-functional challenge.

Non-tariff trade barriers can take a variety of forms. We will confine ourselves here to a brief description of state-imposed restrictions on the number of business licenses and permits granted to foreign investors, as well as the quota system in the textile industry, and the impact of these regulations for companies (cf. the case examples on the pages that follow). These observations serve merely to illustrate the potential risks that national or supranational regulations may expose companies to.

Entering regulated markets where the state imposes limitations on a company's options has become a strategic dilemma for many companies. Competition for the restricted number of permitted joint ventures or business licenses has driven spending beyond the realizable benefit in some cases (especially if you include the opportunity costs resulting from transferring know-how to locals). The phenomenon can also be observed in mature markets, e.g., during the auctioning of UMTS licenses in Western European countries. However, many emerging markets typically have more restrictions and licensing requirements. It is hard for companies to free themselves from this dilemma and avoid the collective destruction of value. A potential option is to adopt coordinated behavior towards the regulatory state, such as by forming consortiums with competitors.

External Factors – How the Mercedes Car Group (MCG) Selects Production Locations

Mercedes Car Group (MCG)[*], which includes the Mercedes-Benz, smart, and Maybach brands, built 1.2 million vehicles and recorded EUR 50 billion in revenues in 2004. MCG employs approximately 106,000 staff. Its chief markets are Western Europe, where 67 percent of its vehicles are sold (just under half of those in Germany), and North America, which has a 20 percent share.

Mercedes passenger vehicles are made at six different locations (Sindelfingen, Rastatt, and Bremen [all in Germany], Tuscaloosa [US], Juiz de Fora [BR], and East London [SA]). The company also runs three aggregate and component factories in Germany, as well as seven smaller locations for CKD assembly[37] (six of them in Asia).

To produce its vehicles cost efficiently, expansion into international markets requires different strategies. When a steady demand for more than 100,000 units per year arises, the company feels it is time to establish a full-grade factory abroad. In 1996, for example, the Tuscaloosa factory was opened to supply the North American market.

Where market demand is lower, economies of scale make full local production too expensive.

---

[37] *CKD (completely knocked down) denotes a type of production in vehicle manufacturing where assembly kits are supplied for export to individual countries, rather than complete vehicles.*

[*] *As of October 2007, Mercedes-Benz Cars.*

However, if customs duties are high and there are restrictions on market access, CKD assembly is still one way for the company to produce locally at a competitive cost. Although variable production costs are higher with CKD assembly, fixed costs are substantially lower than those for a full-grade factory. This means that the required unit production volume is smaller. For example, if the customs duty is 50 percent on vehicles and 20 percent on parts and components (i.e., if there is a difference of 30 percentage points between tariffs), local assembly already starts to become cost-effective once around 1,000 vehicles or more are produced (Figure 2.27).

Production in CKD assembly factories uses parts kits that are imported from Germany and have locally procured components added. The CKD parts kits even include bulky parts of the bodywork (Figure 2.28) that are manufactured centrally due to high tool costs. These parts are welded and paint-ed in the CKD assembly factories; final assembly is carried out on simple lines with little automation.

The CKD assembly concept enables OEMs to respond flexibly to changing regulations in various countries. The creation of the Asian Free Trade Area (AFTA) represents a particular challenge for producers and suppliers in this context. To qualify for AFTA, companies in the ASEAN region have to reach a 40-percent share of value added. CKD assembly can be a starting point, and can enable an increasing share of local value added (in house, but also via local sourcing).

When choosing locations and allocating new production lines to individual factories, companies naturally also need to examine other parameters in addition to the costs involved. For its R-Class vehicle, for instance, the company considered the impact of exchange rate fluctuations on product profitability (net present value) (Figure 2.29). When selecting the production site for this product line, the company expected more than half of R-Class vehicles would be sold in the US. Given the expected sales footprint, the currency risk is significantly lower when production is carried out in the US than at a location in Germany. A significant

Fig. 2.27: Diagram for deciding whether to establish CKD* assembly

CONCEPTUAL

Difference in customs duties between vehicle (assembled) and components/parts
Percent

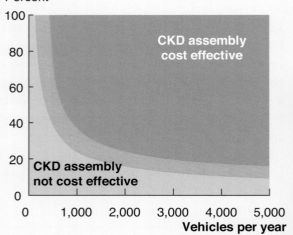

* CKD: completely knocked down
Source: DaimlerChrysler (2005)

Fig. 2.28: Mounting body parts prior to welding using handheld clamps

Source: DaimlerChrysler (2005)

share of components and parts for the R-Class are manufactured in Germany in any case. This share of value added in the eurozone corresponds to roughly the share of unit sales expected there. The balance of sales and value added in both major currency zones nearly eliminates the exchange rate risk. If the final assembly for the R-Class were to be located in Germany, the expected imbalance and hence exchange rate risk would be much higher.

Bottom line: Production close to the market is key to automotive OEMs. Trade barriers and risk considerations are particularly important reasons why CKD assembly opens up new opportunities to enter nascent markets. A clear concept for setting up and supplying CKD assembly factories can help companies to produce cost-effectively even with low volumes.

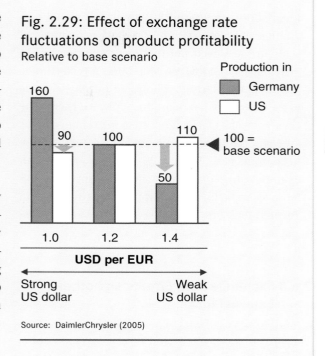

Fig. 2.29: Effect of exchange rate fluctuations on product profitability
Relative to base scenario

Source: DaimlerChrysler (2005)

Radical Change in the Textile Industry: The Consequences of Hampering Globalization

The textile industry demonstrates how radically sectors change once trade barriers fall, and what challenges the companies affected face. Since the **quota system** that was set up in 1974 was abolished on January 1, 2005, global trade flows have been transformed dramatically to favor exports from China. In recent years, tough competition on the Chinese market has led manufacturers such as Fapai Fashion to become highly productive. They have now established reliable processes and are much more quality conscious. They have built up their production capacities and improved productivity so that they can utilize the opportunities presented by the free trade in textiles. Since 1994, 55 percent of all web machines produced worldwide have been installed in China. From January

2004 to January 2005, the volume of exports from China to the US rose by 75 percent, and to Germany by 115 percent.[38] This radical upheaval in a sector where globalization had previously been impeded is set to continue.[39] In addition to low wages,[40] strong regional specialization has been an important driver of success.

In the region around Datang, China, for example, an estimated 9 to 10 billion socks were manufactured in 2004 by approximately 2,500 compa-

---

[38] *Cf. International Textile Manufacturers Federation, Global Trade Information Service (Chinese Ministry of Commerce).*

[39] *Cf. Breuer (2005).*

[40] *Some authors also regard other factors, e.g., subsidies and favorable exchange rates, as reasons for China's immense competitiveness in the textile industry. However, the huge difference in wage costs is almost universally recognized as the main factor; cf. also Lee (2005).*

nies.[41] Because around 30 percent of global production is amassed in one region, economies of scale are achieved not just in production but in distribution and R&D as well. Datang is starting to appeal particularly to wholesale purchasers due to the large number of producers based there. The close network makes it possible to exchange ideas and experience efficiently and develop new products at low cost. The textile industry provides an example of revolutionary change caused by differences in factor costs, but this is taking place in abrupt steps due to changes in state regulation.

This change is not an isolated case. However, the transformation that occurred in other industries, e.g., the manufacture of printed circuit boards in Taiwan or TV sets in China, was more continuous and therefore less noticeable and spectacular.

Bottom line: The formation of clusters in low-cost countries will drive the manufacture of certain product groups almost entirely to these regions.

---

[41] Cf., e.g., Fortune (2004).

2.6.3 Exchange Rate Effects and other External Risks

The current assessment and future development of factors influencing the design of production networks are subject to uncertainty. **Unknown** events with an **unforeseeable** influence on the company, whether natural disasters or terrorist attacks, and **known** events with **unforeseeable** probability, such as exchange rate trends, are difficult to evaluate systematically. They can, however, be countered with diversification strategies.

Events with sufficiently **foreseeable** probability, e.g., automobile claims rates, are easier to quantify. If discrete in nature, they represent real options.[42] Events such as capacity expansion, maintaining overcapacity, being partially unable to fulfill demand, factory closures, or the outsourcing of production volume are regarded as real options on which an entrepreneur can take action. If concrete probabilities can be calculated for external parameters such as exchange rate developments, a company can determine the strategy with the highest expected return and use it to manage risks effectively.

The impact of exchange rate fluctuations is amongst those risks that companies only incur once they globalize their sales and operations. Related issues do not come into play if a company just operates na-

tionally. This means the issue is new and requires management attention and proactive learning. Few companies institutionalize this learning process to familiarize management with the challenges ahead. From subjective observation, most of those who do are based in India and China. It is also important for companies to realize that globalization of the corporate presence – in sales, procurement, and production – can also result in risk diversification, risk avoidance, and the use of arbitrage effects. Globalization can thus reduce risk exposure, especially to catastrophic events.

Uncertainties that normally become relevant to companies only when they start to globalize production and sales include the following:

■ Effects of exchange rate fluctuations

■ Changes in tariff and non-tariff trade barriers

■ Fluctuations in the duration and cost of transportation by sea or air, including customs clearance.

Uncertainties that assume new importance for companies once they start to globalize include:

---

[42] Cf. Cohen (1998), p. 7.

■ Changes in legal regulations, e.g., property law and company law

■ Specific bureaucratic processes such as licensing procedures

■ Uncertainties in the supply chain due to the length and complexity of transportation routes and communication hurdles.

The following section only discusses the effects of **exchange rate fluctuations**, and the special risks to intellectual property of foreign engagements. These examples highlight actions that are also applicable to other fields.

2.6.3.1 Risks Resulting from Exchange Rate Fluctuations

Companies see **exchange rate effects** as an important issue that is perceived and communicated particularly when **unfavorable influences** hold sway. An analysis of 50 corporate communications[43] that were selected on a representative basis shows that companies mention exchange rate effects particularly when they have had a negative impact on profits. Around 80 percent mention exchange rate effects to explain a nominal fall or a smaller than expected rise in revenues and profits. Around 20 percent are neutral or highlight a positive development – these are mostly companies that regularly report the influence of exchange rate fluctuations.

On average, decision makers assign the criterion of exchange rate influence only moderate relevance in making location choices.[44] Managers generally have little perception of their ability to influence exchange rate effects by their choice of production locations and suppliers. Most associate currency risk hedging more with financial instruments. Application of these instruments is often regarded as too laborious, however, or too restrictive on a company's financial scope, due to their impact on the company's credit line.

It is also apparent that the **use of financial instruments** alone does little to reduce exchange rate influences on profits and cash flow in the long term.[45]

The ineffectiveness of financial instruments in reducing the volatility of profits and cash flow is partly because companies often do not have sufficient clarity on their current status. The complex interdependence between the prices of end/intermediate products and exchange rates – and between different currencies – means it is hard to determine which imbalances require hedging and over what period.

However, this does not apply to products that are already covered by purchase contracts detailing a fixed purchase price in the nominal currency, where the risk from exchange rate fluctuations is fixed in terms of both currency and level (transaction exposure). If the lead time for these purchase contracts is long, as is usually the case with aircraft manufacturing, for instance, financial instruments can be used effectively for hedging. Airbus Industrie and its parent company EADS hedge their supply contracts with airlines and finance organizations correspondingly. These contracts are based largely on the US dollar and concluded with an average lead time of around four years (see box for details).[46]

The real impact of exchange rate fluctuations on profits is not just determined by the share of costs and revenue in different currencies and the exchange rate fluctuations. Real exposure and nominal exposure can be quite different. The key reasons for this are fourfold:

■ **Market prices for intermediate products** and services are dependent on exchange rates. This dependence on the exchange rate between the nominal currency and the lead currencies – the US dollar,

---

[43] Balanced sample of European and US companies in the year 2000 when the US dollar was strong and/or the euro was weak, and in 2004 when the US dollar was weak and/or the euro was strong.

[44] ProNet survey: "Exchange rate influences" are ranked 9th out of 16 criteria in terms of their relevance.

[45] Cf. Copeland (1996): Comparison of average fluctuation of profit and cash flow of two corporate groups. The (smaller) group that uses financial instruments to hedge against exchange rate influences (financial hedging) has no fewer fluctuations. The authors conclude that operational hedging is more effective.

[46] Cf. EADS annual reports.

Currency Risks and Competitive Strategy – Airbus Uses Financial Derivatives to Hedge Against Exchange Rate Fluctuations

Large commercial airplanes are sold almost exclusively in US dollars, but more than half of Airbus' costs are tied both nominally and in real terms to currencies outside the US dollar zone. For the year 2004, EADS assumed a real currency imbalance (effective exposure) of around USD 10 billion p.a. for Airbus. The costs and revenues of its main rival Boeing are generated almost entirely in US dollars. A weak euro is therefore a huge benefit to Airbus. Between the years 2000 and 2002, Airbus had a major cost advantage over Boeing due to exchange rate effects, resulting in greater pricing flexibility. The fact that the airplanes ordered are mostly only delivered years after contracts have been concluded is largely irrelevant, given that the use of financial derivatives allows the company to carry the current exchange rate forward into the near future virtually unchanged.[47]

At the end of 2002, Airbus had hedged the advantageous exchange rate of 2000 to 2002 for almost the entire order book (Figure 2.30). Using derivatives trades directly linked to purchase contracts (micro-hedging), the company had hedged roughly USD 43 billion at the end of 2003 at the average rate of USD 0.95 to EUR 1. The impact achieved in 2004 and 2005 was around EUR 2 billion, which was double the net profit for the year. The favorable exchange rate conditions must have helped Airbus to expand its market share. This was less than 20 percent (as a share of purchase orders) in 1995 at a rate of USD 1.46 to EUR 1, but by 2002 the company had achieved a market share

---

Expiring derivatives enabled EADS to absorb a profit impact of approx. EUR 2 billion in 2004 and 2005

Fig. 2.30: Exchange rate hedging at EADS/Airbus

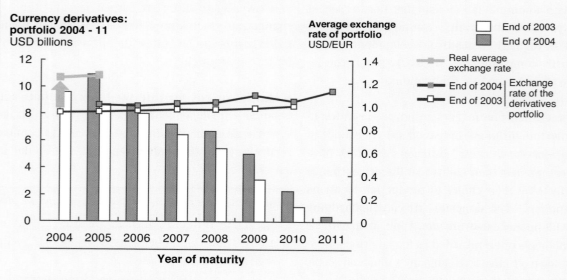

Source: EADS Financial Reports 2003 and 2004

---

[47] The rates for forward sales with a term of up to five years are typically close to the current exchange rate.

of more than 50 percent for the first time ever, at a rate of USD 0.94 to EUR 1.

In 2003 and 2004 the situation changed: the US dollar lost more than 30 percent in value against the euro. When this trend set in, EADS initiated the cost-cutting program "Route '06." The targeted EUR 1.5 billion in savings will be needed just to compensate for the euro's return to a long-term average rate of around USD 1.16 to EUR 1. EADS' use of financial derivatives had bought it an advantage: time in which to react.

Bottom line: Companies should take currency imbalances and exchange rate fluctuations into account when defining their competitive and location strategies, especially in oligopolistic markets. Hedging purchase contracts with financial derivatives can help a company make better use of favorable exchange rate conditions and win time to adjust to a poorer exchange rate situation. Financial derivatives, however, do not mitigate the long-term economic risk that lies in the effective exposure to exchange rate fluctuations.

euro, and yen – differs from one product to the next. Adjustments in local market prices as a result of changes in exchange rates must be taken into consideration to determine the real impact of exchange rates.

■ **Market prices for the company's own products** are dependent on exchange rates:[48] A company may need to adjust the prices of its own products to a new market price level to maintain market share (in the event that the local currency falls in value), or an adjustment may be advantageous because higher prices can be achieved by adjusting to an increased market price level.

■ **Exchange rates are interdependent:** Currencies within closely interlinked economic areas, such as Denmark and the eurozone, fluctuate less than those in currency areas with relatively weak trading relationships. This seems plausible given that trade makes it possible to even out imbalances more quickly. This dependence can be determined via a covariance analysis, and folded into simulations.[49]

■ **Accounting** on the balance sheet date (translation exposure): International companies possess both assets and liabilities in other countries. The value of both is reported in the local currency, not in the balance sheet currency of the parent company. If

exchange rates fluctuate, the balance sheet valuation may change, resulting in book gains or losses.

These interdependencies make it impossible for simple models to sufficiently explain the impact of exchange rate fluctuations on company profits, cash flow, and market capitalization.[50] The following examples examine the dependence of local prices on the exchange rates causing the effective economic exposure. This analysis shows that the argument that exchange rate fluctuations are balanced out by changes in purchasing power[51] holds true for some goods that are traded globally, but not for others.

The dependence of market prices on exchange rates is one reason why nominal and real currency imbalances are not identical. Most companies are not aware of the real imbalance and tend to overestimate it.[52] Determining the actual influence of changes in exchange rates ex ante is difficult, particularly as the change in the price level for intermediate and end products varies (for multiple reasons) depending on the item. The influence of exchange rate fluctuation on prices is hard to isolate and tends not to be suffi-

---

[48] Cf. Hau (1999).

[49] Cf. also Billio (2002) for simulating time sequences of exchange rates.

[50] Cf. Bodnar (2003).

[51] Cf., for example, Madura (2003).

[52] Cf. Copeland (1996).

ciently transparent to companies. To exemplify this, we will analyze the dependence of price levels of selected product groups on the development of exchange rates of the euro or the German mark and the US dollar. Figure 2.31 shows the price indices for dairy products based on local currencies in Germany and the US, as well as the exchange rate ratio (indexed).

We can see that the price indices for dairy products in local currency exhibit relatively little volatility compared to the exchange rate ratio. There was a significant rise in the prices in the US, even though the US dollar was high vis-à-vis the euro/German mark during the period under review, so that a relative drop in US price levels (driven by the exchange rate) would have been expected.

Figure 2.32 shows the time series adjusted for the long-term (linear) trends. This shows that the short-

and mid-term differences between price levels in Germany and the US correlate closely with the exchange rate ratio, i.e., prices in local currency are largely independent of this ratio. This seems plausible because most dairy products are produced locally. The share of transatlantic trade in domestic consumption on either side of the ocean is very small due to low value density, perishability, different customer preferences, and trade restrictions.[53]

In the case of dairy products, the exchange rate effect accounts for 66 percent of absolute fluctuations (i.e., the short- and mid-term differences) in the price index for Germany vs. the US if compared on a US dollar basis. With light fuel oil, by comparison, only 11 percent of the short- and mid-term absolute price difference results from exchange rate fluctuations, i.e., the prices in local currency largely track exchange rate ratios. The impact of exchange rates on the price difference of spot prices for crude oil is

Prices for local products are barely affected by exchange rate fluctuations

Fig. 2.31: Price index for dairy products in Germany and the US, and exchange rate relationship

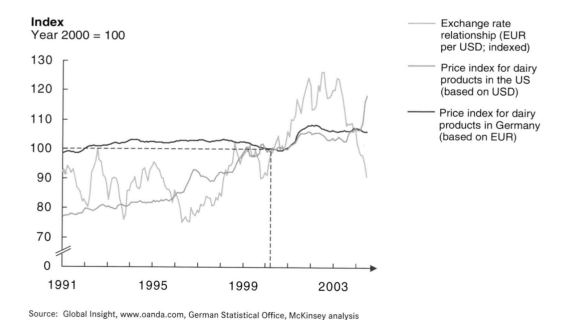

Source: Global Insight, www.oanda.com, German Statistical Office, McKinsey analysis

[53] Database: German Statistical Office (1995).

even less, and is virtually zero when it comes to trading in futures. This yields an almost perfect market for crude oil of the same quality. The crude oil price in euros tracks the crude oil price in US dollars (and vice versa), based on current exchange rates. The exchange rate ratios are almost completely reflected in the nominal price ratios.

Buying crude oil at current market prices in euros and selling it in US dollars thus entails a much lower risk than buying milk (in Europe) in euros and selling it (in the US) in US dollars. Should the euro vs. US dollar exchange rate change, the market price in euros for crude oil would, too. The price of milk in euros would not change, as the cost base for the production of milk is largely euro-based. The company

would therefore have to bear the price difference between the altered purchasing price in US dollars and the unchanged sales price in US dollars.

> **The effective exposure to the US dollar is low with purchases of crude oil, regardless of the nominal currency[54]**

The effective exposure for companies is not just dependent on the nominal currency in which preliminary products are bought and end products sold, but also on **price reactions** for the goods due to exchange rate fluctuations.

If the future structure of revenue and costs is unknown or cannot be forecast, effective hedging with

> **Difference in the US dollar price of local products in different currency zones fluctuates with the exchange rate – so local prices are not impacted by exchange rate fluctuations**

Fig. 2.32: Difference in the prices of dairy products in the US and Germany vs. currency effects

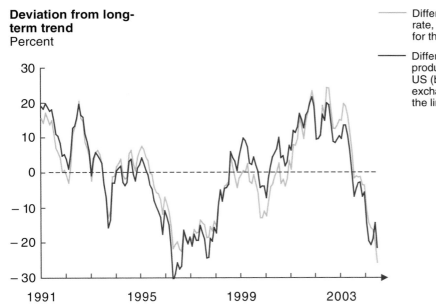

Deviation from long-term trend
Percent

——— Difference in the exchange rate, EUR to USD (adjusted for the linear 12-year trend)

——— Difference in price of dairy products in Germany and the US (based on current exchange rate adjusted for the linear 12-year trend)

Source: Global Insight, German Statistical Office, McKinsey analysis

---

[54] *Assumption: currency risks are hedged against using financial derivatives for trading in commodity futures (e.g., credit sale, i.e., with future delivery date and payment).*

financial instruments is impossible. Applying the principles of the real option theory to the design of global production networks can help to minimize risks and maximize expected profits even under these conditions. Operationally, the uncertainty created by exchange rate fluctuations can be countered via the following **measures:**[55]

■ **Static**

□ **Elimination:** Equalization of the effective exposure by adjusting the effective cost and/or sales footprint

□ **Transfer:** Price adjustment clauses in contracts, price adjustments, or conclusion of contracts in another currency[56]

■ **Dynamic**

□ **Production:** Temporary relocation of production and related wage and salary payments, ancillary wage costs, and indirect costs to a currency zone with a devalued currency, i.e., with lower prices for input factors compared to locations in other currency zones. This kind of relocation can be achieved relatively easily for simple production processes with low fixed costs. Hurdles are higher with increasing capital and know-how intensity.

□ **Sourcing:** Temporary relocation of costs for materials and services to a currency zone with a devalued currency, i.e., where materials prices are lower compared to locations in other currency zones. This requires supplier contracts in different currencies, and suppliers with production in different currency zones.

□ **Sales:** Increased efforts to improve sales in markets where the value of the currency has appreciated – leveraging the lower prices in local currency if necessary to increase competitiveness and thus market share (while passing on part of the exchange rate advantage to the customer). The other side of the coin is to raise prices in markets

with devalued currencies (lower product prices in the currency basket in which costs are incurred), compensating for the decline in profit margin. A drop in sales volume (due to higher prices in local currency) is typically quite acceptable if the move can prevent negative profit margins.

□ **Product selection:** Combined use of the three levers that have already been mentioned by pushing products with attractive production, sourcing, and sales structures.

Figure 2.33 shows the key measures that can be used as part of a global production strategy, together with their potential impact. Ex-post analysis shows the potential benefit of reactive adjustment to cost and sales footprint. Assumptions are made in the simulation about the price elasticity of demand for the end product and the dependence of the market price on exchange rate relationships. The simulation roughly models the situation of a European manufacturer of higher value auto parts.

Various simulations of the impact of exchange rate fluctuations on product margins under different assumptions reveal: a far-reaching, dynamic adjustment of the cost and/or revenue structure typically increases the effective currency exposure. At the same time, rigorously adjusting the operations and sales footprint can also raise the expected profits. Given the high exposure to currency fluctuations, however, standard deviation of profit increases. A far-reaching, dynamic adjustment is therefore risky, particularly if exchange rates fluctuate cyclically.

As a pragmatic solution to this issue and bearing in mind flexibility costs,[57] we suggest defining thresholds for the maximum exposure to each currency. Dynam-

---

[55] Cf. also Boyabatli (2004).

[56] Cf. Min (1991): results of a survey looking into the strategies employed by US companies for sourcing from abroad (incl. payment patterns and flexibility).

[57] Cf. Huchzermeier (1996): Simulative approach to determining the value of real options in product selection (sourcing structure) and in the choice of a company's own production locations, taking into account the interdependencies of exchange rates.

ic adjustments of the operations and sales footprint should take place only within these boundaries. This solution has been implemented in a model described in Chapter 4. Interestingly, the flexibility required to achieve a substantial reduction in profit volatility and increases in expected profit is limited. Shifting 3 to 6 percent of costs in one year is typically sufficient to reap most of the benefits that can be achieved by a dynamic adjustment of a company's footprint.

2.6.3.2 Risks to Intellectual Property

Risks from the violation of industrial property rights such as patents and brand names increase enormously with globalization. So does the wrongful dissemination and use of know-how by business partners, staff, and third parties.

Companies in developing and newly industrialized countries exploit the fact that some states are largely unable and unwilling to enforce intellectual property rights by manufacturing and marketing products illegally. Imports seized by customs authorities in other countries cover the entire spectrum – from DVDs to cell phones, right through to spare parts for cars (Figure 2.34).

The infringement of industrial property rights is often even more important in developing and newly industrialized countries themselves, as the hurdles to marketing such products are lower. The law and its enforcement are often inadequate to protect MNCs' brand names and intellectual property rights. Taking software as an example, there is a clear negative correlation between prosperity and the frequency with

---

Companies can largely avoid negative exchange rate effects

Fig. 2.33: Initiatives* to reduce risks from exchange rate volatility for products with local production/little dependence of local price on exchange rate

■ Can be directly influenced by choice of location

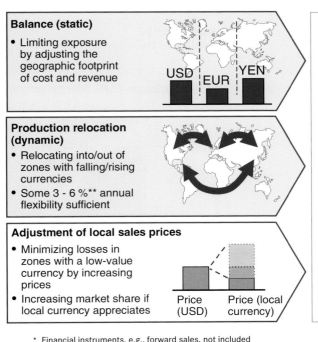

Balance (static)
- Limiting exposure by adjusting the geographic footprint of cost and revenue

USD EUR YEN

Production relocation (dynamic)
- Relocating into/out of zones with falling/rising currencies
- Some 3 - 6 %** annual flexibility sufficient

Adjustment of local sales prices
- Minimizing losses in zones with a low-value currency by increasing prices
- Increasing market share if local currency appreciates

Price (USD)   Price (local currency)

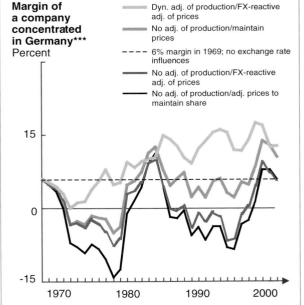

Margin of a company concentrated in Germany* Percent

▬ Dyn. adj. of production/FX-reactive adj. of prices
▬ No adj. of production/maintain prices
- - - 6% margin in 1969; no exchange rate influences
▬ No adj. of production/FX-reactive adj. of prices
▬ No adj. of production/adj. prices to maintain share

15

0

-15

1970   1980   1990   2000

* Financial instruments, e.g., forward sales, not included
** Optimum value depends on the actual development of the exchange rates, price elasticity, and the cost of making production and purchasing more flexible
*** Cost at 65% EUR/25% USD/10% JPY, revenue at 50% USD/30% EUR/20% JPY, annual shift: 3%

Source: McKinsey (ProNet hedging model)

which industrial property rights are violated (Figure 2.35).

While the infringement of industrial property rights is a latent risk that companies need to manage adequately, the dissemination of knowledge in the context of globalizing production represents a specific risk that is more closely associated with internal company decisions. When setting up a new site, a company passes on know-how to local staff, suppliers, public authorities, JV partners, and others, all of whom could misuse it. If trademark rights are violated, a company not only suffers direct economic damage as a result of lost sales: its image is also adversely affected if the counterfeit products are of inferior quality.

Initiatives to prevent, identify, and respond to such problems can help companies minimize the risks re-

sulting from the transfer of knowledge and infringement of industrial property rights. One step that companies frequently have to take is to apply for property rights (patent, brand, utility-model patent, design patent, or copyright) in all relevant countries so that it is subsequently possible for them to assert and enforce claims. The corresponding costs should be taken into account when considering entering a market or region. Companies should also be aware, especially when planning to enter a market in Southeast Asia or China, that existing industrial property rights are vital to minimize the risks from abuse but are in no way sufficient by themselves.

One possible **preventive measure** to hinder the abuse of industrial property rights is the use of counterfeit-proof components and markings (labels, security strips, security labels, holograms, microtags,

---

Infringement of industrial property rights is highest with imports from China and Thailand

Fig. 2.34: Confiscation by German customs authorities

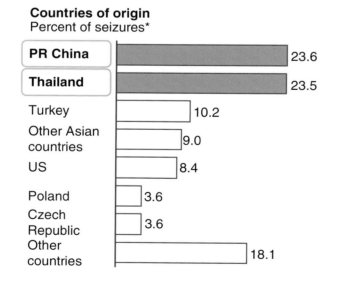

Countries of origin
Percent of seizures*

- PR China — 23.6
- Thailand — 23.5
- Turkey — 10.2
- Other Asian countries — 9.0
- US — 8.4
- Poland — 3.6
- Czech Republic — 3.6
- Other countries — 18.1

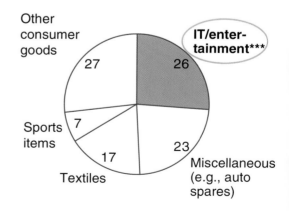

Industries concerned
Percent of seizures by value of goods**

100 % = approx. EUR 145 million

- Other consumer goods — 27
- IT/entertainment*** — 26
- Miscellaneous (e.g., auto spares) — 23
- Textiles — 17
- Sports items — 7

* In 2004, the German Customs Office seized a total of 8,564 items
** Incl. economic damage
*** Software/hardware, image/sound/data carriers

Source: German Customs Office (2004), industrial property rights protection, annual report 2003

etc.). These enable the company's own staff, customers, and customs authorities to identify counterfeit products quickly. The risk of wrongful knowledge transfer can be reduced by producing critical components in countries with a high level of protection, though this does substantially limit the company's scope in choosing locations. Another option is to divide up the manufacture of products. This can reduce the risk significantly, because none of the component suppliers or manufacturing service providers is familiar with or in control of the entire production process. Appropriate options should be provided when defining the product architecture.

Regular visits to relevant trade fairs may also help to identify violations of property rights: this is where around half of all such infringements in mechanical engineering and plant construction are detected.

This is because product and brand counterfeiters depend on broad access to the market, and thus frequently exhibit at relevant fairs. A targeted analysis of the data relating to sales and complaints can alert companies to competitors who are acting wrongfully. Field sales staff should be made particularly aware of the potential risks and indications of counterfeit products in the market, as they are usually the first to encounter them. Collaborating (even with competitors) in professional associations and with dealers can uncover property rights violations faster. A strong incentive for this type of cooperation is that companies involved in illegal activities often counterfeit products belonging to several brands.

One of the most effective **reactive measures** is the immediate seizure of counterfeit products. Customs authorities – particularly in developed countries –

Intellectual property is often poorly protected in LCCs

Fig. 2.35: Technology protection and wage levels

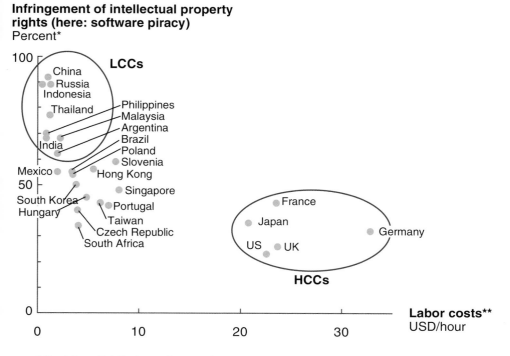

* Illegal share of total business software used
** At long-term average exchange rates (e.g., EUR = USD 1.16)

Source: Business Software Alliance, Global Software Piracy Study 2003, McKinsey/PTW (ProNet analysis)

tend to regard the prosecution of brand and product piracy as an important task. They will independently prosecute over violations, too, but reporting suspicions can greatly facilitate the work of customs authorities in countries involved in production, transit, and distribution. A complaint against an importer and manufacturer in the country of production can prevent the infringement of property rights in the long term. MNCs should also think about complaints in countries other than where they are based or where the infringement occurred, such as the US. This can also improve the chances of an amicable arrangement with the manufacturer or importer, which often represents the most profitable solution.

The expansion of the WTO (World Trade Organization) is committing an increasing number of states – including many Asian members – to effectively combat infringements of property rights that originate in their country. The multilateral TRIPS Agreement (Agreement on Trade-Related Aspects of Intellectual Property Rights) specifies minimum standards for the protection of intellectual property that all WTO member states have to meet. These minimum standards comprise legal guidelines and assurance that proceedings will actually be conducted against violations of property rights that originate in these countries. The WTO provides a committee that settles any disputes relating to the TRIPS Agreement, and imposes trade sanctions if the agreement is breached.

* * *

Companies should be aware that globalizing production and sourcing exacerbates the risks they face but can also have a mitigating effect. Determining country-specific cost of capital and limiting currency exposure are two pragmatic approaches that can help incorporate risk into a holistic, quantitative network assessment. Beyond this, companies need to define concrete action on the most relevant risks. Risks to intellectual property, for instance, can be attenuated by dividing up production and assigning orders to several suppliers, as well as by applying for additional intellectual property rights in relevant markets.

2.7 Handling Migration – Transition Financials

When establishing additional production capacity at a new location abroad, investments and expenses for the production ramp-up have to be financed exclusively from the company's free cash flow or by borrowing – a step that companies often find hard to take. It is also harder to pledge securities for loans in a foreign country. Where existing sites are to be closed, the additional outlay on machinery and buildings can be reduced by transferring and selling existing assets. Especially for companies in Western Europe and Japan, however, restructuring costs offset such income. Relocation expenditure may significantly weaken the economic viability of production network redesign or make certain parts of it unattractive.

2.7.1 Investments

Investments are capital expenditure **for tangible assets** (land, buildings, and equipment) and **intangible assets** (e.g., patents, software). This expenditure is capitalized and not directly expensed. This is important as capital expenditure impacts a company's cash flow immediately, while it only affects the P&L statement via depreciation of the asset value over time. Both financial perspectives are valid in assessing the attractiveness of different production network structures (cf. sections 3.2.3 and 3.3).

The static analysis of a company's economic viability takes into account the costs associated with the use of equipment via depreciation. This perspective also applies to the P&L statement. In principle, assets are depreciated over their expected economic lifetime. Exceptions exist in which either accounting rules or tax laws prescribe depreciation periods that are significantly different from the expected economic lifetime. If this is not the case or the impact of such tax payments is small, the use of the economic lifetime concept seems most appropriate for the purposes of investment analysis.

In contrast, a cash flow analysis considers all cash income and expenditure associated with ongoing

Suspicion of Product Piracy: GM Sues SAIC Subsidiary Chery

At the end of 2003, General Motors (GM) launched a passenger car in China called the Chevrolet (commonly called "Chevy") Spark. It was a slight modification of the Daewoo Matiz,[58] a successful compact car in South Korea since 1998. Six months earlier, however, the Chinese company Chery (a subsidiary of SAIC, a GM joint venture partner) presented an externally similar vehicle at a trade fair: the Chery QQ (see Figure 2.36). According to GM, the Chery QQ was largely an imitation of the Spark: not just the bodywork, but also the optional features, chassis, and powertrain. Allegedly, a door of the Chevy Spark would fit into a Chery QQ without any modifications. It was suspected that internal product data belonging to the joint factory SAIC-GM-Wuling, which manufactures the Chevrolet Spark, had been passed on to Chery.

The Chery QQ was an instant success: the earlier market launch and the lower price compared with the Spark (the QQ had a price tag of RMB 55,000; the Spark cost RMB 67,000) meant that it generated some USD 175 million in revenues by the end of 2003. Spark sales, on the other hand, were very sluggish (only 200 or so registrations in 2003).

In 2004, GM tried to resolve the conflict with Chery in mediation proceedings, conducted via Chinese authorities. These proceedings came to nothing and, at the end of 2004, GM initiated a lawsuit in a Chinese court. The parties reached an undisclosed settlement a year later.

Bottom line: Distinguishing between infringements of property rights and the legitimate use of (transferred) know-how is not always easy. The economic damage from potential misuse may be very high, and compensation may not be legally enforceable.

Fig. 2.36: Product comparison – Chevrolet Spark and Chery QQ

Daewoo Matiz/Chevrolet Spark

Chery QQ

	Daewoo Matiz/Chevrolet Spark	Chery QQ
Manufacturer:	Daewoo, South Korea/SAIC-GM-Wuling, China	Chery, China
Parent company:	GM and SAIC ⟶	SAIC
Market launch:	1998 as Matiz (South Korea), end of 2003 as Spark (China)	Mid-2003 (China)
Initial price:	RMB 67,000	RMB 55,000
Sales in China		
• 2003	200	25,200
• 2004	10,100	49,100
• 2005	26,900	115,000
• 2006*E	36,800	114,400

* Through to the end of Nov; Chery sales do not include successor model QQ6

Source: China Business Info Center, June 20, 2003 (CBIZ.cn), Automotive Resources Asia (2004); China Market News Release 2003 Chevrolet Germany, Motor Information Service (mid)

[58] *Daewoo has belonged to GM since 2001.*

business operations. This dynamic perspective can also be used for assessing the economic attractiveness of investment projects (see section A.3.2: Dynamic investment analysis techniques). Unlike the analysis of a site's profitability, a cash flow-based analysis provides insights into a project's impact on the company's liquidity and allows comparison of alternative investment opportunities by calculating the net present value based on the cash flow impact.

The **sale of assets** that a company no longer needs for its operations because it is relocating its production capacity may help a great deal in financing the production network redesign. Since the prices of land and buildings are generally higher in HCCs than in developing countries, companies can expect the proceeds from sales to make a major contribution to financing the new location. Of course, this only applies if suitable assets are owned at the site being closed. However, it may be advisable for companies to exclude the effects of such income (at least in part) when evaluating location options. The sale of buildings and land often raises hidden reserves, the addition of which is not causally connected to the investment project that is to be evaluated. As a rule, the sale could also take place even if no new location were being planned (e.g., through a sale and lease-back deal). An evaluation based on the assumption that all buildings and land are leased often creates a better basis for comparison.

In accordance with this logic, it is also important to make sure that equipment that has only partially been depreciated does not have any influence on future investment decisions. Write-offs have no impact on cash flow. Given that the procurement of new machinery is already covered in the form of the appropriate investment, the dynamic analysis also maps the economic implications of this.

For a company to make an investment, it requires **liquid funds** that have to be generated by ongoing business activities or raised by loans or capital increases. The restriction of limited liquid funds must be borne in mind when planning the design of the production network. Financial planning for the construction of a new production location should not ignore the fact that income from, for example, the sale of a building no longer required at the old location is often only received after the expenditure has been incurred, e.g., for the construction of a new factory, and that temporary liquidity will therefore be needed. A company's liquidity becomes of specific concern if its performance in the recent past has been poor and has weakened its balance sheet and liquidity. Managers need to keep in mind that the redesign of a production network typically requires significant outlay and only pays off after some years. It is essential to stay ahead and initiate the necessary adjustment early, from a position of strength.

2.7.2 Cost of Production Ramp-Up

The ramp-up of a new production site requires a variety of one-time expenses. These need to be monitored carefully and often allow for significant savings if the process is planned well. They are, however, not the most crucial aspect. The key success factor is ramp-up speed. Potential losses in production volume due to a late or slow ramp-up to full production volume have far greater financial implications. Reaching full production capacity late can have a very negative impact on the economic viability of the production relocation, particularly if products with a short life cycle are involved. If sites are expected to have different ramp-up curves, the financial implications must be included in the location decision.

Ramp-up expenditure typically cannot be capitalized; it counts as operational expenses, and is accounted for as costs in the period in which it is incurred. As such, ramp-up expenditure has an immediate impact on both the profitability of the business as well as its liquidity. Other costs result from initial below-capacity utilization or required duplication of plants. Many of the additional costs only become apparent during construction of the facility abroad, and many companies do not make adequate provision for them in their costings at the start of the decision process.

Applying experience from other companies also proves difficult. We found the structure of ramp-up expenditure markedly different from one company to the next. On average, ramp-up costs run to just over 20 percent of the investment in buildings, plant, and equipment (Figure 2.37).

The extra expenses connected with the start of production are not composed of one dominant block of costs, but a number of different items:

■ **Training expenses** for new employees, who are trained either in the home factory or by expatriates at the new production location. The number of days and the nature of the training required to perform a specific process step and avoid any problems during the start of series production should be determined based on the employee's existing qualifications, the process step involved, and the location of the factory. Involving equipment suppliers and local training institutions can help to minimize expenditure on necessary training initiatives. Even then, a visit by at least a small num-

ber of the workforce to the home or lead factory is still often regarded as an essential success factor for production ramp-up.

■ Additional **inventory** may be needed to maintain service levels, allowing for the downtime of machinery that is being relocated and possibly longer transportation time from the new factory to the customer. Safety stock has to be built up from scratch, especially in the case of investments for expansion at a new location. These investments in inventory must be considered as expenditure and affect a company's liquidity. They also result in costs of tied capital. Building up bridging inventory often requires the use of external production capacity or extra shifts. The costs of products made in this way are therefore higher, and have a causal link with the setup of the new location. The higher than normal production costs typically have to be expensed right away, as inventory should be valued at the standard rate (i.e., typically at normal production costs).

Ramp-up costs correspond to roughly 20% of total investments on average

---

Fig. 2.37: Amount and structure of ramp-up costs

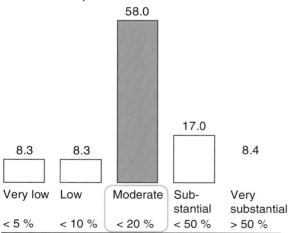

Ramp-up costs compared with investments at the location
Percent of companies interviewed

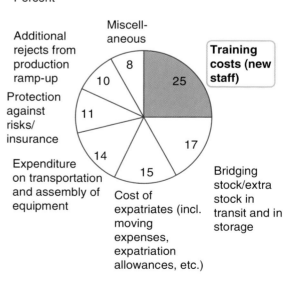

Structure of ramp-up costs
Percent

Source: McKinsey/PTW (ProNet analysis)

- **Spending on expatriates** is country-specific, and therefore estimated separately for each country. In the past, large foreign service allowances, additional bonuses, expense allowances, and cash benefits in kind would be granted specifically for a posting to China. On average, an expatriate in China would receive the following payments on top of normal compensation: roughly 24 percent of the normal compensation as an expatriation bonus, 14 percent as personal expense allowances and cash benefits, and 3 percent additional performance-related bonuses. Added to these are company-paid expenses for the move and school fees, as well as benefits in kind, such as accommodation to Western standards and a company car (often with chauffeur) – in total often an extra 20 to 50 percent or so of base compensation. Additional costs for an expatriate compared to a regular employee based in the home country typically range from USD 80,000 p.a. for a skilled worker to up to USD 220,000 for a factory manager. Since 2003, however, multinationals have become less inclined to make high additional payments given that the market for international skilled workers and executives is now broader-based, particularly in Asia. Even though demand continues to be very high, there is now a substantial pool of expatriates who wish to remain in the relevant countries longer term, as well as a greater willingness on the part of young executives to spend time abroad, even in developing and newly industrialized countries.

- The costs of transporting and installing existing **plant and equipment** typically cannot be capitalized, and also have to be accounted for as start-up costs for the new production location. These include not only the cost of dismantling production machinery and other plant but also the costs of packaging and loading, transportation, and transport insurance for the equipment, as well as the cost of reassembling, commissioning, and performing acceptance tests for individual machines. The entire intercontinental relocation of a machine tool (that roughly fills a 20-foot container) can be expected to cost in the region of USD 22,000.

- The start of series production at a new location also results in **direct additional expenses** due to rejects (materials and cost of the company's own value added), rework, and extra inspection. Companies also have to take into account approval of the production line by their own quality management departments or by the customer, as well as function and fatigue tests on the first series-produced products.

Ramp-up expenses do not sufficiently differ from one country to the next to have a significant influence on the choice of new locations. However, an increase in the start-up expenses above expectations or a substantial delay in production ramp-up can endanger the economic viability of a relocation, especially if the company is targeting a high return on its capital employed compared to, say, return on sales.

2.7.3 Restructuring Costs

Restructuring measures are part of a management program to significantly change either the business segment covered by the company or the way in which this business is conducted.[59] Redesign of the production network usually falls under this definition. Reserves can and may have to be created for the corresponding expenses, which means these activities may depress company profits before the redesign is actually implemented.

Restructuring expenses and the corresponding initiatives include not just expenditure on reducing the workforce but also one-off expenses for terminating or scaling down the operations of individual factories or entire business units. Expected losses on sales of businesses can also be included in the provisions.

Typical examples of restructuring expenses are redundancy payments, redevelopment costs, and penalties for the early termination of contracts, as well as write-offs and losses on the sale of fixed assets. When including these elements in the calculations

---

[59] *Cf. definition according to IAS 37.*

of economic viability, companies should be aware that only the first group constitutes cash expenditure, and is therefore relevant to the dynamic analysis. Figure 2.38 shows the composition of restructuring expenses for a set of manufacturing companies' operations that have been closed or downsized in recent years. These restructuring costs are clearly dominated by redundancy payments for the company's own staff, though the fact that the regional focal point of the survey lies in Western Europe should be taken into account.

Redundancy payments are virtually impossible to quantify in general terms, as the payments differ so greatly. This is even the case within the same legal environment, as Figure 2.38 (right-hand graph) shows using seven examples of restructuring projects in France. The average redundancy payments made to employees varied between EUR 18,000 and EUR 43,000,

with a mean value of EUR 29,000. This difference is not because of the employment structure (particularly the average of the employee's tenure with the company). The way in which negotiations are conducted with staff representatives also has an influence in many countries, even if legislation or established case law is in place. Companies try to prevent walkouts using amicable settlements. It is particularly vital to uphold continuous operations and maintain constant supplies from the existing plant if production is to be shifted, but capacity is not yet sufficient at the new site to take over fully. Customers will expect continuous supplies even during the transition phase.

The comparison of legally prescribed or typical redundancy payments in different countries clearly shows that the level of **redundancy payments** is a **relevant location-related factor** (Figure 2.39). In

---

Redundancy payments make up by far the largest share of restructuring and plant closure costs

Fig. 2.38: Amount and type of restructuring costs

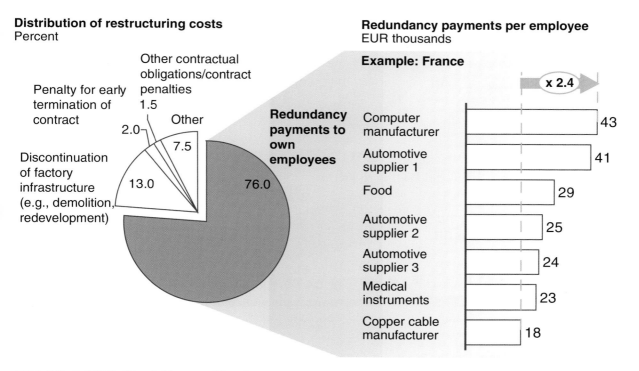

Source: McKinsey/PTW (ProNet analysis), company information

some high-cost countries, the redundancy payments required are so high that investors cannot pull out on a long-term basis. This relative inflexibility has a high price, especially when the possible risks are considered. For example, the typical redundancy payment for an employee in Spain with a company tenure of ten years is ten months' salary, while American employees receive, on average, roughly ten weeks' income. If the different wage level between countries is also taken into account, then the variation in the level of redundancy payments can be as high as a factor of 100. The highest payments are made in Japan and Germany at approximately EUR 65,000 and EUR 39,000 respectively, and the lowest in Russia at a few hundred euros.

In a dynamic investment appraisal, restructuring costs have a similar effect to ramp-up costs: both fac-

tors hamper change in the network due to expenditure that is incurred from expanding capacity, setting up new locations, reducing capacity, and closing locations. However, both cost categories represent one-off expenses and are only pertinent during transition from the existing structure to the target network structure. They are no longer relevant after that.

* * *

This chapter has outlined multiple parameters that influence location choices. These mostly reflect features of countries or regions. We found these parameters to be the most relevant in typical situations, basing this assessment on numerous interviews with managers and a large number of quantitative analyses conducted in the context of case studies and real location decisions. It is important to remember,

Huge disparities between redundancy payments paid in different countries

Fig. 2.39: Income levels and amount of compensation

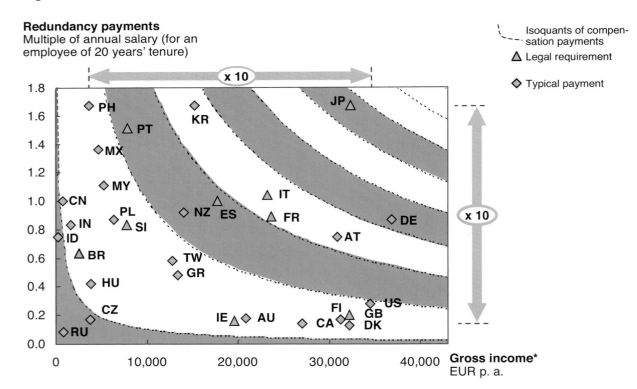

* Including employee-paid share of social security contributions, pretax

Source: UBS Prices and Earnings Report 2003; various reports and company information; McKinsey/PTW (ProNet analysis)

however, that these location-related factors can only deliver conclusions about the attractiveness of a location for production in the context of the products and manufacturing processes considered. Not all parameters discussed in this chapter will be relevant in a specific situation. Identifying which matter most is a crucial objective in the early phases of a location selection process. The insights gained will be important in interpreting and synthesizing quantitative analyses such as investment appraisals.

Further reading

Das, G.: *India Unbound: The Social and Economic Revolution from Independence to the Global Information Age.* New York: Anchor Books, 2002.

Drewry: *The Annual Container Market Review and Forecast 2004/2005.* London: Drewry Shipping Consultants, Ltd., 2004.

Gupta, S. D. (ed.): *Dynamics of Globalization and Development.* Boston: Kluwer Academic Publishers, 1997.

Gutenberg, E.: *Grundlagen der Betriebswirtschaftslehre. Vol. 1: Die Produktion.* Berlin et al, Springer, 1965.

Huchzermeier, A. and M. A. Cohen: "Valuing Operational Flexibility under Exchange Rate Risk" in *Operations Research* (Special Issue on New Directions in Operations Management), 44, January-February, 1996, p. 100 - 113.

ILO: *Yearbook of Labour Statistics.* Geneva: International Labour Organization (ILO), 2004 (also see: http://laborsta.ilo.org/).

IMD: *World Competitiveness Yearbook 2003.* Lausanne: IMD, 2003.

Lewis, W. W.: *The Power of Productivity: Wealth, Poverty, and the Threat to World Stability.* Chicago: The University of Chicago Press, 2004.

MacCormack, A. D., L. J. Newman III, and D. B. Rosenfeld: "The New Dynamics of Global Manufacturing Site Location" in *Sloan Management Review*, Summer Issue 1994, p. 69 - 80.

TOBIAS MEYER

3 Investments Abroad: Using the Right Evaluation Techniques

Summary

When evaluating investments in new production facilities abroad, companies need to select an approach that best matches their information needs and objectives. The various models available have very different scopes of analysis, complexity, and performance. Selecting the right technique is key to capturing the full potential of network reconfiguration and limiting the complexity of redesign to the level the organization can handle.

Our discussion first focuses on three base models with varying levels of detail. For small and mid-sized enterprises (SMEs), a simple process of sequentially excluding options may be most suitable – the knockout model. Portfolio analysis is a particularly useful tool for diversified groups developing a global site strategy. Corporations with highly integrated manufacturing structures, on the other hand, may find the strategic location concept more appropriate.

A wide range of methods can be used to evaluate the location options identified, combining investment analysis techniques, various indicators and perspectives, and IT implementation. Companies can tailor the approach they take towards evaluating global production sites to their specific requirements. However, our analysis showed that three elements are crucial. First, the approach has to be action-oriented. While it needs to function as a tool for determining the target structure, it should also lead to specific steps towards implementation. Second, it must cover significant parts of the value chain. Suppliers and customers should typically be included in the analysis, particularly if materials costs and other upstream value-added stages account for a vast share of total manufacturing costs. Third, with any technique applied the ratio of costs to benefits has to be right. Realistically mapping economic and operational conditions can take a great deal of effort and time. The level of detail targeted should be strictly geared to the additional insights and accuracy a more granular analysis can provide – a focus that is all too often neglected.

Key questions, Chapter 3

- What are appropriate approaches toward production site location for an SME, the board of a diversified corporate group and the head of a vertically integrated multinational company or business unit of such an MNC?

- How can companies reliably evaluate the ROI of production sites abroad?

- What aspects should the evaluation take into account, and what different analyses and perspectives are required to create sufficient transparency and a solid basis for decision?

- How relevant are qualitative criteria in selecting production sites?

- How can the long-term cost impact of a location decision be evaluated? How can the ROI of specific relocation initiatives be calculated?

- What calculation methods and other tools are useful for conducting investment evaluations?

- How should ROI analyses be synthesized and presented to provide top management with the most effective possible support in their decision making?

Weaknesses in strategic planning are a key reason why companies fail to fully capture the anticipated potential of network reconfiguration. Savings opportunities from factor cost advantages and operational improvement are often overestimated – not because the calculations are misguided in themselves, but because extra costs at the interfaces with other companies and in-house corporate functions as well as the migration costs are not factored in adequately. Decisions are made without sufficient insight into the real implications.

Assessing the economic attractiveness of location options is not a simple task. A seemingly infinite number of potentially relevant factors have to be correctly evaluated and aggregated. This generally requires a sequential approach, moving from a fairly large number of possible options to a shorter list and ultimately to the final solution (Figure 3.1). Various investment analysis techniques are available for determining the ROI and/or cost reduction potential of every potential initiative.

It is vital to select the right scope of analysis and perspective, depending on the purpose of reconfiguration. A focus on direct production costs alone is generally too narrow. Savings in this arena of 10 to 15 percent are not as attractive as they first appear if one-off costs for site construction, relocation, and ramp-up are high. Failure to consider synergies between locations in the current network – between production and R&D, for example – and economies of scale can lead to additional costs that should be included in the ROI analysis, at least implicitly. It is also usually not enough just to look at the present value of net cash flows (NPV[1]) and payback time of individual actions. Location decisions have to be viewed against the backdrop of competition on the market due to their long-term character. This chapter presents a number of models and techniques that can be used for evaluating new locations depending on a company's situation.

---

[1] *Discounted cash flow (DCF) and present value of net cash flows (net present value: NPV) are used synonymously in this chapter.*

3.1 Basic Models

The best model for a company to use when adjusting its global footprint is determined primarily by its present constellation and objectives. These two aspects usually indicate how many locations, products, and processes should be investigated and at what level of detail (Figure 3.2).

- A **simple knockout process** can be used to choose a production location for a clearly defined set of products or production volume, reducing a given number of options step by step. The analysis is mostly limited to the company's own in-house production – interfaces with suppliers, customers, and other areas of the company are only incorporated implicitly. The approach is therefore particularly suitable for SMEs with a small number of locations

and a limited number of interfaces (internally and with suppliers and customers).

- **Portfolio analysis** is appropriate for groups and corporations with a number of different, barely overlapping business units (BUs). The objective is prioritization, i.e., to identify those BUs with the greatest potential for globalization.

- The **strategic location concept** is an integrated concept that encompasses multiple products, production steps, and locations, taking into account all relevant parameters and interactions. It allows optimization of the entire production network from an overall perspective.

At present, most companies – including relatively large MNCs – favor the simple knockout process, and

---

The process for choosing global locations should increasingly narrow down the number of options analyzed

---

Fig. 3.1: Scope of analysis and selection process

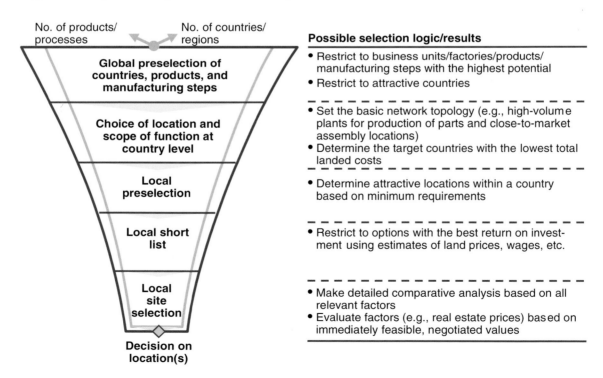

Source: McKinsey/PTW (ProNet analysis)

require some six months on average to complete it.[2] The other two methods can be viewed as an extension of this conventional approach[3], because the decision has to be further detailed on a local level (cities, municipalities, building plots) in both the portfolio analysis and integrated approach. This again requires a sequential method in which – as with the simple knockout process – the solution space is narrowed down step by step based on different sets of criteria. Decision makers should be aware of the advantages and disadvantages of the different methods. We examine each of them in greater detail in 3.1.1 - 3.1.3.

3.1.1 The Simple Knockout Process

A new production location is determined by a pragmatic knockout process: a (usually large) number of

options are evaluated in increasing detail, and gradually reduced by excluding those that prove unattractive (Figure 3.3). Many companies fix the product volume and manufacturing technology at the very beginning. This approach allows the company to focus quickly on the most appealing options.

A sweeping classification as "attractive" or "unattractive" should be avoided, however. Very different requirements apply for locations intended to supply and help open up local markets abroad, as against solely for exporting to existing markets. Locations may well be attractive as markets but not for production, and vice versa.

Given its simplicity, the knockout process cannot take into account most of the interdependencies between locations, products, and production steps. It

The model should be chosen according to the goal

Fig. 3.2: Models for location selection

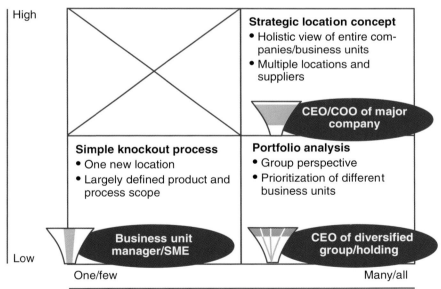

Scope of analysis: number of factories, products, and manufacturing steps considered in parallel

Source: McKinsey/PTW (ProNet analysis)

---

[2] *ProNet survey (Question B.5).*

[3] *Eversheim (1996).*

therefore makes sense to limit its use to a narrow set of countries, manufacturing steps, and products[4] right from the start. Usually the simple knockout process for the selection of production locations includes the following steps:

1. Global preselection: The first step is to examine which countries and regions might be considered for production based on some minimum requirements. This review is performed against the backdrop of the predefined products, production volumes, and production steps with their specific capital intensity, know-how sensitivity, and manufacturing complexity. An option is eliminated if it does not fulfill all criteria. The criteria should be selected on a company-specific basis and also include "soft" (qualitative)

factors.[5] Factors such as knowledge intensity of the production processes may rule out an option from the outset.

Global preselection should be carried out by experienced decision makers, since estimates often have to be used for soft criteria such as "sufficient political stability" or "required proximity to the market." If performed according to these guidelines, preselection can achieve a clear geographic focus,[6] greatly reducing the cost and complexity of location selection without affecting its quality.

2. Target region/country: Selection of the target region or country should include an ROI analysis. This should cover an estimate of the cost position of the

---

The knockout method can be carried out quickly, but does not cover network synergies

Fig. 3.3: Scope of the knockout method

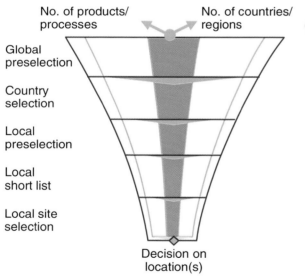

No. of products/ processes — No. of countries/ regions

- Global preselection
- Country selection
- Local preselection
- Local short list
- Local site selection

Decision on location(s)

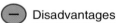 **+ Advantages**

- Simplicity
- Limited effort required
- Quick realization
- Various evaluation techniques can be combined at different levels

− Disadvantages

- No guarantee of correct prioritization
- Synergy losses between factories
- Higher materials costs due to lack of coordination with Purchasing
- Higher transaction costs (transportation, warehouses, customs duties)
- Synergy losses due to uncoordinated choice of manufacturing technology
- Suboptimal use of economies of scale

Source: McKinsey/PTW (ProNet analysis)

---

[4] Cf. also Hack (1999).

[5] Cf. Zheng (2002); the importance of soft location factors is especially high at the preselection stage, while the assessment of alternative locations is more heavily determined by quantitative (hard) factors.

[6] Cf. Eversheim (1996), Fig. 9–35.

products that may be manufactured there, the required investment, and the migration costs.

This stage may also involve additional selection by knockout criteria using, for example, maximum transport time to the relevant markets, or a cap on average labor costs. The result should be the definition of one target region with a limited number of alternatives. These then go on to the next stage.

3. Local preselection: Local factors are increasingly used to determine attractive locations in the target country or target region. These differ to some extent from the global factors. Criteria such as the attractiveness of a region for expatriates or transport links become more important, while others such as customs duties and taxation that apply to the entire country take a back seat. The outcome is a number of options that meet the basic requirements but need further scrutiny.

4. Local short list: The local short list narrows the choice down further via quantitative analysis such as an investment analysis assuming different factor costs and prices of land, buildings, and equipment. The objective is to define the three to five best options. Discussions and negotiations are initiated on these, with the owners of the land, for instance, allowing more precise assumptions to be folded into the ROI analysis.

Parallel negotiations are advisable, partly because of the greater transparency on the costs of the various options and their prospects of realization. Companies often confine themselves too early to one option, harming their negotiating position, since walking away leads to higher opportunity costs due to the time lost. This is not such an issue with parallel negotiations.

5. Investment proposal and decision: One of the remaining locations is now selected using a comparative ROI analysis. The assumptions used are based on negotiated, directly implementable figures, such as the purchase price or rent/lease costs of buildings and land. Possible selection indicators are

NPV, ROI, or payback time. The decision makers should be able to compare at least two evaluated options with an initial scenario (base case or "do nothing" scenario) before making a binding decision.

The solution space resulting from the previous step is analyzed and evaluated in greater detail in every phase of the above process. The relevant parameters are modified if plans from an earlier phase prove impossible or clearly disadvantageous.

It is not unusual for inaccuracies and inconsistencies to creep in during the selection process since higher-level evaluations use aggregated (and therefore approximate) parameters, making an iterative approach advisable. Hierarchical planning systems[7] integrate iteration from the start and provide defined feedback mechanisms.

The principles of such hierarchical planning also apply to the choice of location and design of production networks. If, for example, there are no adequately equipped industrial buildings in a region at the budgeted costs, the assumptions in the superordinate plans have to be adjusted accordingly.

The costs or abstract benefits[8] (usually in the form of an index value) of the remaining location options are compared with each other to reach a final conclusion and select a defined physical location as the new production site. The use of non-monetary values (such as an index) allows the capture of relevant qualitative criteria, but alignment with the ROI analysis may be difficult. Some methods[9] will lead to decisions that fall outside the quantitative framework that is an obligatory element or even backbone of the investment approval process in most companies today. Although the comparison of location options using indices and utility value generates a certain transparency on their comparative benefits, these evaluation methods are therefore ultimately of limited use. They fail to provide a synthesized, quan-

---

[7] Cf. Drexl (1994).

[8] Cf. Eversheim (1996), pp. 9–42 to 9–52.

[9] Cf. e.g., Peren (1998), pp. 71 ff.

tified assessment of economic attractiveness and ROI.

3.1.2 Portfolio Analysis

Portfolio analysis is suitable for prioritizing and assessing the attractiveness of location options for different business segments. It is particularly suitable for diversified groups to evaluate the potential of network reconfiguration and to set objectives for the business units (Figure 3.4). The selection process for the prioritized business segments and the evaluation methods used may vary (cf. also sections 3.2 and 3.4).

Portfolio analysis is divided up into three phases:

1. Segment production activities and define the evaluation criteria: Segmentation should be based on the least interdependent characteristic. If, for example, product lines are manufactured largely independently of one another, production activities should be segmented by product line. If the locations oper-

ate largely independently of one another and have only a limited number of logistical connections, the analysis should be conducted per location.

2. Evaluate the segments: The second step involves assessing the optimization potential of the defined segments, i.e., the absolute and relative value of relocating the activities in question. This makes use of investment analysis methods (comparative cost analysis – relative and absolute, NPV/ROI), other suitable techniques (such as cost-benefit analysis based on an index), or a combination of both.

3. Compare and define action needs: Figure 3.5 is a comparative analysis of all a company's segments (defined here as production facilities). In this case, the greatest need for action would be in the top right-hand quadrant of the matrix, particularly for plants represented by a large orange circle, since this stands for a negative profit contribution accompanied by high relocation potential (long-term cost reduction potential and NPV).

Portfolio analysis is the CEO's prioritization tool

Fig. 3.4: Scope of portfolio analysis

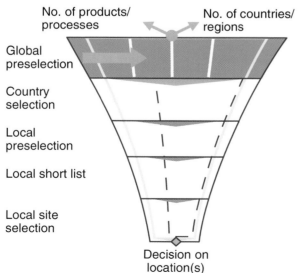

 Advantages

- Analysis of the entire company
- Application of the same evaluation methods for all BUs
- Relatively simple
- Helps organization to focus on biggest opportunities

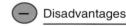 Disadvantages

- Synergy losses between factories
- Higher materials costs due to lack of coordination with Purchasing
- Higher transaction costs (transportation, warehouses, customs duties)
- Synergy losses due to uncoordinated choice of manufacturing technology
- Suboptimal use of economies of scale

Source: McKinsey/PTW (ProNet analysis)

Assessing the optimization potential from global production by location, product, or business unit provides a valuable decision-making basis for top management to prioritize forthcoming activities.

Classifying business segments – by production facility, in the example – is useful in multiple ways. The diagram maps the four main dimensions of the analysis:

■ The size of the circles shows the absolute NPV of a move. This ensures that larger business segments with a greater impact on overall profitability are given more attention.

■ The color represents the current profitability of the business segment. Business segments with a negative return on investment are colored orange. This signals problematic areas in a group where opti-

mization of the location structure can contribute to a turnaround.

■ The horizontal position of the business segments shows the long-term effect of a redesign of the location structure. In the present case, the savings in operating expenses for materials, manufacturing, and logistics are used as an indicator.

■ The vertical position describes the (short-term) attractiveness of the action. This takes into account the savings from relocating production, as well as implementation expenditure (both expenses for the new location and any restructuring of existing sites). There is a certain correlation between the long-term effect (measured as total landed costs) and short-term attractiveness (measured as NPV). However, the more a company has to spend on investments and one-off expenses when changing

Portfolio analysis identifies the business segments where reconfiguration can deliver the highest potential

Fig. 3.5: Portfolio analysis of production processes

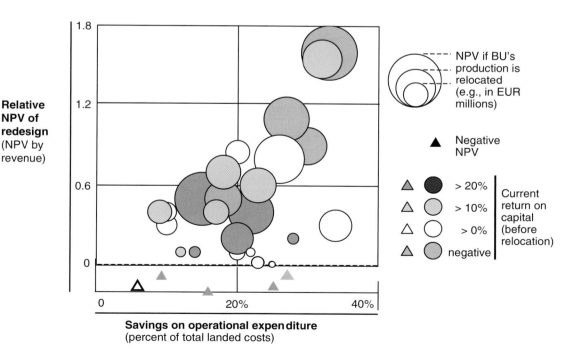

Source: McKinsey

its location structure for individual business segments, the less significant this correlation becomes.

3.1.3 Strategic Location Concept

In designing entire production networks, it is not as easy to limit the scope of analysis as when selecting largely stand-alone production locations. Normally the detailed planning for every factory also has an impact on other parts of the production network. The interactions between different stages of production, products, and corporate functions are high, particularly with complex series-produced products – automobiles and machinery, for example. Failure to consider these dependencies can lead to a significant increase in inventory, customs duties, transportation costs, and insufficient capture of economies of scale and synergies between functions and product segments (Figure 3.6).

The following interactions should be considered in integrated production structures:

■ The analysis has to include the **entire manufacturing value chain** to be able to capture the supply relationships between the individual manufacturing steps and the shared fixed costs, e.g., for production lines, in sufficient detail.

■ The **materials flows** between locations have to be captured to be able to accurately determine the transaction costs, e.g., transport costs, inventory costs, customs duties, and economies of scale.

■ Interactions between locations on issues of **manufacturing technology and/or product design** and variants have to be taken into account to be able to define production technology and processes that are aligned with the location characteristics such as labor costs and qualifications. Related costs and one-off expenses have to be allocated correctly within the network.

■ Finally, the **lead time, service level restrictions, and demand volatility** have to be captured to determine constraints for the location selection with-

An integrated approach can only be usefully applied in the context of a major transformation program

Fig. 3.6: Scope of the integrated approach

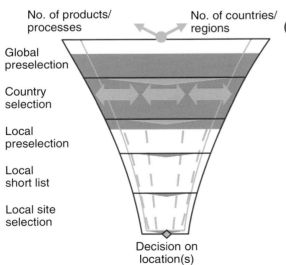

Source: McKinsey/PTW (ProNet analysis)

in the network. One reason is because the reliability and flexibility of supply chains have to satisfy constantly increasing requirements. Another is because advancing standardization and platform building are opening up ever greater economies of scale in upstream manufacturing.

The integrated analysis of several value-added stages provides an important decision-making tool in deciding on the vertical integration of each location and how to minimize risks.

In addition to the knockout criteria described in the previous section, all these interactions have to be incorporated into the decision-making process. Failing to do this risks only partially capturing the potential. The developing network may demonstrate considerable inefficiencies, particularly at the interfaces between new and existing locations and corporate functions. Planned savings may be wholly or partially swallowed by higher logistics and overhead costs, customs duties, negative economies of scale, and untapped cross-functional/cross-production synergies.

When building complex global networks, companies therefore need a strategic location concept that can be applied within the sequential selection process (Figure 3.7). This complements conventional location planning in two ways: It ensures that not only countries and regions, but also the scope of the products and manufacturing steps analyzed are successively narrowed down – starting out from a portfolio analysis where the entire production scope is examined for relocation potential. It also ensures that interactions between existing sites are adequately incorporated.

The strategic location concept is the key element in an integrated approach

Fig. 3.7: Integrated approach and importance of the strategic location concept

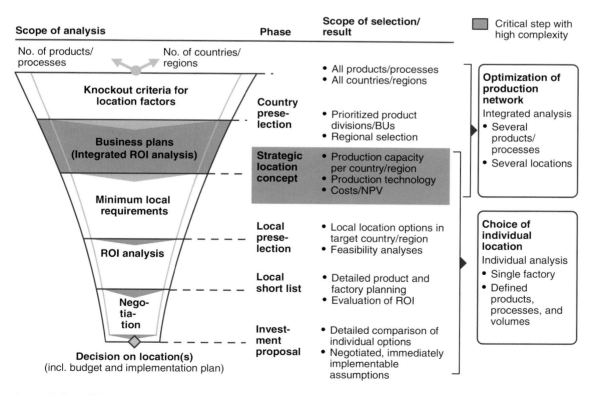

Source: McKinsey/PTW (ProNet analysis)

Companies should consider the spectrum of different products when selecting locations and take into account interactions between sites

The **manufacturing steps of suppliers** who define critical product features are also regarded as a component of the production network. They therefore need to be included in the analysis, regardless of the actual geographic flexibility of the existing suppliers (initially). Often this flexibility can be created by coordinating the choice of location with the supplier or by building up new suppliers at the target location. This interaction has to be driven by the purchasing department: another good reason to ensure it is intensely involved. But even if supplier locations do not shift, the changes in logistics costs and delivery times are relevant to the efficiency of the total system and should be factored into decisions on the company's own locations.

The network-wide analysis can also be used as an opportunity to optimize the **level of vertical integration** and possibly adjust it (temporarily or permanently). The company's own core competencies may play a role, as well as ideas on how to minimize restructuring costs at existing locations, or bridge the period required to build up suppliers via temporary insourcing. Outsourcing internationally may be the most appropriate option if a company lacks the resources or scale to build its own production abroad.

Key elements of a strategic location concept are the economically optimum **target structure** and a **migration plan** with the most important implementation steps. This may include expanding an existing factory (in the target region if available) and/or the setup of a new one. Actions for individual locations are then derived from this master plan. This is accompanied by a shift in organizational accountability for larger corporations. If overall production network optimization is driven by the COO or the corporate development department, planning for the individual location may be assigned to a dedicated project team

under the future plant manager or – for existing locations – may be handled within the regional organization.

Much higher savings can be achieved via **integrated analysis** than by isolated consideration of individual production steps. This particularly applies to multi-stage manufacturing value chains and a complex location structure (characterized by a high share of supplies for a manufacturing location from other plants and know-how-intensive manufacturing processes and technology). The high level of complexity of such networks calls for special tools (e.g., computerized optimization models) to achieve maximum effectiveness. This sophistication is highly demanding and is the main disadvantage of the integrated approach. It can be extremely costly to develop and continuously monitor complex strategies, policies, and controlling systems. Business systems optimized in line with theoretical assumptions may also prove not to be robust in reality – and are therefore neither more effective nor efficient than existing systems. Against this backdrop, an integrated approach is only advisable in the context of fundamental redesign of the production network, with highly integrated production structures. Management should also pay close attention to ensure that the hidden costs of complexity of a global production network are adequately reflected and that sufficient experience and talent are available to implement such a strategy.

The strategic location concept outlined here is discussed in greater depth in Chapter 4. Since the initial and the last two phases of the approach (see Figures 3.6 and 3.7) correspond to a large extent to the conventional approach, we focus on the extension – the integrated evaluation of process and location factors on a country level – in greater detail there.

3.2 Dimensions of the Analysis

One can look at investment opportunities from different perspectives. Assessing the return using a short-term focus of only three years is one possibility; a longer time horizon of 10 or even 30 years is another. Both perspectives are valid. Which is most suitable depends on the investment and investor.

The economic attractiveness of an investment opportunity can be assessed by a wide range of criteria using different analytical perspectives (see Table 3.1). A company should choose a set of indicators and dimensions depending on its specific circumstances and objectives. It is however advisable to largely stick to the same metrics – at least for comparable types of investment. This allows management to familiarize itself with the approach and indicators, enabling better comparison of different investment proposals. Table 3.1 shows nine of the most relevant dimensions that should be defined when compiling an investment proposal.

3.2.1 Narrow Versus Extended Functional Scope

In analyzing production locations, the narrowest scope is a focus on direct operating costs, i.e., the cost of labor and machinery. This inevitably leads to a distorted result, as it does not include a large share of costs, such as materials costs. The costs incurred at interfaces with other functions, such as R&D, may be relevant enough to be taken into account. Interdependence with some, such as finance, may be minuscule. Those functions should be excluded (or relegated to a more general category).

Synergies and economies of scope between production and development are typically relevant, especially in innovative, fast-moving industries. They can be included by defining an opportunity cost rate incurred by all potential locations at which none of the relevant functions are co-located with manufacturing processes (i.e., where no synergies can be tapped as a result). In other cases, dependence may be so great that selected manufacturing steps or products have to be produced at locations with R&D departments. This must then be taken into account as a constraint in the network design. The enhancing effect of local knowledge clusters on staff effectiveness can be another relevant fact to consider. So should synergies with other companies, if these are signif-

icant (they may, for example, enable suppliers to produce in larger batch sizes, and therefore at lower cost).

The inclusion of **improvement potential** at existing locations also extends the analysis scope. Folding an optimized configuration of existing locations into the evaluation of options is best practice. Obviously assumptions on future productivity and improvement potential at the existing site need to be realistic: the program that would be required to achieve these improvements must be clearly defined. The expenditure needed to implement the program and leaner production processes[10] must also be considered (cf. section 7.3). Streamlining existing locations can influence the timing and sequencing of relocation steps. Early improvements are beneficial even if production still ultimately has to be shifted as restructuring costs fall (cf. section 4.1.5).

3.2.2 Tactical Versus Strategic Choice of Location

The decision to set up or close production locations is always strategic. A long-term international commitment has considerable implications for the company and generally requires substantial expenditure. Closures likewise entail significant expense and impact corporate performance.

The allocation of products and mid-term capacity planning for sites, in contrast, is of a tactical nature. The criteria considered here are largely separate from those relevant to the strategic choice of location: free capacity and directly applicable product and process know-how play a much greater role.

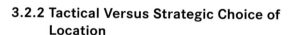

Tactical and strategic location planning can deliver conflicting results

Decision makers should be aware of the latent conflict between tactical and strategic location planning. While strategic planning aims at long-term orientation to the lowest costs (materials, manufacturing, and logistics costs), a tactical planning perspective

---

[10] See, e.g., Drew (2004) for the concept of lean manufacturing.

Table 3.1: Dimensions of an ROI analysis

Dimension	Values	
Scope along functions	**Non-production functions** ■ Sales/distribution ■ Finance ■ R&D management ■ …	**Production in a broader sense** ■ Production capacity per product per factory ■ Transport/inventory management ■ Procurement planning (in collaboration with purchasing)
Planning horizons	**Tactical** ■ Capacity adjustments ■ Time horizon: > 1 year and < 3 years	**Strategic** ■ Opening/closure of locations ■ Target value: present value of net inflows
Time perspective and corresponding financial indicator	**Static/single-period** ■ More long-term, steady-state conditions ■ Target value: costs or profit	**Dynamic/several periods** ■ Development over time ■ Target value: present value of net inflows
Scope of analysis	**Single-stage** ■ One manufacturing step ■ Decoupled view of several manufacturing steps	**Multi-stage** ■ Integrated analysis of several manufacturing steps ■ Higher complexity due to interdependencies, e.g., inventory ■ Mapping of dependencies, e.g., using bills of materials
Type of analysis	**Qualitative** ■ Nominally or ordinally scaled characteristics[11], e.g., "good" infrastructure ■ Knockout criteria, analysis of locations along strengths and weaknesses	**Quantitative** ■ Metrically scaled attributes, e.g., sales, production volumes ■ Landed cost analysis, NPV calculation of redesign of network configuration
Granularity	**Macro-environment** ■ Analysis of groups of issues/summary indicators ■ Example: high political stability	**Micro-environment** ■ Analysis/evaluation based on detailed individual factors ■ Example: costs per effective hour worked
Uncertainty	**Deterministic/certainty** ■ Parameters are regarded as certain	**Stochastic/risk** ■ Parameters are subject to uncertainty
Problem-solving precision	**Exact** ■ The solution definitely represents the global optimum	**Heuristic** ■ The solution only represents the global optimum by chance
Problem-solving method	**Simulation** ■ Result specified ■ Achievement of target value(s) ■ Extension via stochastic choice of profit parameters[12]	**Optimization** ■ Determination of the result by optimizing the target value(s) using an algorithm

[11] Cf. Hartung (1999), pp. 10 ff.

[12] Cf. Domschke (1998), p. 7.

is very different. Much of the costs are regarded as fixed, such as for existing equipment and staff. The stronger orientation towards marginal cost logic is not misconceived for short-term optimization but may potentially conflict with longer-term objectives.

3.2.3 Static Versus Dynamic Analysis

Static methods are based on the assumption of constant parameter values, such as sales volumes or factor costs. They consider a "steady-state" condition. Forecasts for a specific point in time are often used to gear the method to the future. The forecast parameter values are considered to represent an average for the entire period of time within the scope of the analysis. Time differences between events – between the erection of a hall and the start-up of production, for example (corresponding to the time difference between outflows and inflows of funds) – are largely disregarded.

Dynamic analysis makes it possible to include changes in operational and economic parameters over time, such as:

- Discontinuous cash flows, e.g., due to investments/divestments

- Discontinuous cash flows and profit contributions due to one-off expenditure, e.g., for restructuring, and extraordinary income, e.g., from selling assets at more than book value

- Changes in factor costs, market volumes, market requirements, e.g., delivery times, and product characteristics, e.g., value density, over time.

The most suitable financial indicators for a static analysis are cost or profit (and the related quotients)[13]. If dynamic analysis is used, it is more appropriate to switch to a cash-flow-oriented perspective and to apply corresponding investment analysis techniques.

A static and thus more long-term view is more appropriate in mature, steadily developing industries. A dynamic view based on NPV calculation is essen-

tial when assessing investments in highly volatile markets.[14]

> **Detailed, long-term planning of the location structure only makes sense in a relatively stable environment**

Companies often use **combined techniques** to perform sophisticated analyses of the financial implications of foreign investments.[15] A long-term cost comparison technique can deliver information about the attainable cost position of selected products, while dynamic analysis of the cash flow effects serves as the basis for calculating the NPV and payback time of a specific project. The use of different techniques is useful particularly where these build on the same base data. Different methods may be helpful both during the evaluation process, e.g., for preselecting options based on a simpler technique, and the final evaluation, e.g., using a cost comparison, payback, and NPV analysis, all in parallel.

3.2.4 Single-Stage Versus Multi-Stage Supply Chain

The focus on only one manufacturing step, such as the assembly of a component, greatly simplifies the evaluation of a location, but also leads to inaccuracies in evaluating total network costs.

A one-stage manufacturing process is easy to model and makes homogeneous demands on the production location. Delivery relationships between production locations do not need mapping. Delivery times and service levels can be calculated easily, assuming sufficient availability of intermediate products and raw materials.

Analyzing location options for just one manufacturing step in the supply chain is only worthwhile if interactions within the network are very low and the step accounts for a high share of the end product's

---

[13] *Cf. Fig. 3.1 and Perridon (1999).*

[14] *Cf. Mintzberg (1999), particularly pp. 396 ff.*

[15] *Cf. Thommen (1998), p. 551.*

total value added. This applies to the manufacture of simple garments, for instance, but not the assembly of automotive components. Particularly when customs duties, transportation, and inventory costs are in the same order of magnitude as the value added of the manufacturing step, it is essential to perform an analysis along several steps. Considering the multi-stage nature of the supply chain can minimize the transaction costs between the manufacturing steps.

3.2.5 Qualitative Versus Quantitative Evaluation

Quantitative techniques for assessing location options range from simple cost comparison analyses to complex NPV calculations and the comprehensive analysis of total landed costs across multiple markets and production steps.

> **Transparent separation of quantitative and qualitative factors is essential**

Qualitative analysis uses criteria that cannot be measured on a metric scale – corruption and crime levels, for example – and which are therefore hard to translate into economic terms. To incorporate qualitative factors into an investment analysis, companies can consider methods that help to quantify these soft factors[16], such as political stability – but this is no easy matter.

Including qualitative criteria in the quantitative analysis means converting them into metrically or ordinally scaled attributes in a meaningful way. This can be done for the criterion "political and economic stability," for example, by defining a country-specific cost of capital rate that reflects the specific investment risk. This risk can be approximated by making a historical analysis of the default risks of investments in unstable countries. The results are then used as the basis for determining an increase in the internal rate of return (IRR) required to compensate for the higher default risk.

It is crucial that the relevant quantitative and quantifiable factors are clearly delineated from those that

are not quantifiable, and thus cannot be included in the resulting indicator, such as the ROI. It is important to provide decision makers with sufficient transparency on the scope of analysis the indicator value comprises. This is paramount to reading the information correctly and drawing the appropriate conclusions.

Qualitative evaluation criteria can be considered in the decision-making process analysis in three other ways:

- **Using checklists:** Minimum or fixed requirements are laid down for a number of criteria. Countries, regions, or cities that do not fulfill a criterion are excluded as candidates.

- **Generating indices or utility values:** A metric interpretation of the nominally or ordinally scaled factors is required to do this. The (weighted) average of different attribute rankings for each factor is aggregated in an index.

- **Making comparisons with a requirements profile:** The gaps between current and target status are aggregated into an index for each location.

Figure 3.8 gives an overview of typical checklist techniques for selecting locations via knockout criteria. The insights they offer are clearly very limited. They do not provide any firm basis for estimating the suitability of a location or the economic attractiveness of an investment there. Purely qualitative methods should therefore only be used for preselection. Detailed analysis should then be used for a sufficient number of options to ensure that all the potentially attractive configurations are examined.

3.2.6 Deterministic Versus Stochastic Perspective

A **deterministic view** posits that the assumptions are certain to materialize. Location decisions however often require a long-term perspective and thus entail

---

[16] Cf., e.g., Harding (1988), p. 24 ff.

a certain degree of uncertainty when it comes to the assessment of input factors and market conditions.

Stochastic parameters can be used to explicitly consider risk. Capturing them in a quantitative analysis and incorporating likelihood distributions in the ROI, however, is a very complex matter. For the purposes of selecting production locations, most companies should confine themselves to analyzing several (deterministic) scenarios with varying input factor values. Often more general conclusions can be drawn on this basis, including an approximate assessment of the most relevant risks.

3.2.7 Simulation Versus Optimization

Problem-solving techniques are enablers. They have significant impact on the scope of the analysis: the

more powerful the enabling techniques, the broader and more comprehensive the scope of analysis can be. The problem-solving technique selected can therefore have significant impact on the type and content of options discussed. It can even influence the ultimate proposal – the location concept itself.

When a **simulation** technique is used, the output values, e.g., production volumes per product and location, are actually specified, and only the target indicator value, e.g., production costs, is calculated. The solution is improved by performing a comparative analysis, i.e., changing the output values and calculating the target indicator value for these scenarios. The global optimum, i.e., best possible solution, basically remains unknown. Defining the scenarios correctly is therefore of great importance, as they all represent potential solutions being considered for

---

Simple knockout processes are generally only suitable for preselection

Fig. 3.8: Simple methods of location selection – examples

Example 1

"Mathematics for location decisions"		
Criteria	**Lower limit**	**Upper limit**
• Distance from the customer (km)	0	1,529
• Monthly income in manufacturing ind. (EUR)	1,674	4,250
• Hourly labor costs in important industries (EUR)	8.9	28.2
• Productivity (EUR p.a.)	20,121	115,235
• Corporation tax (%)	12.5	40
• Distance from highway (km)	0	1,529
• Distance from international airport (km)	0	1,529
• Crime (indexed)	15	26.4
• Corruption (indexed)	4.2	9.7
• Electricity price (EUR per KWh)	0.037	0.097
• Gas price (EUR)	5.73	10.44
• Distance from suppliers (km)	0	1,657
• Share of employees in production (%)	0.0149	0.1422
• Gross value added in manuf. by purchasing power	71	28,214
• Unemployment rate within 30 km radius (%)	1.15	29.2
• Employment rate within 30 km radius (%)	0.0034	0.2522
• Growth opportunities	2	40
• Birth rate (per 1,000 inhabitants)	4.1	17.5
• Economic development aid (indexed)	3	1
• ...	...	...

Example 2

"Decision support for SMEs"		Country rating				
Criteria	**Weighting***	**E**	**D**	**C**	**B**	**A**
• Political and economic stability						
• Infrastructure within country						
• Currency situation and exchange rates						
• Inflation rate						
• Personnel costs						
• Employee qualifications and work ethic						
• Employee availability						
• Real estate prices (construction/leasing costs)						
• Energy costs						
• Environmental requirements						
• Legal framework and legal safeguards						
• Foreign trade regulations, customs duties, borders						
• Red tape, administrative efficiency						
• Profit transfer conditions, taxes						
• Labor law/trade unions						
• Ownership structures/ guarantees						

Enter how far fulfilled by alternatives (like school grades) from A = "Very good" to E = "Poor"

Multiply weighting by fulfillment. Next, total fulfillment levels per alternative, and total the weightings. Then divide each sum for the alternatives by the sum of the weightings. The lower the arithmetical value of an alternative, the better it meets the requirements

Source: Based on Jacob (2006)

* 1 = "Not so important" to 3 = "Very important"

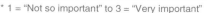

evaluation. Scenario definition can therefore be a highly controversial subject in network reconfiguration projects.

Output values do not need to be defined if an **optimization** algorithm is used. You only have to determine the input parameter values and constraints. In practice, these can be derived from a sound fact base, and management can easily check their plausibility. Applying an optimization approach requires the following elements: the target function (which for practical purposes is similar to the formula for calculating the target indicator value), boundary conditions (e.g., that deliveries have to match predefined sales volumes per country), and the input parameter values typically defined and mapped in a formal process model (such as labor costs).

A variety of **operations research methods** are available to implement optimization models.[17] Optimization methods have great advantages, particularly for dealing with complex planning tasks. Defining the output values manually – as required when using a simulation approach – can be very difficult, time consuming, and often encourages adherence to the status quo. Optimization models can therefore be of great benefit to the organization when preparing for and making complex decisions. The initial effort of setting them up and learning how to extract the benefits typically pays off.

3.3 Specific Tools and Analysis Methods

Various tools and methods are available for supporting the location selection process. While simple schemes can be appropriate for preselecting sites, detailed guidelines and the support of special software packages may be needed for the quantitative analysis of a larger production network. This section describes some tools and methods relevant when selecting production locations abroad. The list is by no means exhaustive, and some are only used in combination with other techniques. The combination needs to be defined depending on the scope and complexity of the analysis and various characteristics of

the planning issue to be resolved. Examples of the latter are the type of boundary constraints and nature of the input factors (such as the number of parameters for which values need to be modeled).

3.3.1 Investment Analysis

Various analysis methods are suitable for assessing the financial attractiveness of location options, including static and dynamic techniques. The appendix contains a detailed description of these different methods and explains when they are applicable.

This section discusses five general points that need attention when performing investment analyses to ensure viable results:

■ Choose an appropriate **target indicator**

■ Ensure **consistency** with P&L standards and cash flow statements

■ **Define** the costs/outputs, expenditure/income, and outflows/inflows considered in the location decision

■ Ensure clear **presentation** of results

■ Select indicators familiar to **management**.

1. Choice of target indicator: Which investment analysis method is most suitable in a specific case? Management with a long-term focus might be most interested in a comparison of total project costs over a time period of ten years or more. A management team facing cash constraints would be more short-term oriented and likely request a cash flow analysis that is part of the overall investment proposal. Functional responsibilities also play a role. The HR manager is probably most interested in the impact of production network redesign on staff requirements and labor costs per site, whereas the procurement manager will want to know the impact on suppliers.

---

[17] Cf., e.g., Domschke (1998).

All these information requirements can be met by defining adequate indicators and calculating their values for redesign scenarios. Some of these indicators will be more suitable for a holistic assessment than others, but all are valid.

When using an optimization technique, however, one indicator must be defined as the target indicator. Other parameters can be used to define boundary constraints, but they will be of a different nature. While boundary conditions define minimum requirements, the target indicator is the parameter around which the network design is optimized. One example is total landed costs. The target indicator has to fulfill two main requirements:

■ Appropriately reflect the company's **strategic objective**. This applies both to the type of indicator (whether costs, revenues, profits, net cash flow, etc.) and the scope for which it is defined. The terminal asset value, for example, should generally

not be included as a cash inflow in the NPV (as the target indicator) if the site is to remain a going concern beyond the period of analysis. If this is the assumption, it is more appropriate to use an NPV as target indicator that includes an infinite series of expected net cash inflows from the plant.

■ Accommodate decision makers' **multidimensional objectives.** To do this, companies use various financial indicators (such as costs, NPV, or cash flow impact), broadening their scope to incorporate other factors. The opportunity costs of lost orders can be applied to evaluate the service level, for example. A country-specific cost of capital rate can take into account a country's political risks. The target indicator will then also reflect these criteria.

2. Consistency with P&L and cash flow statements: Investment analysis techniques build on different elements of a company's P&L or cash flow statement. Investment analyses have to adhere strictly to one

---

Comparative cost analysis versus NPV

Fig. 3.9: Methods of financial analysis

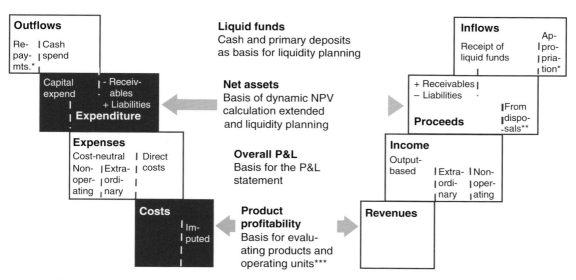

* Outflow: repayment of debt or withdrawal of equity/profits. Inflows: borrowing or increase in capital contributions
** Pure exchange of assets, e.g., via the sale of fixed assets at book value. Disposals above/below book value lead to an overlap with the performance parameters, e.g., by earning extraordinary income
*** Imputed costs and revenues may occur for a variety of reasons, e.g., to bridge differences between accounting standards (e.g., depreciation of assets over legally defined depreciation periods), and actual economies (e.g., depreciation of assets dependent on actual use/wear)

Source: McKinsey/PTW (ProNet analysis) based on Perridon (1999), p. 7

level of the financial accounting scheme (Figure 3.9). If, for example, an NPV analysis is performed based on cash flow effects, non-cash charges need to be disregarded, such as write-offs or book gains from revaluation.[18]

3. Definition of the scope of analysis: Decisions on what elements of costs/outputs, expenditure/income, and outflows/inflows to include in the investment analysis are often no simple matter. While the costs of physical relocation entail cash outflows and can be directly allocated to the individual initiative, the costs of production start-up and opportunity costs of the relocation often cannot be allocated unequivocally (Figure 3.10). Also, only a certain share of this expenditure has a direct impact on cash flow. Which elements should be included in the target indicator and (as a result) in assessment of the ROI needs careful consideration.

4. Transparent evaluation and presentation: The importance of making the components and structure of the financial analyses transparent for decision makers is often underestimated. Production network design has a complex impact on other company functions. These interactions cannot be fully quantified or mapped and evaluated as an integrated whole. It is therefore essential for decision makers to understand the exact scope of the analysis and the range of factors the target indicator represents. The qualitative and quantitative factors that cannot be included in the investment analysis should be clearly indicated (cost impact of network redesign on the R&D department, perhaps). Management should also be supplied with background data for checking the plausibility of key assumptions, such as labor cost rates per location.

5. Relevant indicators from a management perspective: Decision makers' preferred indicators vary when assessing an investment (Figure 3.11). Most senior managers consider a fairly large set of indicators before making a decision. The less clear the advantages of the project, the more indicators tend to be required. A striking factor here is that payback period is considered more relevant as an indicator than NPV, as the former implicitly also makes a statement on the risk of the initiative.

The impact of location decisions on a company's financial performance can be substantial: accurate projections are vital. Relevant financial indicators include

---

The ROI analysis should take into account all relocation expenses incurred due to relocation

Fig. 3.10: One-off expenditure when relocating production

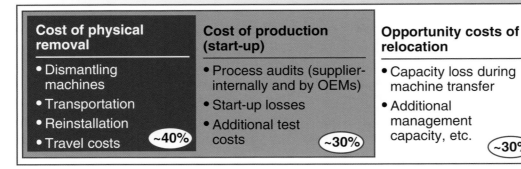

Source: McKinsey/PTW (ProNet analysis)

---

[18] Cf. also Coenenberg (1997), p. 38.

the effects on capital expenditure, income, EBITA, and cash flow. Impact on the cost and profitability of key products/product lines also needs to be transparent. Particularly important is usually the financial impact of production network redesign over time, for the typical time periods – usually calendar years and quarters – used by the financial community. Any causes of major change in the value of key indicators (such as lower labor costs, one-off expenses, hikes or cuts in logistics costs) should be transparent.

3.3.2 Decision Support Systems

The evaluation of production network configurations becomes a complex task when the scope of analysis is broad. Simplification entails lower accuracy. Mastering this complexity is therefore a key success factor in integrated strategic location planning.

Decision support systems offer a potential solution. Suitable IT systems allow the quantitative analysis of complex networks with a uniform approach. This makes them the logical successor to simple management tools such as cost accounting or SWOT analysis. However, these IT-based management tools are not yet universally applied. The following sections explain the use of decision support systems for choosing production locations and outline recent trends. Despite the rapid progress in operations research methods, there are still considerable obstacles to their use.

3.3.2.1 Operations Research Methods

A large number of operations research techniques for location selection have existed since the 1970s.[19] With a few exceptions however they are not yet wide-

---

Payback time is the most important indicator for decision makers when making location decisions

Fig. 3.11: Indicators for evaluating location options

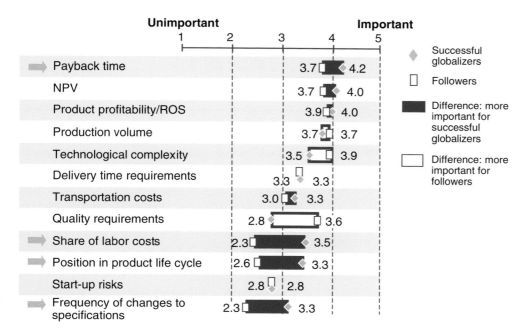

Question A9: "What indicators does your company use to decide which products to move to another location?"

Source: McKinsey/PTW (ProNet analysis)

---

[19] Cf. Vidal (1997) and Bhutta (2004).

ly used in practice.[20] This is largely due to two weaknesses:[21]

- **Lack of realism:** Academic approaches do not sufficiently take into account parameters that are highly relevant in practice, such as the impact of the number of variants, customs duties, delivery time restrictions, and safety stock.

- **Process and output are not well presented:** Most of the academic literature gives no tangible illustrations of the process models, assumptions, and results in its presentation of optimization methods. Often, no use is made of the common management tools for communicating complex issues and solutions. Most publications address optimization methods on a more theoretical or conceptual basis (e.g., solution algorithms), and only a few papers describe practical applications that offer proven value.

The approaches of the 1970s and early 1980s were too simple, partly due to the limited technical capabilities at that time – business issues could not be represented realistically. However, some companies began investigating quantitative optimization methods for location planning.[22] Approaches emerged in the mid 1980s that were still very circumscribed[23] given the magnitude of the issue (the number of locations, products, and manufacturing steps). Greater efforts were made to apply these models to more complex structures[24] and cross-company production networks beginning in the 1990s. They are becoming increasingly realistic, taking into account factors such as customs duties and fixed costs.

However, most approaches and relevant parameters still have considerable gaps. They do not consider expenditure for training, production start-up, or factory closures, for example. Only a few approaches[25] use detailed process models with realistic factor input volumes and include the specific complexity of manufacturing processes (such as requirements on staff education standards).

3.3.2.2 New Capabilities Enable the Use of Decision Support Tools

More recently, decision support systems have been gaining in importance in many areas of strategic, tactical, and operational management. A key reason is the increased complexity of business processes. Another important factor is greater data availability.

> **Better data availability is increasing the use of computer-aided systems in the strategic decision-making process**

Thirty years ago the purchasing behavior of retail customers was mainly assessed based on surveys and experience. Little data was readily available and collected on an ongoing basis. Nowadays plenty of data is gathered at various points in the supply chain – directly at the point of sale, for instance (by recording details of the merchandise purchased at the checkout, etc.).

Such data allows strategic insights such as the price elasticity of demand, which can now be assessed based on sound empirical data. Many factors previously regarded as soft can now be incorporated in quantitative analyses. IT systems in operations make

---

[20] Cf. Cohen (1998).

[21] Cf. Vos (1996): Beschreibung und Kritik von Ansätzen (Schwerpunkt Operations Research); Translation: Description and critique of approaches (focus: Operations Research): (1) The prevailing academic approaches are too theoretical, and practitioners cannot understand/implement them. (2) Factors of influence with proven practical relevance are ignored, since they cannot be integrated into the established models.

[22] Cf. Breitman (1987): This describes the development of a decision support system that goes back to a 1973 initiative of General Motors.

[23] Cf., e.g., Haug (1992): Model with one manufacturing step, one end product, and two periods.

[24] Cf. Arntzen (1995): Application of a hybrid integers model in several projects for Digital Equipment Corporation (DEC). Cf. also Kirka (1995): application of a linear dynamic multi-product model with several resources for one SME.

[25] Cf. Paquet (2003): Dynamic optimization model based on a detailed production process model selecting production locations in the US. The approach contains new, pioneering elements but not the specific factors of international location choices.

it possible to capture machine and other operating data that can provide quantitative backing for tactical and strategic decisions.

Structured modeling of the entire production chain allows more exact evaluation of cost drivers and complexity

---

To put such a wide range of information to use requires a robust structure and rigorous parameter definitions. A model of the company's (or BU's) business processes can be an enabler, allowing the deployment of a decision support systems. This requires recording, parameterizing and mapping the business process structure in a formal model. Such models are not limited to strategic planning. Very similar models are used for decision support systems in tactical production planning (i.e., scheduling the individual assembly steps and completing materials and personnel requirements planning). Materials requirements planning based on bills of materials is another widely used application.

3.3.2.3 How to Decide which Technique to Use

The growing use of operations research methods, particularly optimization methods, for choosing locations is not just because these systems are now more widely available: The complexity of location planning has also spiraled. The greater complexity has two main drivers:

- The **growing number of relevant factors** influencing the choice of location increases the complexity of evaluation and is a direct consequence of globalization. Companies are acting in a more heterogeneous environment when they expand activities beyond their own borders. Differences in factor costs, productivity, and transaction costs are considerable and have to be carefully analyzed before a decision is made.

- **Increasing fragmentation of the value chain:** Companies in industrialized countries – especially large manufacturers – have increasingly been focusing on ever smaller sections of the produc-

tion chain. This trend became apparent in the 1980s and 1990s, and is still continuing. Specialization of production facilities has also grown. The number of cross-factory and cross-company interfaces has dramatically increased as a result, ratcheting up the complexity of location choices.

Methods based purely on manual scenario planning and the use of simulation techniques have their limits. Preparation and analysis of the scenarios is often too complex. Too many scenarios would have to be prepared to ensure comprehensive assessment of the options available. This frequently becomes evident when completely new constellations emerge towards the end of a project phase that are superior to those already examined. The shift in direction and detailed analysis of these new scenarios often causes considerable project delays – though this is the lesser evil compared with implementing a poor solution for lack of time or patience. Mathematical modeling is more effective in dealing with complex network optimization problems. Framing the business issue as a mathematical model enables the use of commercial optimization software to solve it.[26] This approach avoids lengthy software programming while still allowing a degree of tailoring that most software packages for strategic supply chain management do not cater for.

The problem-solving method of choice depends on the specific network structure, the type of input parameters, and the precision required. With deterministic parameters, you can use linear programming and mixed integer programming. Mathematical methods often fail in practice when the parameters are stochastic. Even if theoretically possible, stochastic parameters increase the complexity of the model and the computing time required so dramatically that they are not viable for practical analyses.

3.3.3 Sensitivity Analyses

Sensitivity analyses show how changes in the values of key input parameters will affect financial per-

---

[26] *Cf., e.g., ILOG (2005), Aksen (1998), Lustig (2001).*

formance and ROI (e.g., the profit impact of a 10 percent change in key input parameter values).

Standard scenario analysis can be used for sensitivity analyses, but its applicability is limited. Evaluating the impact of a large number of factors using manual scenario preparation is very laborious. When automated, this analysis technique can be more powerful for calculating the sensitivity of the target indicator to changes in the values of input parameters.

When linear or hybrid integer optimization methods are used, the influence of individual factors can be examined by calculating the shadow costs. These represent the change in the target indicator value if the input parameters are marginally altered.

Leading companies are also increasingly using methods that evaluate goal conflicts quantitatively. Assessing the costs of achieving one additional percentage point of service level can be useful, for example, in deciding on an appropriate target value for this indicator. As an example, a production footprint balanced across multiple currency zones will reduce the risk of currency fluctuations. It is interesting for management to understand how a reduction in currency risks impacts total costs. Such an analysis provides transparency and helps management make sensible trade-off decisions that strike the balance between conflicting goals (e.g., achieving the lowest costs and minimizing risk).

3.3.4 Central Guidelines and Templates

Central corporate finance departments should propose guidelines and templates for assessing investments. These can help managers structure the analyses effectively and make it easier for decision makers to understand the investment proposal.

In practice, guidelines on the scope of analysis and approach are often too broad. In contrast, guidelines on the formal approval process and the departments to be involved may well be very detailed. This blurs the meaning of the financial indicators and assessment results, making the decisions harder.

To give an example: Customs duties are often buried in the manufacturing costs category. However, customs duties can vary hugely depending on the choice of location. If this is the case, it is important to itemize them separately. This provides management with a better picture of the actual manufacturing costs ex works, and draws attention to the customs duties issue should related costs be high. These costs can often be influenced by (for example) changing the customs classification of parts via design modifications. This may offer great optimization potential – and managers' attention is simply never drawn to the fact. Central guidelines can help overcome these deficits. These are usually cross-functional topics. Proactive steering can prevent each function only focusing on its area of responsibility – an inherent trend in any organization.

Top management should also deliberately encourage the organization to consider uncomfortable issues, such as developing location-specific manufacturing technology and processes. This is generally not a key concern of the R&D or quality management departments: They are more interested in continuing to use proven and reliable (if more expensive) technology. Appropriate guidelines help to force the issue.

3.4 Survey Results and Case Studies

We conducted a survey with more than 50 companies and examined 15 real-life location decisions in detail. The following section summarizes our main findings, after which we discuss six case studies in greater detail.

3.4.1 How Companies Prepare Investment Decisions

The quantitative assessment of ROI is a central element in the decision-making process across all the companies we surveyed. An investment decision was never taken solely based on qualitative criteria in any of the cases we examined, though the quantitative analysis may be supplemented by qualitative elements. **Holding** companies generally formalize the investment analysis more, and define the evaluation technique to be used. They usually choose an NPV technique, used in conjunction with simple methods of risk assessment. At **SMEs**, a simple investment analysis with a relatively narrow scope is often the only instrument used. The input parameter values used in the analysis, such as the investment volume or start-up costs, are mostly determined by simple auxiliary calculations or just estimated.

Surprisingly, both large corporations and SMEs very frequently use the **modified NPV technique** (comparing differences in outflows against a base scenario) to evaluate streamlining investments. This is common even with investments for expanding sales (rather than just cost-cutting): The costs of scenarios with identical sales volume assumptions are compared and used to determine the most suitable location structure. Overall, the majority of companies we surveyed apply relatively simple tools and processes and a relatively narrow focus of analysis (Figure 3.12). Decision-making in this field does not appear

to be very sophisticated compared to other corporate capabilities and processes.

> **Conventional approaches to location selection pay insufficient attention to interdependencies. The consequences are often unexpectedly high costs and operational difficulties**

Only around one in four use standardized methods to choose locations or evaluate investments abroad. And only about 20 percent consider four or more potential locations in the selection process. One-third of companies perform an ROI analysis for just one location – they do not make any systematic examination of possible alternatives. Only around ten percent consult intensively with their suppliers during the decision-making process: Two thirds do not consult them at all. This pragmatism may partly be because network reconfiguration is a relatively rare event for most corporations. As a result, they neglect to build up special skills in this arena.

Most companies take a very pragmatic approach

Fig. 3.12: Method and scope of analysis for selecting locations
Percent

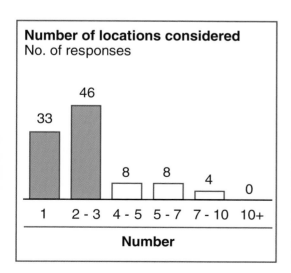

Number of locations considered
No. of responses

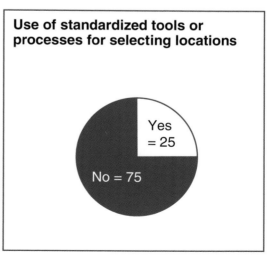

Use of standardized tools or processes for selecting locations

Question C7: "Does your company use a standardized tool or process to evaluate the attractiveness of potential production locations?" (based on 52 interviews)
Question G3: "How many countries are typically considered as potential new production locations?" (based on 27 interviews)

Source: McKinsey/PTW (ProNet analysis) (ProNet survey)

This background reveals why the results of network redesign are so often unsatisfactory (cf. section 1.4 for the results of specific industry sectors).

3.4.2 Case Studies

The following case studies describe six different approaches and evaluation techniques used by companies we analyzed. The cases are sanitized and disclose no company-specific information. The examples each represent only one (generally the key) step in the overall selection process (Figure 3.13). The focus here is on selecting production locations at the level of continents or countries. However, the same methods can often be employed in the preceding and succeeding stages, too. Chapter 6 goes beyond the economic assessment of the investment that is the subject of this chapter to address the topics of local site selection and production ramp-up.

Case Study 1: Country Preselection

A highly diversified European mechanical engineering group with annual revenues of approximately EUR 2 billion and 21 production locations is reviewing its location structure and wants to ensure the use of uniform standards. Corporate management wishes to limit exposure to country-specific risks, and has developed a perspective on the minimum level of political and social stability required to consider countries potential production locations. In the first step of the selection process, a standardized matrix for preselecting countries based on qualitative

---

The relevance of the different parameters changes depending on level of analysis

Fig. 3.13: Scope of analysis and relevance of input parameters   ▉ Focus of the case examples

Most relevant parameters

Global preselection of countries, products, and manufacturing steps
- Political stability/access
- Geographic position/transportation costs and times
- Minimum requirements concerning the market, infrastructure, or costs

Choice of location and scope of function at country level
- Labor costs and other factor costs
- Size and growth of market, customer requirements
- Logistics costs (incl. customs duties)
- Taxes and subsidies
- Availability of skilled workers and know-how

Local preselection (approx. 10 - 30)
- Local labor costs, staff availability, and qualifications
- Geographic position and transport links

Local short list (approx. 3 - 5)
- Local labor costs, staff availability, and qualifications
- Prices of land and buildings
- Availability of subsidies

Local site selection
- (Detailed comparative analysis based on all relevant factors)

Decision on location(s)

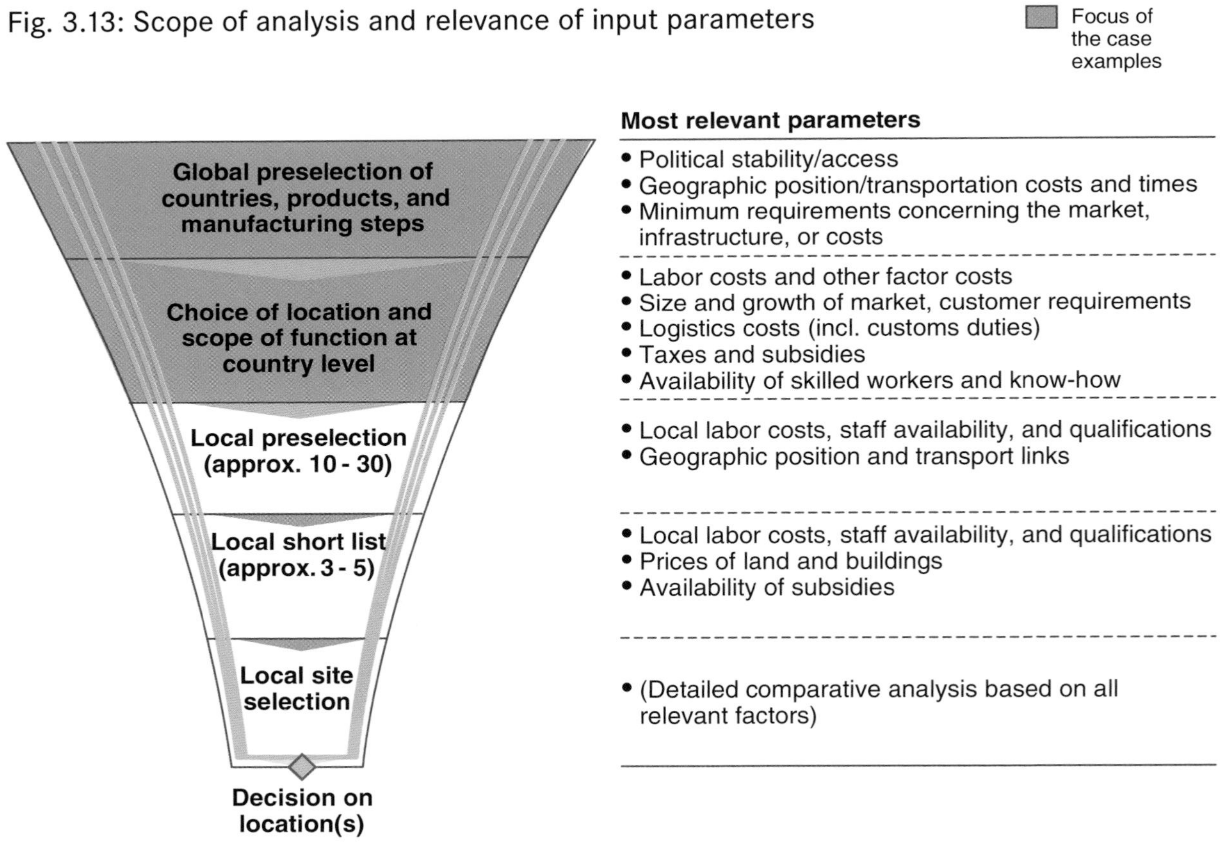

Source: McKinsey/PTW (ProNet analysis)

criteria is used throughout the group as a result. These two criteria are:

■ **Political stability:** Threat to operating processes and the value of the investment from war, social unrest, international sanctions, corruption, and other crime, as well as political intervention.

■ **Economic stability:** Threat to operating processes and the value of the investment from hyperinflation, loss of purchasing power or declining attractiveness of the local market.

A corporate department classifies the countries centrally using an evaluation matrix (Figure 3.14). Uniform evaluation standards are used for the entire company, ensuring that the group's risk preferences have already been included when the investment decision is processed. This rating has several implications.

A category "E" rating, for example, excludes the possibility of all in-house production activities in the country. A "C" or less on economic stability means that only exportable goods will be produced in the country, as the grade indicates insufficient confidence in continuous local demand.

The rating also determines the cost of capital rate to be used in the investment analysis. The use of higher cost of capital rates to value investments in countries with less political or economic stability assumes a higher risk of default.

The concept of incorporating qualitative parameters in the form of a location-specific imputed rate of return can be extended to other factors. This will also cover the expected loss of capital associated with increased external risks. Macroeconomic indicators and survey results can be used for classifi-

---

Preselection by knockout criteria can also be used to determine the cost of capital rate

Fig. 3.14: Simple matrix for country preselection – classification of suitability for production processes   SCHEMATIC

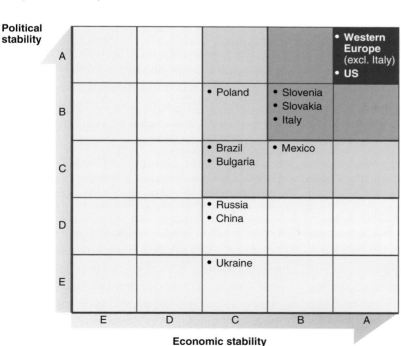

Source: EIU (2004), McKinsey/PTW (ProNet analysis)

cation, limiting the effort needed to collect primary data.

Bottom line: Uniform standards for evaluating country-specific investment risks can be applied using a simple matrix for preselection and setting country-specific cost of capital rates.

Case Study 2: Reviewing Global Location Structure

A North American manufacturer of industrial machinery and vehicles with revenues of over USD 20 billion currently operates around 70 factories. Of these, around 30 are primarily concerned with close-to-market assembly of end products. The other locations produce parts and components. The degree of vertical integration is relatively high and the total value-added share of the OEM is around 40 percent (compared with automotive OEMs, for example, where the figure is typically around 25 percent).

The company wishes to supply markets more cost-efficiently based on a systematic review of its global location structure. To do this, it needs to determine the most suitable continental regions for a defined scope of production. It decides to use static comparative cost analysis.

The analysis focuses on the product assembly sites close to markets and a limited share of parts and components production. These are under direct control, and management is willing to consider and rapidly implement changes. The redesign remit also includes changes in manufacturing technology and processes in these plants. Suppliers' sites and the majority of parts and components production are not considered. The analysis leaves out a large share of the network, so there is a risk that higher costs will be incurred at the interfaces, and the economic impact of relocation will be diminished right from the start. Delivery times and service levels are generally to be kept at the current levels, though opportunities to relax them will also be considered.

The **cost elements** in the scope of analysis are:

■ Variable costs of production, particularly labor costs

■ Fixed costs of production, e.g., administration costs per factory

■ Logistics costs (transport costs, inventory in transit, and safety inventory)

■ Customs duties.

These elements are aggregated to total landed costs – the target indicator representing the efficiency of a network configuration. The project team used a simple spreadsheet analysis to compare a large number of possible location structures using the target indicator for comparison. Scenarios represent different environmental conditions that determine the values for input parameters such as market demand, labor costs, etc. Each individual scenario has to be defined, entered manually, and analyzed separately. Out of the six scenarios, the one with the lowest costs is then examined in greater detail and refined.

The detailed scenario analysis includes a sensitivity analysis for a few key input parameters, e.g., exchange rates of some low-wage country currencies against the US dollar. A calculation is made of the relative impact of changes to these parameters on total landed costs. An index value for qualitative criteria, such as political risks, is also generated for each network configuration. This provides management with additional information and a different perspective on the characteristics of the different network setups.

The process puts several viable strategic options up for decision – various structures with assigned production volumes, indicating the cost savings they would capture. So it is certainly practicable. However, this example also illustrates the limits of manual approaches. Although the scope of analysis was restricted to the assembly plants, the number of possible location/product combinations is barely manageable. As a result, some of the underlying assumptions had to be overly simplified. The cost structure, for instance, is detailed only for entire product lines, not for individual production steps and products, and the geography is confined to continental regions (such as Western Europe), without considering specific countries, states, or provinces.

The result of this location selection process therefore requires further detailing before it can be implemented.

Bottom line: The approach features advanced elements, such as a sensitivity analysis. Products, process, and locations characteristics and know-how were only considered on a high level. The strategic plan that resulted was not sufficiently concrete for direct implementation. But it provided directional guidance and a sense for the size of the opportunity.

Case Study 3:
Selecting Production Facilities for Specific Products

A North American manufacturer of medical products with a turnover of more than USD 1 billion per annum wants to determine the potential savings from optimizing its location structure at the level of individual products and plants. A team analyzes scenarios that vary the allocation of production volumes to existing and potential sites using a static comparative cost analysis. A spreadsheet-based approach is chosen to compare total landed costs for the different scenarios.

The following elements are incorporated in the indicator 'total landed costs':

- Direct manufacturing costs (assuming the same manufacturing process and labor and capital productivity)

- Customs duties, transport costs, and inventory in transit (but excluding safety inventory)

- Materials costs including the change in this cost factor depending on the location selection and pro-

curement strategies (based on the cost structure of the parts and components).

The evaluation is partially automated. An algorithm implemented in a simple decision-support tool assigns production volumes and transportation automatically after relevant markets and potential sites have been selected. This enables the project team to analyze location configurations faster and ensures more accurate results, particularly for transportation and manufacturing costs.

Bottom line: This approach allows faster analysis than with purely manual scenario planning. However, it does not include the fixed costs incurred at every location (e.g., from providing machines for a specific manufacturing step). Another drawback is that it only details in-house production in multiple discrete production steps. Outsourced production is not modeled at this level of granularity. Also, the approach does not adequately map all the network effects and interdependencies. Overall, however, the approach allows fairly precise evaluation of concrete options.

Case Study 4:
Analyzing the Value of an Optimized Location Setup

A European conglomerate with a focus on mechanical engineering and revenues of over EUR 2 billion p.a. aims to optimize its location structure. It has already preselected countries with acceptable conditions. It now plans to assess each of its BUs to reveal the potential of network reconfiguration.

As is often the case with highly diversified companies, both the number of products or product segments and the number of (legacy) production locations is high. This makes the analysis too complex – it would be better to include fewer products and locations in the analysis. The resources and capabilities required to handle the complexity become too high, particularly if management is only asking for prioritization of the opportunity.

The pragmatic analytical logic shown in Figure 3.15 allows a rough-cut evaluation with limited effort. The approach is applied uniformly to all business units to focus the actual redesign effort that follows on the BUs with the highest potential.

The approach takes a dynamic perspective, but makes shortcuts. The calculation of operating cost reductions assumes that production will be completely relocated in one step (and not gradually ramped up over time). As a result, only a ballpark estimate is made, using the value that will only be reached once the relocation is complete. The additional cash outflows that accompany the redesign (additional investments and one-off expenses) are also merely approximated. Interactions between the BUs and the impact of different manufacturing technologies are not explicitly included. Using these assumptions, the NPV of relocating each BU's production can be calculated with a simplified formula that assumes an extraordinary cash outflow at the beginning of the relocation effort and then constant net savings.

Bottom line: This method is useful as an element in the strategic location planning of diversified groups. It allows quick analysis of the BUs. The analysis can then be further detailed within the same framework using the same indicator – NPV – yielding consistent and comparable results. A downside of the approach is that it neglects the interdependencies between the products and plants of different BUs. The method should therefore only be used if both vertical and horizontal interdependencies between the production of different BUs are limited.

Many elements have to be taken into account to make reliable estimates of the NPV of production relocation projects

Fig. 3.15: Diagram of a dynamic investment analysis   *REAL-LIFE EXAMPLE*

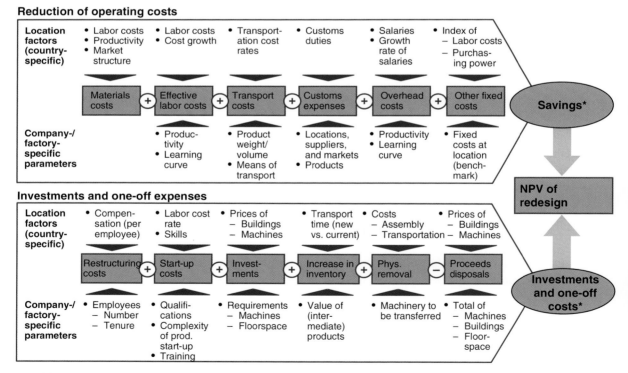

* If cash expenses

Source: McKinsey/PTW (ProNet analysis)

Case Study 5:
Allocating Products to Existing Plants

A North American white goods manufacturer with revenues of over USD 10 billion p.a. and around 40 production locations worldwide wants to identify the optimum production location for a defined product volume. The location will pursue both volume and cost goals, i.e., expansion and streamlining investments are equally possible.

Several scenarios are analyzed that differ by the allocation of products to locations and also take possible capacity adjustments into account. The current distribution of production volume to existing factories is defined in two base cases with different total volumes. This ensures that consistent underlying assumptions on demand volume will be applied to the base case and alternative scenarios, even when simulating investments to expand overall production.

Modified NPV is the most appropriate technique in this constellation. The options are evaluated over ten years based on NPV, without factoring in the infinite value or possible residual values of investments. The method is implemented in a spreadsheet calculation and allows the aggregation of individual product allocations into overall scenarios. The downside: It does not explicitly include interdependencies between products.

Bottom line: The technique delivers granular, practicable results. Capacity per product, location, and year are all analyzed, so the initiatives can be directly implemented and incorporated in central investment, sourcing, and production planning as well as site planning. A weakness is the lack of guidelines to automatically calculate one-off expenses and net investments – these have to be determined by the user instead.

Case Study 6:
Setting up One New Factory

A production facility belonging to a corporate group with around 240 employees is run operationally as a more or less autonomous, mid-sized enterprise. Be-

cause of high cost pressure on the company's technologically mature products, they wish to move production to a location with lower labor costs. They decide to use a dynamic ROI analysis to create transparency on the financial implications of a move and provide a framework for evaluating different location options.

The method they choose is a dynamic investment analysis based on outflows or outflow differences. This is used to compare three alternative options with the base case – a typical approach. The structure of the calculation for each scenario is simple and pragmatic (Figure 3.16). Input factors are mostly estimated directly or worked out using simple auxiliary calculations. The analysis is largely focused on the costs of the facility's own value added. Interfaces with other business units and functions are not evaluated, while interfaces with suppliers and customers are only considered for their impact on transportation costs. Other knock-on effects such as inventory implications are not included.

Bottom line: This kind of simple investment analysis may be helpful if the scope being examined is very limited. The danger is higher costs than expected at the interfaces because the analysis does not cover network effects.

* * *

Production network redesign is increasingly emerging as a key method of sustainably improving competitiveness. It is vital that companies are familiar with the most appropriate approaches, analysis techniques, and tools to develop and implement an effective globalization strategy.

However, existing approaches only partially fulfill the criteria for making strategic choices that we have highlighted in this chapter. Many have a vital flaw: They fail to adequately consider network effects – the interdependencies between products and plants. Chapter 4 describes a new approach: The ProNet network design approach (supported by a tailored optimization model as a decision-support tool) is particularly suited to major corporations with highly integrated production network structures.

Fig. 3.16: Real-life example: pragmatic investment evaluation for production relocation
(EUR millions p. a.)

	2004	2005	2006	2007	2008	2009	2010	2011	2012	2013	Total
Cost difference vs. base case											
Personnel costs	-0.4	1.9	4.2	8.6	8.7	8.8	8.8	8.8	8.8	8.7	66.9
Transportation costs	0.0	-0.3	-0.6	-0.7	-0.7	-0.7	-0.7	-0.7	-0.7	-0.7	-5.8
Investments and sale of assets											
Land and buildings	-7	-8.5									-15.5
Machinery and plant	-0.9	-1.7	-0.2								-2.8
Proceeds from old factory			1.5	15							16.5
Start-up, closure, transfer costs											
Training of new staff	-2.6	-2.4	-0.9								-5.9
Subcontracted production/support		-0.6	-0.6	-0.3							-1.5
Severance payments	-1.1	-2.1	-2.3	-1.3							-6.8
Bonus/retention payments	-2.4	-1.6	-1.0								-5.0
Machine transfer		-0.4	-0.2								-0.6
Project costs/provision		-0.9	-1.5								-2.4
Nominal net cash flow surplus	-14.4	-16.7	-1.6	21.3	8.0	8.1	8.1	8.1	8.1	8.0	**37.1**
Discounted (WACC 11%)	-14.4	-15.0	-1.3	15.6	5.3	4.8	4.3	3.9	3.5	3.1	**9.9**

Number of employees on annual average

		2004	2005	2006	2007	2008	2009	2010	2011	2012	2013
Factory A	Skilled workers	270	180	100	0	0	0	0	0	0	0
	White-collar workers	65	45	30	0	0	0	0	0	0	0
Factory B	Skilled workers	50	200	320	320	320	320	320	320	320	320
	White-collar workers	20	70	75	75	75	75	75	75	75	75

Assumptions (costs per employee in EUR '000 in 2004)			**Assumptions** (costs per employee in EUR '000)		
Labor costs location A	Skilled worker	30	Bonuses location A	Skilled worker	6
Growth rate 3% p.a.	White-collar w.	36		White-collar w.	12
Labor costs location B	Skilled worker	6	Compensation payments	Skilled worker	20
Growth rate 10% p.a.	White-collar w.	12		White-collar w.	22

Source: Company information, McKinsey/PTW (ProNet analysis)

Building a Global Business in Ten Years – Investment Decisions at Deutsche Post World Net

With revenues of EUR 60.545 billion and 520,000 employees (2006), Deutsche Post World Net (DPWN) is the largest logistics company worldwide. Under the brand names Deutsche Post and DHL, the company offers just about every possible logistics service – from delivering letters to shipping goods by container. DPWN has followed an aggressive strategy of internationalization since the mid 1990s. Today, some 60 percent of revenues are generated outside Germany, with an upward trend.

Production locations are the primary hubs for logistics companies, sorting centers, depots, and warehouses where letters, packages, express shipments, and pallets are processed and directed. In recent years, DPWN decided to make a number of large-scale investments, including building air logistics centers in Leipzig, Germany, Hong Kong, and Wilmington, Ohio. The company follows a central policy when making investment decisions, ensuring consistent evaluation across corporate divisions and projects (Figure 3.17).

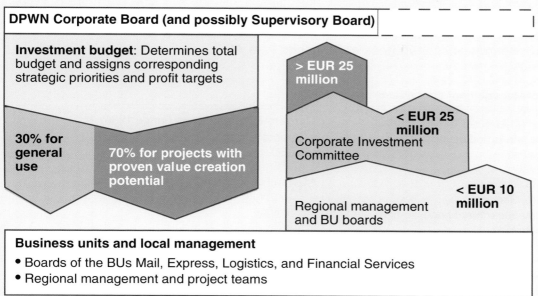

Fig. 3.17: Investment planning and decision-making process at Deutsche Post AG

SCHEMATIC

DPWN Corporate Board (and possibly Supervisory Board)

Investment budget: Determines total budget and assigns corresponding strategic priorities and profit targets

> EUR 25 million

30% for general use

70% for projects with proven value creation potential

< EUR 25 million
Corporate Investment Committee

< EUR 10 million
Regional management and BU boards

Business units and local management
- Boards of the BUs Mail, Express, Logistics, and Financial Services
- Regional management and project teams

Source: Deutsche Post AG (2005)

Creating Transparency – Use of Investment Proposals Templates

DPWN essentially uses NPV to assess planned investments. It values savings as positive cash flow differences when making investments to streamline its operations. These figures are calculated only for the duration of the anticipated project; infinite values are not generally taken into account. All future expenditure that becomes binding due to investment decisions, e.g., long-term rental agreements, is capitalized. This is performed using the marginal interest rate on borrowings, which is lower than the rate (weighted average cost of capital – WACC) used to discount earnings and savings. This approach ensures the consistent valuation of purchase and rental options, while at the same time using a conservative calculation technique (in line with commercial prudence).

DPWN develops a pessimistic, neutral (most probable), and optimistic evaluation of the competitive landscape for each investment project. This profitability analysis is structured in three categories: project costs, revenue impact, and cost savings. Each category is further differentiated: within revenue impact, for example, expected impact on volume and on price are itemized separately. A spreadsheet calculation assesses profitability and makes sure that the guidelines are applied consistently throughout the Group.

Making Decisions – Committees and Processes

The board determines the total funds available for investment and their distribution among the business units top-down. The business units are required to invest 70 percent of their allotted funds in specific projects and to present these projects to the relevant committees.

Investment projects are prepared by the business units. Certain criteria may also require that these proposals are cleared with the central units Purchasing, Real Estate, IT, or Finance/Leasing. This procedure ensures that the expertise of these operational departments flows into the decision, and identifies synergies among the plans of different BUs. Having central committees review the proposals also ensures that the projects support DPWN's strategic objectives (Figure 3.18).

Fig. 3.18: Board-level committees at Deutsche Post AG

Board of Management committees

Investments and Procurement
- Corporate Board members:
  - Express
  - Logistics
  - CFO
  - Corp. Services
- Central department heads:
  - Controlling
  - Purchasing

Mergers & Acquisitions (since 1996)
- Corporate Board members:
  - CEO
  - CFO
- Central department heads:
  - Corporate Strategy
  - Finance
  - Law

Source: Deutsche Post AG (2005)

Making Decisions on Capital Investments – a Discussion with Dr. Edgar Ernst (former CFO, Deutsche Post AG) on his Experiences with Foreign Investments

Dr. Edgar Ernst was a member of the Board of Management of Deutsche Bundespost Postdienst and subsequently Deutsche Post AG from 1992 to 2007, making him one of the most senior DAX-company CFOs. Dr. Ernst talked to the author about his practical experiences with foreign investments.

Source: Deutsche Post AG

Dr. Ernst, companies only have limited resources available for investment. How does DPWN set its priorities?

We start out with more ideas than we have capital at our disposal. The first step is to determine how large the total investment volume should be. The credit rating that we would like to earn plays a role in this decision. We don't just consider financial commitments but also pensions and other obligations. We meet annually to set priorities. On average, this resulted in a volume of around EUR 1.8 billion. That's around EUR 7 million per working day.

How is this top-down planning implemented?

The Board members discuss and agree on the top-down plan in their BUs, implementing the Board's strategic priorities. The operations side (in other words, the unit boards and regional management) takes the lead in bottom-up planning. In the logistics business, most capital investments are in buildings, often for sorting facilities. This means Purchasing and Real Estate need to be included as well.

Initially, the integration of central functions was often viewed with some skepticism. However, it is increasingly becoming clear that we have gained in effectiveness in this area, too. Take the project in Wilmington, Ohio, where we built a hub to extend our air network. In retrospect, the project staff on site admit we were right. Thanks to the expertise of Purchasing and applying the standards we had negotiated centrally with Siemens Dematic as a supplier, we were able to procure the sorting technology for 15 percent less than the original estimate.

At DPWN, the Group executive committee is responsible for all projects with investments of over EUR 10 million. Projects of more than EUR 25 million are further discussed by the entire Group board. What is the rationale for such limits?

We carried out an ABC analysis for investment projects and arrived at this breakdown. Of course, it is bound to be arbitrary to a certain extent. But our goal is to achieve the right focus and that requires a practicable filter.

How do you ensure that investment projects will produce an adequate return?

All in all, the process matters more than the figures for a specific project, because it is the right process that often makes it possible to achieve quality. We have gone a long way in many areas; we could still be better at project controlling, so we are working on that. Our experience had been that once an investment decision was made, we wouldn't hear about it again for a while. Management attention wasn't being focused where it was needed early enough. To counteract this, we now usually ask for reports on large projects at three-monthly intervals. These reports illustrate aspects of the project both quantitatively and qualitatively, and the overall status is shown using a traffic-light coding system of red, amber, and green.

Dr. Ernst, hundreds if not thousands of proposals have crossed your desk. Which information and indicators do you look at first when considering a proposal?

First, I want to understand the topic, so I read the Management Summary. Then I take a look at the figures. I am less interested in one specific figure than in how the economics will develop over time. This includes capex, i.e., the investments, and expenses, that is the costs that will affect cash flow, and impact (on revenues and costs), especially for the first three years – and this as absolute figures. It is also interesting to see what share of expenditure is already covered by provisions.

In most cooperative efforts, DPWN has immediately acquired a capital share in the partner. What is the thinking behind this strategy?

Our strategy is always to hold a majority stake in a company, subsequently acquire 100 percent of the equity, and then integrate it. This strategy cannot always be implemented immediately, however. We only gradually acquired Blue Dart, the leading express delivery service in India, for instance. Initially, we also only acquired 50 percent of Securicor SOE in the UK, a company with a courier, express, and parcel delivery network due to tax considerations.

However, we basically always try to take over the majority, at least in the longer term. In Spain, for example, we used put and call options to successively acquire the majority share in the logistics company Guipuzcoana. Having control over the hubs in a network is critical in this industry. Building a reliable system is impossible if your partner can just get up and go. Also, rebranding and operational integration are generally only possible if you acquire all the equity in a company.

We do not consider financial investments desirable. They also have no external impact: One reason is that revenues cannot be consolidated. We consider financial investments only when there are clear advantages or legal arguments against a majority holding. We only have a minority interest in Sinotrans, for example. This allows us to have a seat on the board that also supervises our joint venture with Sinotrans. It also gives us access to key information.

What do you think of this strategy of majority interest and integration with hindsight?

It has certainly been the right approach. Take the example of DHL: There, we gradually achieved 100 percent ownership. The value enhancement program STAR would not have been possible otherwise.

Deutsche Post World Net is expanding rapidly abroad. Lifeguards in Sydney advertise for DHL, minibuses with the DHL logo are a familiar sight in Delhi. DPWN has indicated that it intends to continue growing, particularly in Asia. How do you plan to achieve this?

Our share of the market to and from Asia is around 27 percent, and within Asia this figure is 40 percent. One reason why we are focusing on investment abroad is because we are reaching our limits within Germany. We invested massively here in the 1990s and created new operational platforms. As the decision of the German antitrust authorities on Trans-o-flex shows, we are limited in terms of acquisition here.

I would like to make more acquisitions in Asia, but there are not enough attractive targets. In China, for example, we want to link cities with overland transport and offer the corresponding services. We have to develop all this ourselves because there simply isn't anyone of substantial scale doing this already whom we could acquire.

What standards do you apply to address the specific risks in low-wage countries, and what experiences have you gained there?

We generally use the discounted cash flow method for evaluations. We also expect a higher return on investments in countries with higher risks.

We have also gained direct experience with IT centers in low-wage countries with DHL, most recently due to the transfer of our European IT center from London to Prague. That worked well because in Prague we could attract well-trained staff, many of whom are even trilingual. The availability of skilled staff is a critical issue for us at many low-wage locations, which may not be the case for companies needing simple assembly work. Savings also don't just come from labor costs. Rents in Prague are also considerably lower than in London.

Transferring complex processes does not always go as smoothly as in the example I mentioned. A few years ago, I was in charge of a project to transfer certain IT services to India. It didn't work; the communication and coordination among employees were simply not adequate. The only really valuable thing about that project was the experience we gained.

Further reading

Owen, S. H. and M. S. Daskin: "Strategic Facility Location: A Review" in *European Journal of Operational Research*, Vol. 111 (1998), p. 423 - 447.

Perridon, L. and M. Steiner: *Finanzwirtschaft der Unternehmung*. 10th Edition. Munich: Verlag Vahlen, 1999.

Vanderbeck, E. J. (2005). *Principles of Cost Accounting*. 13th International Student Edition, Mason, Ohio, Thomson/South-Western.

Weygandt, J. J., D. E. Kieso, and P. D. Kimmel: *Financial Accounting with Annual Report*. 5th Edition. New York: John Wiley, 2005.

Appendix: Investment Analysis Techniques

A.3.1 Static Investment Analysis Techniques

Static techniques assume a constant level of trading and thus ignore time differences (e.g., between outpayments and inpayments). The business system is analyzed assuming a steady-state, stable condition.

(1) Comparative Cost Analysis

This technique compares different location configurations based on their costs. The crucial factor here is the definition of the elements of cost and performance accounting that are included. Irregular expenditure, e.g., investments, is not captured direct, but via costs and expenses, e.g., depreciation. Typically,[27] expenses for staff, materials, machine maintenance (e.g., external maintenance services), rent, and depreciation are factored in, as well as imputed costs, e.g., the cost of capital. For international location choices, transaction expenditure and costs determined by the product flows between sites are also relevant. These are transportation costs, customs duties, and the opportunity costs of additional inventory.

Comparative cost analysis is suitable for comparing several location configurations assuming comparable boundary conditions. In particular, identical sales volumes and market structure have to be assumed.

Static comparative cost analysis has weaknesses in two areas. First, average values have to be applied that reflect the genuine course of events inaccurately when development is along a trend curve. Second, an analysis based on costs or expenses (i.e., only one side of the profit and loss account) merely determines the economic attractiveness of an investment relative to other options that assume the same sales volumes. It cannot provide insights into the marginal return on the investment, specifically compared to a "do nothing" option. The return on investment of a project can only be valued in absolute terms by comparing both expenditure and income with a continuation of business in the status quo setup, without the investment. A comparative cost analysis cannot achieve this.

(2) Profit Comparison Method

The profit comparison method differs from comparative cost analysis to the extent that it considers both sides of the profit and loss account (e.g., expenditure and income or costs and revenues).

The profit comparison method is particularly suitable for selecting a location when a cross-functional decision is needed. For example, entering the US

---

[27] *Cf. Perridon (1999), p. 40.*

market for a company based in Germany might only make sense in direct connection with production in the US. Decisions on market entry (income/revenues) and production location (expenses/costs) would thus be linked. The prerequisite for applying the method usefully is that it must be possible to allocate expenses and income or costs and revenues to the specific investment decision.

(3) Average Return Method

In contrast to the profit comparison method, the average return method also integrates the capital employed and shows the relationship between this investment and its return, measured as annual net cash inflow[28]. It is advisable to use capital employed (i.e., including investment financed by debt) in this calculation, and not only equity. The financing decision (i.e., the mix of equity and debt) includes a risk-return trade-off that is generally independent of the investment's economic attractiveness. While this trade-off decision is important for the financial investor, it is of little importance from an operations perspective.

The average return method is marred by the same problems as methods (1) and (2) above. Using the return on capital employed as an evaluation criterion appears to make sense particularly where the company's opportunities for refinancing are limited. For companies with easy access to equity capital (e.g., from issuing shares) or debt (e.g., due to high solvency and a high equity/debt ratio), the suitability of return on capital employed as the only target value appears questionable. Being fixated on a maximum return on capital could, for example, lead to the company's confining itself to currently profitable niches that are not sustainable on their own in the long term.

(4) Payback Period (Static)

The payback period in static analysis is the quotient of the initial investment and the expected average surplus income generated by the investment. The payback period is highly relevant as an indicator of

the return on investment projects and the associated risk. There is also a dynamic form of this analysis technique.

A.3.2 Dynamic Investment Analysis Techniques

(5) Payback Period (Dynamic)

The payback period in a dynamic analysis is the time at which the cumulative inflow surpluses generated by the investment compensate for the initial investment.

The payback time gives indications of both the return on and the risk of a project. A fairly short payback period means a lower dependence on events in the more distant future, for which forecasts are generally subject to higher uncertainty.

(6) Net Present Value (NPV) Method

The NPV method determines the NPV C of the cash flows associated with the investment project to be evaluated. The net inflows for a period are discounted to the date of the analysis and cumulated.

$$NPV = \sum_{t=0}^{T} \frac{\left(C_t^I - C_t^O\right)}{\left(1+r\right)^t} \cong \sum_{t=1}^{T} \frac{C_t}{\left(1+r\right)^t} - C_0 \qquad (A.2.1)$$

C_t^I: Cash inflow in period t, i.e., cash transactions, settlements of accounts payable by customers, inflows from divestitures, etc.

C_t^O: Cash outflow in period t, i.e., expenses with direct cash impact, capex, etc.

C_t: Net cash flow in period t

C_0: Initial investment (including one-off expenses)

r: Discount rate

(7) Internal Rate of Return Method

The internal rate of return represents the return on the capital employed over the period of the invest-

---

[28] Cf. Perridon (1999), p. 51.

ment project. It is implicitly assumed that inflows of funds can be invested at the same rate.

The calculation of the rate of return r is difficult because it involves an equation to the nth degree. It can, however, be resolved with sufficient accuracy for typical parameters using several iterations of Newton's method of approximation, or by interpolation of estimates of r.

$$0 = \sum_{t=0}^{T} \frac{(C_t^I - C_t^O)}{(1+r)^t} \qquad \text{(A.2.2)}$$

r: Internal rate of return

(8) Annuity Method

The annuity method is not only helpful for evaluating investments, but also for profit planning and budgeting. The annuity represents the constant surplus inflows induced by the investments, i.e., the inflow of funds after taking into account capital repayments and interest payments on an investment hypothetically financed entirely by debt at a rate of interest z.

$$A = \left(\sum_{t=0}^{T} \frac{(C_t^I - C_t^O)}{(1+r)^t} \right) \times \frac{(1+r)^T \times r}{(1+r)^T - 1} \qquad \text{(A.2.3)}$$

A: Annuity, i.e., a constant sum that is available as part of the free cash flow after repayment of capital and interest payment in each period[29].

(9) Modified Net Present Value (NPV) Method to Evaluate Streamlining Investments

The modified NPV method to evaluate streamlining investments is similar to comparative cost analysis to the extent that only outflows (costs or expenses in comparative cost analysis) are considered, and thus issues of production and sales are largely kept separate. The method is only suitable for the relative evaluation of alternatives, whereby the structural status quo can be used as the basis for comparison, i.e., the continuation of the current production structure. Similarly, it is possible to perform a *ceteris paribus*

analysis by assuming constant market-side parameters, i.e., constant sales volumes and corresponding inflows in all alternative scenarios. In practical terms, this boundary condition is often implemented by stipulating a uniform demand profile for all scenarios considered, i.e., the same unit volumes to be delivered per market. The assumption of identical unit volumes per period and market is an important prerequisite for maintaining the relative comparability of scenarios when applying this method.

$$NPV^* = \sum_{t=0}^{T} \frac{(C_t^B - C_t^O)}{(1+r)^t} \cong \sum_{t=1}^{T} \frac{C_t^{net}}{(1+r)^t} - C_0 \quad \text{(A.2.4)}$$

NPV^*: NPV of the differences in cash flows

C_t^B: Outflows in the base scenario in the period t, i.e., expenses with direct cash impact, reinvestments, etc., while assuming the structural status quo remains in place

$C_t^O A_t$: Outflows in the period t, i.e., expenses with direct cash impact, investments, reinvestments, etc., assuming the streamlining investment is made

C_t^{net}: Net cash flow/savings (surplus inflows resulting from the project)

C_0: Initial investment (including one-off expenditure)

r: Discount rate

---

[29] *Cf. Perridon (1999), p. 67.*

TOBIAS MEYER, FRANK JACOB

4 Network Design: Optimizing the Global Production Footprint

Summary

An optimized global production footprint can give a company a strategic edge by delivering long-term savings of around 20 percent of total landed costs.[1] Savings can even exceed 40 percent for companies with a legacy of fragmented sites in high-wage countries. The globalization of a company's production and sourcing also creates a platform for entry into new markets. However, planning and execution are still a feat – on average, the companies we surveyed had only achieved production cost reductions of 13 percent in recent relocations.

To get the full benefit of global production, companies must adopt an integrated perspective that extends across the value chain and covers multiple input factors from labor costs and productivity, materials, energy, and logistics through to customs, taxes, and exchange rates. Changes to product design and process technology should also be explored. As these elements can dramatically alter the economics, companies also need a new quantitative approach that does justice to the many factors involved.

Management should also take a fresh look at existing locations. Operational improvements, such as lean production methods, can make existing sites more competitive. These opportunities need to be weighed in order to compare sites fairly; they can also point the way to a more gradual transition to a new footprint and lower restructuring costs. A clear transition plan to the target structure needs to factor in the speed and sequence of migration to optimize net present value and return on investment.

Companies have to be proactive to maintain their competitiveness. The challenge is to move from an incremental to an integrated approach folding production into a global operations and growth strategy that is able to react dynamically to market changes and is regularly reviewed by top management.

---

[1] Total landed costs include manufacturing, materials, and logistics costs, customs, and duties up to delivery of the product to the customer.

Key questions, Chapter 4

- How can companies identify when a review of their location structure is needed and for which product segments or business units is this issue most pressing?

- What is the best approach for a comprehensive redesign of the production network? What fundamental principles need to be observed?

- Why is it important for a company to align its choice of locations with the globalization of procurement, and how can this be achieved?

- How can a company map its production processes to realistically evaluate the costs of production at all (potential) locations?

- What is the best way to generate a strategic concept for the globalization of production and an optimized global footprint?

- How can product mix and capacity per location be determined? How can production technology options be included in the decision framework?

- How can make-or-buy decisions and the potential restructuring of existing locations be integrated into a location concept?

- What are the pros and cons of a typical production network pattern?

- How can migration towards a target setup be structured so that it is financially viable and operationally feasible?

To date, new manufacturing locations have netted cost savings of only 13 percent on average[2] – an astonishingly low figure, particularly given that differences in labor costs between existing high-wage and new sites can amount to a factor of ten. 13 percent savings in cost of goods manufactured (ex works) can barely make up for the increase in logistics costs, inventory, cost of capital, and the additional cost for management to coordinate the location ramp-up and integrate the location into the corporate network. Under these conditions, exporting from the new location to existing or new markets (other than the country where the new site is situated) generally has little upside.

> **Companies often fall far short of their cost reduction aspirations when reconfiguring their global footprint: They act incrementally and too slowly**

One of the main reasons for the relatively low savings, as noted in Chapter 3, is the use of conventional location planning techniques for global, multi-staged production value chains. This often results in decisions that lead to higher interface costs not accurately accounted for in the original basis for decision. These costs include higher transportation and other logistics costs, higher management costs, and lower management productivity associated with expensive expatriates and travel, slow ramp-ups, and initially low productivity and high error rates at new locations. Conventional methodology used to evaluate locations does not include such factors, and so companies tend to ignore these additional costs. As a result, locations abroad frequently fall short of expectations. They do not achieve the targeted cost position and their performance is more of a hurdle than an enabler to opening up new markets.

This chapter describes a new methodology for designing a global production footprint. The resulting **strategic location concept** can serve as a master plan for globalizing manufacturing and sourcing.

---

[2] Cf. results of the ProNet survey.

The approach is based on analysis of the strengths and weaknesses of existing methods as described in Chapter 3. The concept incorporates the experience of numerous managers we interviewed about their decision-making and success criteria and has been used and refined in multiple projects.

4.1 Holistic Approach

This methodology goes beyond conventional approaches: It covers a company's entire production network, viewing location selection and strategic procurement as one integrated task. An array of factors are analyzed, from labor costs through to customs duties, including their interdependencies wherever relevant. A key goal is to minimize total production network costs.

This approach draws a clear distinction between determining the target structure (greenfield perspective) and optimizing the NPV of migration based on the existing network (brownfield perspective). The greenfield perspective reveals the ideal target structure, i.e., the production network with the minimum total landed costs for supplying all relevant markets. The cost position of a **greenfield network** is an interesting benchmark. Beyond showing the total savings potential for the company, it also allows conclusions to be drawn on the structural cost position of a potential fast-moving competitor, and the threat it would represent. The **brownfield perspective** takes into account existing facilities and other assets, and represents a plan for how to transition from the status quo to the target structure. The brownfield perspective is the more realistic approach for management. It takes into account the investment needs and costs of changing the current setup, and helps develop a view on the speed and sequencing needed to optimize NPV.

Realizing these benefits means investing a good deal of effort in analysis and evaluation. This is worthwhile for companies with upwards of several hundred employees and sufficient will to change. Without this critical mass, it makes more sense to choose one of the simpler approaches described in Chapter 3.

4.1.1 Principles for Redesign

A location strategy clearly has to be tailored to a company's industry and specific competitive situation to make a sustainable contribution to corporate performance. Nevertheless, a few universal principles also apply. They are based on both an analysis of successful global companies and the experience of decision makers from our survey.[3] The case studies at the end of this chapter further support these principles' validity.

> **Integrated optimization of the total production and supply chain has a much higher cost impact**

Comparative analysis of (a) the *stand-alone optimization* of individual manufacturing steps (such as final assembly) and of (b) the *integrated optimization* of the entire process chain reveals the very different potential of each approach. On average, **integrated optimization** achieves nearly *twice* the savings attainable from an isolated choice of location for individual steps (calculated using total landed costs).

The relevance of the labor cost factor also increases substantially with integrated optimization. The impact of labor costs runs throughout the entire value chain (Figure 4.1). Labor costs also affect material costs – often a large share of overall costs, which can be reduced by relocating the procurement base. If individual manufacturing steps are considered in isolation, the relevance of labor costs is only roughly the same as that of transport costs or customs duties.

> **Significantly higher cost impact can be achieved in several manufacturing steps if the production technology is adapted to the new location**

When redesigning production networks, the share of savings realized by **manufacturing technology** geared specifically to the location is dependent on the industry and product concerned. It may be un-

---

[3] *Cf. ProNet survey; cf. also Abele (2005).*

economical to use different processes at different locations in some industries, such as semiconductor manufacturing, because of the high process development costs. However, significant impact can usually be achieved with location-specific processes in the manufacture of components and the assembly of simple consumer goods. The four case examples (cf. section 4.3) show that companies do need to consider extending their technology portfolio. Between 5 and 80 percent of savings in those examples depend on the use of alternative manufacturing technology and processes. (The analysis covered differing degrees of automation and alternative manufacturing technology both with and without a change in product design, while maintaining exactly the same product functionality.) For some companies, such an extension will lead to only minor adjustment of the

automation of materials flow and work piece handling. Others will find that using alternative manufacturing technology and processes makes expansion or change of the production footprint both more economically viable and easier to manage.

The location structure should be redesigned proactively and – if possible – ahead of globalization of the relevant market

Substantial relocation of production capacity in an industry segment causes a permanent change in the **industry cost curve**. Competitive conditions and the pricing strategies of industry players change as a result. The development of prices and revenues gives the first incumbent to expand production into low-wage locations a strategic edge – a first-mover advantage. This

Analyzing the overall supply chain increases the relevance of labor costs, and the savings that can be captured

Fig. 4.1: Sensitivity analysis
Percent

EXAMPLE: AUTOMOBILE GEARBOX

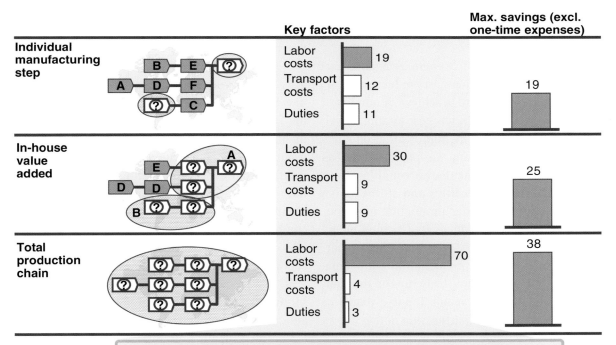

Cost disadvantage from failing to consider these factors during optimization (labor/transport costs and customs duties)

Source: ProNet Value Chain Optimizer v9

company will have a know-how lead over local rivals in low-wage countries, at least in continuously developing industries. It will be able to compete with competitors that have factories in highly developed industrialized countries both by developing and manufacturing high-tech products in its home-based factories and cost-effectively manufacturing simpler standard products at low-wage locations. This ability to play both the high-tech as well as the low-cost game and shift products and technology between locations according to their maturity and complexity is of utmost importance for almost every manufacturing company today.

Many companies that are late in venturing abroad to where production is cheaper and close to the market are forced out of the **mass segment**. This means they jeopardize their opportunities to open up price-sensitive markets in fast-growing developing and newly industrializing countries. In addition, R&D spend has to provide a faster payback since product innovations can only be used in premium products – they cannot be rolled into simpler products at a later stage. However, a successful presence in the more price-sensitive volume segments is increasingly becoming a prerequisite for overall corporate success.

In mechanical and automotive engineering, this participation in high-volume, low-cost production is particularly important but hard to achieve. A number of factors make production in low-cost locations challenging: complex technology, significant economies of scale for both factories and equipment, as well as brand image risks. However, there are multiple ways to circumvent these difficulties and still leverage the advantages of a broader production footprint. Porsche, for example, produces its Cayenne model in a partnership with Volkswagen, enabling it to participate in low-cost mass production in a low-wage location (the Porsche Cayenne body runs on the manufacturing line of Volkswagen's Touareg model in Bratislava, Slovakia). The partnership option allows Porsche to use scale and location advantages for cost-effective production while avoiding high investments, fixed costs and the considerable risks that running a low-cost plant alone would pose to this small, high-end car manufacturer.

With an NPV-maximizing production network redesign, significantly **negative financial performance and cash flow effects** are to be expected during the first two to three years after starting implementation. A company can only meet this challenge by taking action before its competitive position has been eroded. Timely production footprint redesign is vital for many companies to safeguard their long-term success – or, for some, their survival.

4.1.2 Approach for Generating a Strategic Location Concept

The approach illustrated in Figure 4.2 is based on the principles described above, and has been successfully used multiple times. It helps to manage the complexity of redesigning a global production network: a process that should not be underestimated. A vast number of options need to be evaluated using many different criteria. The approach ensures systematic planning and focus on the elements and parameters that have the greatest impact on financial and operations performance. The step-by-step explanations that follow explain the analyses typically needed to prepare a comprehensive, accurate location concept as a basis for decision.

This section provides an overview of the five modules that create the strategic location concept. Module 1 essentially determines the urgency of a fundamental production network redesign and lays the groundwork for comparing the economics of various location configurations. Modules 2–4 are preliminary to generating the location concept. They guide planners to build a model of the current production configuration, test that model, and generate and assess different scenarios. The strategic location concept is developed from this information in Module 5.

Module 1: Identify strategic objectives and urgency. The need to redesign the network has to flow out of the company's strategic objectives. These objectives include market share targets in the various market segments and regions, technological aspirations (implied by a shaper vs. follower strategy), and whether a "first-mover" or "ready-made nest" strategy should be pursued when selecting new locations. We describe five indicators that deliver results for both lines of inquiry in parallel. A natural concomitant of these analyses is to reveal the urgency of fundamental footprint redesign (section 4.1.3).

Module 2: Model existing production. The second module falls into three subsections. The first is to segment the product and process portfolio (4.1.4.1). The aim is to group production steps that are similar in their cost structure and complexity. An example would be a welding process, including surface preparation, positioning of parts, and the actual welding, but excluding cleaning or further processing of the part – processes that are different in their cost structure and complexity, and can be separated from the welding process itself. The resulting manufacturing process model provides the underlying structure necessary for assessing the economics of location configurations. Getting this step right and develop-

ing a sensible model of the manufacturing process as a whole is critical. It is particularly important to find the right level of detail, neither modeling more production steps than necessary nor making the structure too coarse. The second task is to collect the process parameters for each production step, such as the amount of energy or time required (4.1.4.2). The third is to check the validity of the process model (section 4.1.4.3). This is done by comparing the cost of goods manufactured calculated using the model (assuming actual production quantities and factor costs, etc.) with actual costs (base case).

Module 3: Assess the potential of current locations. The third module identifies the improvement potential of existing production locations (section 4.1.5), provided this area has not yet been sufficiently explored. If existing plants have already been optimized and management does not see significant further improvement potential, this module is not

---

An integrated approach to production footprint redesign

Fig. 4.2: Integrated globalization strategy – overview

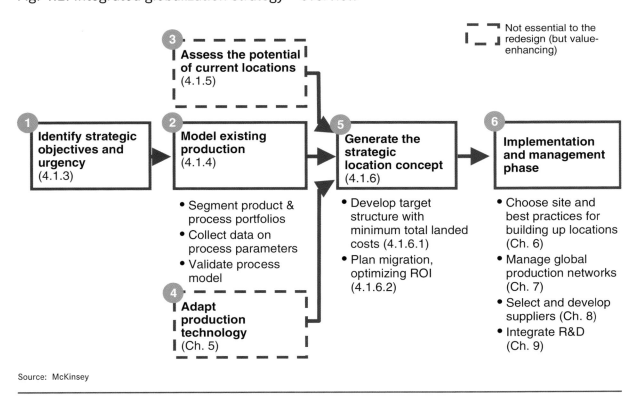

Source: McKinsey

required. However, improving productivity in existing plants using lean manufacturing methods (which help to reduce inventory, increase labor and machine productivity, and quality, but often require a high level of worker qualification and experience) can often be a viable alternative to relocating to sites with lower factor costs. Companies must also consider the value that existing sites contribute to the total production network, which often goes beyond the locations' actual manufacturing role, including economies of scope from combining manufacturing with R&D.

Module 4: Adapt production technology. The fourth module, like Module 3, is not strictly required for deciding on a new production footprint but can significantly improve the concept's economics and implementability. It examines whether alternative production technologies and product designs might be more cost-effective in other locations, with different factor cost structures or smaller production volumes. We will not be discussing this in a more detailed section below as the next chapter is entirely concerned with this theme (see Chapter 5 "production technology: Adapting to local requirements").

Module 5: Develop the strategic location concept. It is important to note that Modules 2, 3, and 4 are not sequential: They each mine indicators, parameters, and data that are then fed into calculations for Module 5. This key module consists of two stages: developing the target structure (greenfield: the ideal structure based on total landed costs and assuming no constraints), and migration planning (brownfield: transferring from the status quo to a realistic target setup).

The first stage (section 4.1.6.1) develops the **ideal target structure** using a cost comparison method, considering questions such as:

■ How many production locations are needed?

■ Where should which manufacturing steps for which product be based?

■ What are the **implications** for the company's own factories, and for those belonging to suppliers?

■ What would the **cost position** of an optimal production network be?

The second stage (section 4.1.6.2) develops a **migration plan** that optimizes return on investment (ROI). The focus is on optimizing the ROI of the overall network redesign by analyzing:

■ When should capacity be created or reduced for which manufacturing processes in which locations?

■ How do trends involving factor costs, sales volumes, and other relevant factors affect the structure of the production network and the cost position?

■ What are the **financial implications** of (for example) the capital requirements for investments, spending on restructuring, and start-up costs?

A final module, implementation and management, is only briefly mentioned in this chapter, as it is the subject of several others later (Chapters 6, 7, 8, and 9).

4.1.3 Identifying Strategic Objectives and Urgency

Five indicators determine the urgency of production network redesign and help identify strategic threats that the company needs to address. Depending on how sophisticated the company's strategic planning process already is, these can either be used for a one-off analysis, or to create a tailored system of indicators for continuously tracking the competitive environment. A distinction is made between leading indicators that identify opportunities and risks in advance, coinciding indicators that reveal the current market and competitive position, and lagging indicators ("after-the-event" analysis).

1. New Markets and Revenue Shifts (Leading Indicator)

There are various possible reasons for a change in revenues in regional markets:

■ **In the short term**, market share is determined by the attractiveness of the current product portfolio and the launch of new products. The size of the total market is relatively stable.

■ **In the mid-term**, market size fluctuates frequently in line with the economic cycle or the accurence of major innovations. The resulting effects on order volume should be – as far as possible – incorporated in plant capacity planning.

■ **In the long term**, markets alter structurally, leading to changes in the relative and absolute size of markets, the industry cost curve, and prices. These changes are of primary interest to companies when making strategic location choices.

Rapidly developing economies pass through phases in which the demand for specific product segments grows very strongly until the saturation point is reached. In India and China, for example, the production of bicycles for domestic use experienced dramatic growth in the 1960s and 1970s but has hardly grown since the start of the 1990s. The market is saturated, and bicycles are increasingly being replaced by motorbikes and automobiles (section 2.2.2).

Companies have to detect change of this kind early and win market shares during the expansion phase. As explained in Chapter 2, decision makers rate close-to-market production as a key success factor for supplying the market flexibly and cost-effectively. Companies also have to act on this, which means recognizing structural shifts in markets early and initiating adequate changes in the production network in time to capture the market opportunity.

The implications of structural change on a company's sales are often very transparent. Figure 4.3 is a snapshot of how clearly the figures can indicate problems in the making (or opportunities, if appropriate action is taken). It shows how a printing press manufacturer's sales outside Europe have risen over just five years. This has not yet been accompanied by any change in the production network.

2. Skills and Clusters (Leading Indicator)

Three factors need constant tracking in the know-how arena: education (as a metric of the general skills of local staff), technology shifts of competitors, and the development of clusters. The general standard of education is relevant to all industries, while specialized technical expertise and knowledge clusters are usually industry-specific.

Shifts in the market require an adequate response – but not all companies act

---

Fig. 4.3: Regional distribution of contracts and staff

Percent

EXAMPLE: PRINTING PRESS MANUFACTURER

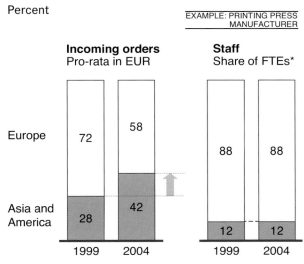

* Full-time equivalents

Source: Company data, McKinsey analysis

The standard of education is an important indicator of a location's future attractiveness and is increasingly developing to the advantage of developing countries

---

The importance of **education** and research to successful economic development is now recognized in virtually all nations. A high standard of education is no longer a unique selling proposition of Western industrial nations and Japan and will become an even smaller plus point in selecting locations in 10 years' time. China, India, the countries of Southeast Asia, and parts of Southwest Asia are catching up particularly fast. China has increased its spending on education from 2 percent of GDP in the 1990s to almost 3.5 percent. India has traditionally given high priority to education: At an average of around 4 to 5 percent of GDP, its spending relative to GDP is on the same scale as that of industrialized nations. Saudi Arabia invests over 9 percent of GDP. It is trying to develop areas with robust value added to safeguard high living standards in future without being solely dependent on the export of natural resources, particularly oil.

A large share of graduates in low-wage countries are not qualified for employment at international companies, and this will continue to pose problems. These countries are providing targeted support for particularly high achievers, enabling them to educate a small but increasing number of highly qualified graduates to international standards. However, for the foreseeable future the demand for talent with a skill set fitting the needs of international companies will far outstrip supply in nearly all low-wage locations. Monitoring local capabilities and securing early access to suitable talent pools is therefore vital.

A technology shift by competitors is a strong indicator of local skills

An important element in determining local skills is **proof of feasibility** by a competitor. Is a rival installing a new production technology in a low-wage country for the first time? If so, is there evidence of a capability shift in that region (as well as the competitor's business acumen)?

Competitors' use of the local supplier base is also particularly important. The nature of parts that competitors subcontract in low-wage countries provides information about the skills of companies based there. Companies should consider taking advantage of the supplier-building work carried out by competitors (bearing in mind, of course, the risks of disseminating proprietary know-how). This data may indicate the best timing for subcontracting parts to suppliers in the region.

Clusters are becoming ever more important, particularly in dynamic industries

Comparative advantages can develop in favor of both low-wage and high-wage locations. High-wage locations have an advantage for more mature but continuously developing industries if they play a lead role in that particular segment. If a company wishes to participate fully in the advantages of such a knowledge cluster, it often has to include a share of its production there. Geographical concentration of production is on the increase in some industry segments, and is virtually obligatory in stable, homogenous markets.[4]

Singapore and Malaysia complement each other, for example, and offer a good business environment for producers of consumer electronics. While Singapore has know-how in the manufacture of semiconductors as well as product design and marketing, Malaysia is a favorable location for manufacturing simple parts cheaply and for assembling and packaging the equipment. Although the region only has a 0.6 percent share of global GDP, approximately 10 percent of consumer electronics are made there, whether cell phones, printers, scanners, modems, or games consoles. Companies also choose this location to hedge themselves and avoid excessive dependence on China as a manufacturing location. Taiwan is another example. Though small compared to the rest of the world (1 percent of global GDP), it is going to great lengths to build a powerful position in software and biotechnology alongside its existing strengths. These are semiconductor components, where its share of world production is already roughly 11 percent[5], and LCDs, where it is responsible for roughly 55 percent of global production of LCD panels and monitors.[6] Taiwan is likely to intensify its knowledge base in areas that can generate lucrative synergies with these fields.

Companies should track cluster development proactively: The know-how and staff available there make them attractive, and they usually also have a good

---

[4] Cf. Porter (1998).

[5] On the basis of wafer starts per month.

[6] Cf. Schulz (2004).

supplier structure. Locating production within a relevant industry cluster particularly helps to curb startup, expatriate, and material costs. The reliable supply of materials, production machinery availability, and worker productivity are other positive effects. And local proximity to competitors promotes the exchange of best practices in production. High-tech companies should not be the only ones to consider cluster effects. Even in more traditional fields – from automotive supply to textiles – new centers of know-how are emerging that specialize in one subsegment and can, therefore, exist alongside established clusters.

3. Revenue and Cost Shares per Region (Coinciding Indicator)

Is the relationship between production, procurement, and sales volumes unbalanced? This signals a need to consider reconfiguration, especially if it applies to goods with relatively low value density. Long distances from assembly plants to the market raise costs while reducing supply chain flexibility.

A major imbalance between market and production structures harbors risks

Fig. 4.4: Regional balance of production and sales
Percent
EXAMPLE: AUTOMOTIVE SUPPLIER

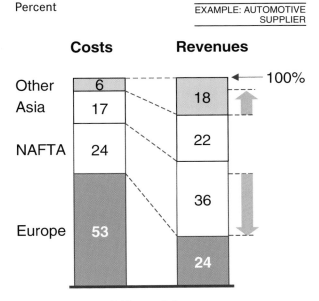

Source: Company data, McKinsey analysis

A severe imbalance between currency zones can be another strong driver for redesign. An imbalance leads to higher risks from exchange rate fluctuations, which can have a severe impact on profits and cash flow. Exchange rate fluctuations with a profit impact of several percent of revenue – sometimes higher than the industry profit margin – are not unusual. The recent rise of the euro against the US dollar and its impact on exporting manufacturers with a Europe-centric production footprint such as Porsche and Airbus is only one example. Similarly, Figure 4.4 shows an automotive supplier's distribution of production and sales across three continents. The imbalance means significant risk from exchange rate fluctuation: The rise in the value of the euro has a detrimental impact on the supplier's competitive position.

A geographical imbalance is particularly critical for simple goods if the share of product volume produced in high-wage regions is greater than the share of sales generated there. This gives new market players in low-wage countries a structural advantage, based on lower factor costs, lower logistics costs, and greater market proximity.

4. Change in Competitors' Location Structure (Coinciding Indicator)

The globalization of production tends to depress competitors' costs and can change industry cost curves, leading to significant structural discontinuity. Companies should track these shifts and determine the implications for the competitive position of their individual business units. Two areas should be considered for investigation:

- The current production footprint of **existing competitors** and changes to it as a result of relocation to expand and substitute their portfolio

- The growth (revenues, customer base, skills) and global presence of companies from low-wage countries, i.e., potentially **new competitors**.

Figure 4.5 shows the location-related differences in labor costs in the production functions of two competing companies. Company A has a much weaker structural cost position. Only around 10 percent of its production staff work in low-wage countries (compared with around 40 percent at company B), while its labor costs in high-wage countries are approximately USD 7 per hour higher. The cumulative cost disadvantage amounts to around USD 2.5 billion per annum. For company A to achieve the same financial performance as company B, it would have to compensate for this disadvantage by substantially higher productivity or higher prices.

Companies should know the cost structure of existing and potential competitors and track their development regularly. Action must be taken to compensate where a structural cost disadvantage exists or is expected. Given that companies are increasingly adopting each other's best practices, this cannot usually be achieved simply by improving productivity. A company also needs to align itself to the market to

make the best possible use of its specific skills, adapting its location structure both for growth and substitution if this is not possible (Figure 4.6).

A change in the cost curve of existing competitors and market entry by new competitors has consequences that extend beyond the direct competitive relationship. Globalization can lead to fundamental price changes, especially in mature industries and oligopolistic markets (Figure 4.7). This is especially true if the cost advantage of producing at new locations – steel production in Brazil, for example – is substantial. This will often be due to low labor costs and the proximity to raw materials and the market, while production at existing locations is capital-intensive and, therefore, has a high share of fixed costs. In a situation like this, established competitors will go on producing even when market prices are below their own full costs of production, thus contributing to a sustained fall in prices.

Two-thirds of the labor cost disadvantage of company A is due to a more limited presence in low-wage countries

Fig. 4.5: Comparison of labor cost structures in production

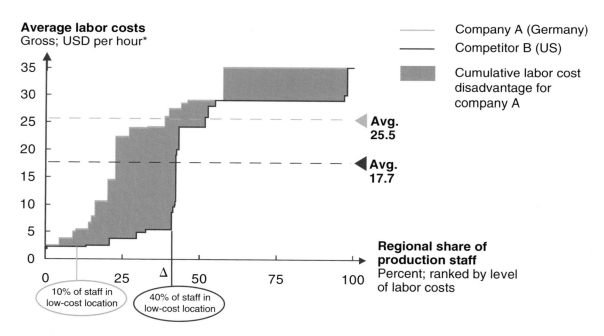

* Identical average labor costs assumed per employee per country for both companies

Source: McKinsey/PTW (data based on annual reports and company information)

Companies may use new locations to expand or substitute existing sites, depending on the situation on their home market

Fig. 4.6: Change in share of production in different regions
By number of staff, index 1999 = 100

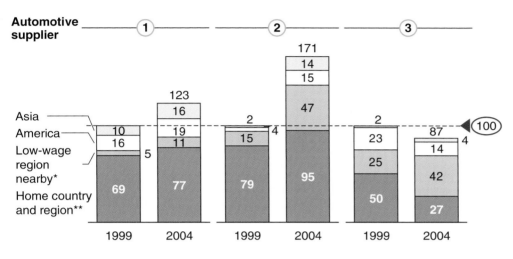

* E.g., Eastern Europe for Western European companies; (1 and 2) Mexico for US company (3)
** E.g., Western Europe (companies 1 and 2) or US (company 3)

Source: McKinsey (data based on annual reports and company information)

Competitors from low-wage countries entering the market can have a significant impact on the industry cost curve

Fig. 4.7: Industry cost structure

SCHEMATIC

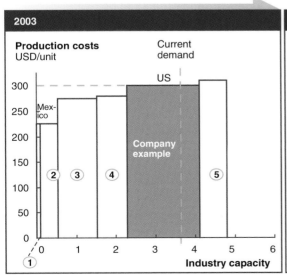
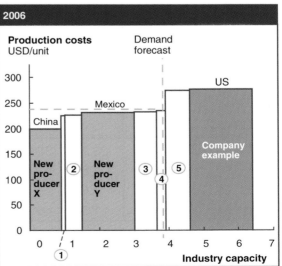

Source: McKinsey

Companies must also keep an eye on players who are still only potentially direct competitors. Asia's rapid economic upturn has brought forth companies that have the skills and resources for international expansion because of their solid business base in their domestic market. Even though these companies only have limited experience in other markets so far (e.g., North America and Western Europe), the mere fact that they are attempting to enter the market threatens the margins of established companies in the segments concerned.

The expansion of companies with headquarters **in low-wage Asian countries** and increasing international presence is very dynamic (Figure 4.8). Their growth rates are mostly well above those of their respective market volumes and are, thus, being achieved by squeezing out local and international competitors. This trend will continue unless incumbents redesign their production network before the relevant markets are fully globalized. Companies with their origins in low-wage countries will have a substantial cost advantage for decades to come, even if there is a relatively sharp rise in the local cost base (wages, for example). If these companies build their skills and improve their productivity, they can exploit this structural cost advantage to gain market share.

5. Rise in Imports from Low-Wage Countries and Significant, Long-Term Drop in Prices (Lagging Indicator)

A historical review shows a correlation between price trends and the share of imports from low-wage countries in domestic consumption. A rise in these imports is accompanied by a drop in prices too great for companies to offset with the usual increases in pro-

Leading companies from low-wage countries are reaching critical mass for international expansion

Fig. 4.8: Expanding companies from low-wage countries

Industry	Company	Revenues, growth*, and profit, USD billions			Comments
Basic materials and construction	Sinopec	72.2	39%	4.35	Asia's largest oil company
	Baosteel	19.9	37%	2.62	Has large international holdings
	China Construction**	9.1	40%****	n.a.	Over 220,000 employees
	Tata Steel**	2.4	25%	0.39	Best-cost steel manufacturer
Electronics	Haier***	12.3	26%	0.22	Internat. market presence/production
	Huawei	5.6	46%	0.47	Over 24,000 employees
	TCL	4.9	42%	0.03	Several internat. JVs and presence
	Lenovo	3.0	15%	0.16	Largest PC manufacturer in China
Service	China Mobile	23.2	22%	5.07	Approx. 200 million customers
	TCS	1.8	15%	0.41	Largest provider in Asia
	Wipro Technologies	1.6	26%	0.33	Large international provider
	Infosys Technologies	1.5	46%	0.42	Large internal national provider

　* Revenues & profit for 2004; revenue growth in 2004 (compared with 2003); non-adjusted exchange rate effects < 3 percentage points
　** Revenues & profit for 2003; revenue growth in 2003 (compared with 2002); non-adjusted exchange rate effects < 3 percentage points
　*** Haier Group, i.e., including the listed subsidiaries Quindas Haier and Haier Electronics; intragroup revenue may not be fully consolidated
　**** Estimate

Source: Bloomberg, OneSource, Internet research, annual reports, McKinsey analysis

ductivity from continuous improvement. This correlation is evident not just for individual product categories but even for entire areas of industry, such as clothing, leather goods, or automobile components (Figure 4.9).

Companies are usually able to compensate to a large extent for increases of 1 to 5 percent per annum in factor costs relating to their own value added by raising productivity. If there is a substantial fall in nominal prices, however, productivity improvements are not normally sufficient to prevent a decline in margins. For a company to improve the nominal cost situation, it would have to increase physical productivity well above the rise in factor prices. This is hard to achieve because labor unions tend to align their wage demands to gains in productivity.

Companies need to take proactive steps when a substantial, sustained price decline is expected. For those whose production systems are more or less stretched to the limit, preemptive improvement of the structural cost position is one possible solution, as is a divestment or unique positioning strategy. It takes several years to implement all these approaches. If a company decides to improve its structural cost position, it has to keep in mind that the effects must be achieved before the onset of a substantial price decline. This timing is the only way to achieve payback from expenditure on production relocation. If the location structure is redesigned late, the company will not achieve any improvement in its returns, as the reduction in costs will be offset by the decline in revenues per unit. Nevertheless, the fall in margins will still be slower than for companies that persist with

---

The rise in imports from low-wage countries is accompanied by declining prices

---

Fig. 4.9: Trend in prices and imports from low-wage countries, 1998 - 2003*
Price indices for industry segments

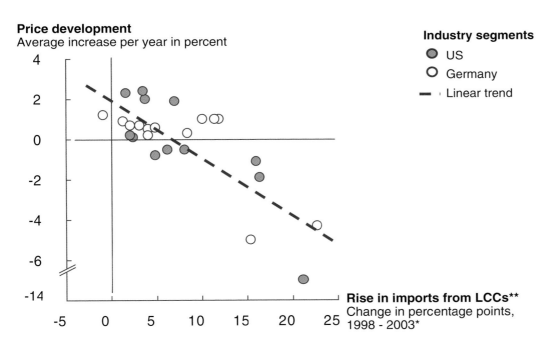

* For Germany: 1997 - 2002
** As "gross LCC penetration: LCC imports/(LCC imports + domestic value added)"; LCC = low-cost country

Source: Eurostat, US Bureau of Labor Statistics, Federal Office of Statistics, McKinsey analysis

the status quo. Analysis shows that incumbents from high-wage countries can only survive successfully in mass markets with global production if they revise their competitive strategies very early and implement measures to redesign them.

Improving competitive position by changing your production footprint is not an easy task. The results of the ProNet survey, in fact, show[7] that the correlation between company profitability and share of production in low-wage countries is not statistically significant. Companies that find themselves competing primarily on price can only achieve a positive impact if they are among the first to move to more cost-effective countries, and if they adapt the location structure dynamically to changing requirements and constraints.

4.1.4 Modeling Existing Production

The second vital batch of data needed for an integrated optimization model is to model existing production. The resulting **production process model** forms the basis for a quantitative assessment of the production network. It includes the main economic and technological features of the production network, using a manageable number of parameters.

This production process model concentrates on mapping the input-output relationships for parts and other input factors. The result is similar to a product's bill of materials (BoM), except that the process model contains additional parameters and information regarding the manufacturing process, such as its complexity. This distinguishes the process model used for optimizing the production footprint from approaches in tactical supply chain management and methods for operational planning and control. These focus more heavily on the design of operational processes such as the shop floor layout and materials flow, scheduling such as the planning of machine setup, and the flow of information.

There are three key steps to creating this model:

- Step 1: **Segmentation of the product portfolio**, first horizontally, i.e., along product lines, products, and variants, and then vertically, i.e., along finished products, components, parts, and raw materials. The aim behind this is to group production steps that are similar in cost structure and complexity. The result is then transposed into a **process model** that shows the relationships between the different manufacturing steps.

- Step 2: The next step is to collect the **process parameters** for each production step. This is usually done in workshops involving experts from the relevant functions, i.e., production, controlling, and product and process development. A parameterized process description means input factor quantities for each manufacturing step and product can be collected faster. However, this still does not achieve sufficient accuracy.[8]

- Step 3: The **process model is validated** (to create the **base case**). This involves using the process model to reproduce the status quo. The parameters generated by the model (total costs, cost structure, number of staff, etc.) are compared with the actual values. If they deviate too much, the process parameters have to be adjusted.

The approach for creating the production process model using these steps is explained below. Pointers are provided on how to make the modeling process as effective as possible, so that the conclusions obtained provide valuable support for the decision-making process. It is important to find the right balance when deciding on how detailed to make the analysis. Overly demanding aspirations are not advisable. Companies may find that the additional accuracy achieved is only minor, and the extensive time and resources spent on modeling could have been put to better use.

---

[7] Cf. Abele (2005), Appendix 2.

[8] Cf. Veloso (2001).

4.1.4.1 Segmenting Product and Process Portfolios

Companies tend to have a very large number of end products, some of which only account for a miniscule share of sales or only differ slightly from other product variants. Analyzing the return on investment of production locations and the allocation of products in excessive detail does not make sense, because there is a disproportionate relationship between the effort involved and precision gained. It is, therefore, advisable to first of all select **a small number of representative products** or areas that account for a substantial share of sales or value added. An ABC analysis[9] (analysis of the consumption value per product/part) and clustering of the product portfolio into a small number of product families may be helpful.

Production process models: as aggregated as possible, as detailed as necessary

With **horizontal segmentation**, end products, components, and parts should be selected whose manufacturing is to be explicitly mapped in the model. Similar products, components, and parts should be aggregated into groups. The cost structure (especially capital and labor intensity) and technological constraints (particularly manufacturing complexity and machinery) should be largely homogeneous within these groups. Products, components, and parts with little influence on the core characteristics of the end product (e.g., C parts according to the ABC analysis) should also be disregarded. The effort required to manufacture such

Structured presentation of products and components is the first step in modeling production processes

Fig. 4.10: Core products and rest of product portfolio   <u>SCHEMATIC</u>

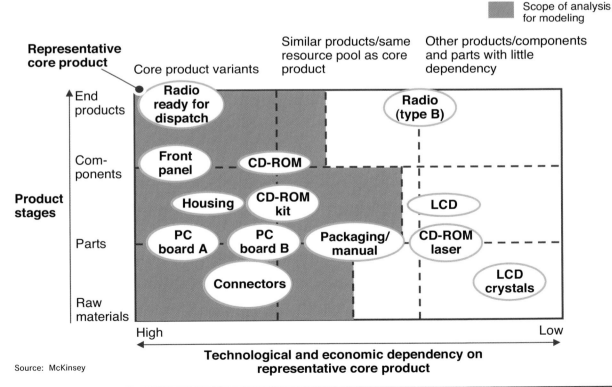

Source: McKinsey

---

[9] Cf., e.g., Silver (1998), p. 32.

parts should either be added to other processes or incorporated as a lump sum.

With **vertical segmentation**, the main manufacturing steps should serve as guidelines. The main categories of the manufacturing process[10] – forming, shaping, joining, machining, coating, and assembly – normally display significantly different cost structures and varying levels of complexity. Particularly with assembly, it may be necessary to consider several stages separately. Manufacturing steps that are inherently connected for economic or technical reasons should not be modeled separately, even if they have different cost structures. Work piece handling and tool changing, for example, are intrinsically connected with a machining process. The manufacture of sand cores for a casting process is unlikely to be feasible at a location different from the casting process itself, because cores are hard to transport. Deburring

castings is a different matter. It is quite conceivable to separate casting from deburring, assigning the two processes to different locations.

When using the product structure as a guideline for segmentation, companies should keep in mind that bills of materials (BoM) are usually too detailed, and that manufacturing steps have to be aggregated into individual BoM headings to obtain a practicable production process level.

Figure 4.10 illustrates the method of segmenting the product and process portfolio using the example of a consumer electronics company.

The segmented structure is then translated into a process model that describes the relationships between the individual manufacturing steps. The process model defines the level of detail of the opti-

The production process model maps the supply chain relationships

Fig. 4.11: Manufacturing steps – example

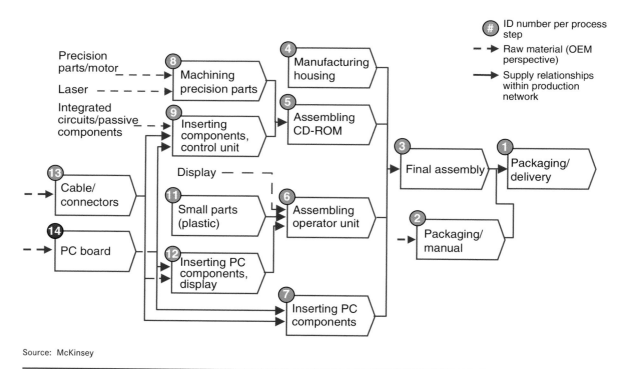

Source: McKinsey

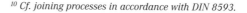

[10] Cf. joining processes in accordance with DIN 8593.

mization. Activities aggregated into one process step will always have to be carried out at the same location. The production process model – as shown in Figure 4.11 – serves as an important basis for defining and illustrating the relationships between the manufacturing steps.

These relationships, i.e., the aggregated product structure, should be represented in a formalized way so that they can be run in the optimization model. With the selected optimization approach, this is done by defining demand coefficients. These coefficients define the secondary demand for the upstream parts and components based on the primary demand for the end product. The coefficients are incorporated into the equations to determine the secondary materials demand, the materials flows, and the consumption of resources. Mapping the supply relationships between manufacturing steps in line with a general BoM structure (Figure 4.12) enables a component to be processed further in several successive steps (diverging BoM structure). Likewise, several parts can be included in one manufacturing step, as is typically the case with assembly (converging BoM structure). In the example shown in Figure 4.12 (referring to a case study in the aerospace industry – also note the process model in Figure 4.26), manufacturing step 11 generates demand for manufacturing step 10, while manufacturing step 10 generates demand for the intermediate products of manufacturing steps 3, 6, and 7, etc.

4.1.4.2 Collecting Data on the Process Parameters

The structure of the process model supplies the framework for **collecting data on the process parameters**. The process parameters describe manufacturing such that all relevant decision criteria for the choice of location included in the model are actually mapped.

Detailed knowledge and discussion of the **manufacturing technology** to be used are normally required to determine process parameters. This should be taken into account in selecting the workshop participants and estimating the time required for gathering data. Important raw data for determining the process parameters include:

The demand coefficients formally capture the supply relationship

Fig. 4.12: Manufacture of a component – input-output matrix   <u>EXAMPLE: MANUFACTURING</u>

		Production process (predecessor)										
		1	2	3	4	5	6	7	8	9	10	11
Production process (successor)	1	0	0	0	0	0	0	0	0	0	0	0
	2	1	0	0	0	0	0	0	0	0	0	0
	3	0.55	0.45	0	1.5	0	0	0	0	0	0	0
	4	0	0	0	0	0	0	0	0	0	0	0
	5	0	0	0	0	0	0	0	0	0	0	0
	6	1	0	0	0	5	0	0	0	3	0	0
	7	0	0	0	0	0	0.2	0	0	0	0	0
	8	0	0	0	0	0	0	0	0	0	0	0
	9	0	0	0	0	0	0	0	1	0	0	0
	10	0	0	19	0	0	3	16	0	0	0	0
	11	0	0	0	0	0	0	0	0	0	1	0

Source: McKinsey

■ Cycle times

■ Operating times

■ The number of work pieces processed in parallel

■ Prices of machines and tools

■ Installation costs

■ Time required for maintenance

■ Depreciation periods

■ Space required for production equipment (machine tools, chucks, tool shops, assembly units, etc.), inventory (raw materials, work in progress, and finished products), the materials flow systems, and other operating resources

■ Resources required for ancillary processes.

Sometimes, it may be necessary to make a rough sketch of the component with production engineers and estimate the key parameters (if the work pieces are new, the planned manufacturing processes have never been used, or the historical data for existing processes is poor).

In addition to determining the factor input volumes per production process, qualitative criteria should also be captured in the form of constraints. For location factors relevant to a number of different industries, e.g., quality of the infrastructure, it is advisable to use the indicators available.[11] Customized indicators may be necessary if requirements are industry- or process-specific. An indicator to determine availability of skilled workers in a region, for example, might be the value added of the relevant industry segment as a share of the region's total value added. Figure 4.13 illustrates definition of the process parameters for one specific application of the optimization model.

4.1.4.3 Validating the Process Model

The next stage is to validate the process model, i.e., check it to ensure that it presents a sufficiently accurate map of reality by **simulating the existing network structure in the model**. The output parameters of the optimization model that describe the configuration of the production network are set in advance. The following parameters should be transferred from the real location structure into the model to develop a strategic location concept:

■ Production volumes per process step, process type, and location

■ Selection of the locations used in reality and exclusion of locations that are not used

■ Transport volumes, i.e., allocation of markets to production facilities.

It should be remembered that optimization is carried out as a comparison of different location configurations using the same process model. It is vitally important that deviations and inaccuracies are eliminated from the model during validation. The stakes are clearly too high to allow any margin of error.

Deviations in the absolute level of aggregated cost of goods manufactured per product or the costs of the total network should be below 10 percent. The same applies to variances in the number of direct employees in production at existing locations. It is often difficult to achieve higher accuracy, partly because it is almost impossible to allocate and model overhead costs precisely without disproportionate effort.

The scope of costs and expenses captured must be carefully scrutinized when making the comparison. While the capital costs should be included when optimizing the global production network, these should not be included in a location's direct expenditure. When comparing the cost of goods manufactured per

---

[11] Cf., e.g., IMD (2003).

product, you have to ensure that assumptions on the internal rate of return, for example, are aligned.

4.1.5 Assessing the Potential of Current Locations

When analyzing the costs of potential new locations, it is typical to perform an evaluation for a point in time in the future. This makes good sense, and should apply equally to existing locations. The analysis should take into account not just expected values for factor costs and sales volumes, but also the expected future productivity of the location.

Productivity can, of course, often be considerably improved in the mid- to long term. Experience shows that the cost advantages of this can vary widely. The potential of home factories is often particularly high if manufacturing processes have not changed fundamentally since their introduction, and if the company does not yet use a system of continuous improvement, with possible savings of 20 to 30 percent of direct manufacturing costs. The potential at companies with high operational excellence is distinctly less. Benchmarking the production of a few select products or individual manufacturing steps can help to estimate the potential more successfully and create a basis for sound location planning assumptions.

Efficiency improvement programs sometimes incur considerable one-off expenditure, especially with widespread organizations that have numerous loca-

Definition of the process factors must be adapted to the respective requirements

Fig. 4.13: Model description of production processes

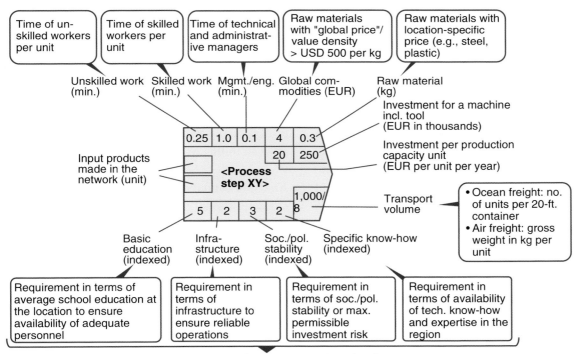

Minimum requirements of processes on location factors
(implementation via standardized indicator (for example); cf. WEF, 2004)

Source: McKinsey

tions. They usually require the full-time deployment of approximately 1 percent of the workforce in production during the project phase to achieve extensive and sustainable impact. The recruitment of specialists and training of local employees require a lead time of roughly one year. As a result, it may make sense to model the implementation of an efficiency improvement program as a separate option when optimizing the production network. The migration from the existing plant structure to a more productive one should be evaluated with one-off expenditure in the same way as starting up production at new locations.

An efficiency improvement program of this kind may also lead to a continuous reduction in production staff, which could lead to lower one-off expenditure when all or part of production is subsequently relocated. Particularly at locations with strong social security systems, such as Japan, France, or Germany, restructuring costs and especially severance payments are a substantial hurdle for transferring capacity to new locations.

Companies reviewing the potential of existing locations should also consider possible economies of scope of their home production with other corporate functions, suppliers, and customers. **Economies of scope** can contribute to higher productivity at the home plant and influence the entire production network via the transfer of best practices. Economies of scope with production can also lead to greater effectiveness in other functions. For example, if developers have direct and frequent exchanges with skilled production staff, this may make it easier for them to design made-to-manufacture products that can be ramped up fast for series production. The real potential of home production may lie in these qualitative synergies. It is crucial to ensure that the evaluation of economies of scope is performed neutrally and supplies a fact-based estimate of whether they can truly be created. For example, it is often possible to check whether any processes and management methods have been transferred to other locations in the past, leading to sustained improvement.

4.1.6 Generating the Strategic Location Concept

The steps described so far were largely preparatory. The individual factors are brought together when the target structure is developed and a migration plan is defined.

4.1.6.1 Developing the Target Structure

For the greenfield work that is now required, it is advisable to analyze different environmental scenarios to examine the sensitivity of the solution to changes in the selected input factors. Conflicting goals can also be made transparent in quantitative terms: How will costs increase, for example, if service level requirements are raised?

The typical steps for generating the target structure are as follows:

- **Blueprint:** The optimization model independently defines the lowest-cost location structure under the constraints and cost functions. The result of the first optimization run therefore usually infringes upon the implicit restrictions or preferences of top management – e.g., not setting up capital-intensive manufacturing steps in countries with low legal stability. Such restrictions or additional (opportunity) costs have to be integrated into the model after an initial discussion.

- **Analysis of goal conflicts:** The next round of discussions should center around selected conflicting goals. This includes a comparison of total cost with delivery time restrictions, maximum currency imbalances, and the separation of critical manufacturing steps from other corporate functions (such as central R&D at the main home-based factory).

- **Detailed design:** The detailed design of the target production network structure should allow decision makers to validate or at least check the plausibility of critical assumptions. Corresponding indicators should deal with parameters that decision makers are familiar with, such as employee figures per location, cost structures (e.g., labor costs, transportation and depreciation as a share of total costs), and production volumes. In addition, it should be illustrated how implementation of the target structure will impact specific corporate functions (particularly production, purchasing, and possibly logistics) and the individual companies in the production network. This will trigger a discussion of the implementation hurdles and provide incentives to proactively tackle the problems that will be encountered in implementation further downstream. Although these discussions generally lead to the preparation of several detailed designs for the target production network structure, they ensure easier and faster implementation of the concept. Such scenarios should also be part of the basis for decision and interim reviews with management. This helps to inform management of relevant trade-offs and align the project team and management around the final proposal, which also needs to include a migration scenario and ROI analysis.

4.1.6.2 Planning Migration and Optimizing ROI

Two steps are required to ensure targeted implementation of the location concept by the functional departments, country organizations, and project teams:

- **Migration planning**, i.e., determining and scheduling the steps from the starting position to the target structure helps make implementation of the concept feasible, since it defines concrete tasks and deadlines.

- The organization's overall and individual **skills** have to be compared with the needs of the relocation projects. In the event of gaps, either the organization's skills should be enhanced or the complexity of the project reduced (also Chapter 6.1). The allocation of the structured tasks to project teams for implementation ensures accountability.

Companies can plan migration and optimize the return on investment, i.e., the NPV of the project, in two ways:

- **Pragmatic migration:** What is needed to move from the existing network structure to the desired status? The specific actions are planned in line with the optimal structure in a "roll back the future" operation. The actions are scheduled as early as possible, factoring in operational restrictions, whether the limited availability of management capacity to establish locations abroad, for example, or the lifecycle status of product lines to be relocated.

Expected future savings are estimated in line with the planned implementation progress that emerges. Cash flows resulting from savings, additional investments,[12] proceeds from the sale of assets, changes in inventory, and one-off cash expenses are captured in a simple NPV calculation. This allows evaluation of the project using the relevant indicators.

- **Optimized migration:** Pragmatic migration assumes that rapid implementation will be the most cost-effective. This is not necessarily the case if investment requirements and one-off expenses are high. Adapting the speed and sequence of implementation can significantly improve the project's NPV if annual savings are not significantly above the one-off expenses.

If very fast payback or a high internal rate of return is required, some steps may even be entirely uneconomical. Dynamic analysis may reveal that a different location configuration offers a higher return because, although it delivers lower savings, less expenditure is required for implementation. Migration needs to be adjusted under these circumstances by adding the dimension of time to the model and changing the target function. While optimization of the model structure is based on the cost comparison method, it is advisable to use the cash-flow-based NPV method when dynamic analysis is used.[13]

---

[12] *Additional investments here are investments that go beyond the reinvestment budget. Reinvestments are taken to mean those investments in the means of production that would be needed to maintain operations at the original location.*

[13] *Cf. section 3.2.3.*

The following rule of thumb applies: **Optimizing migration** of the production network in terms of speed and sequence of the projects delivers significant impact compared to pragmatic migration if the ratio of additional investments and one-off expenditure compared to static savings is over 3:1. This 3:1 ratio of the total extra spend on the migration vis-à-vis savings roughly corresponds to a payback time of around four years. An increase in the NPV of the network redesign by more than 10 percent or a difference of several years in the scheduling of the relocation of several production processes is regarded as a significant effect here.

When the ratio of investments and one-off expenditure to savings is more than 3:1, it normally makes sense to adjust migration

Optimizing migration also appears to make sense if the parameter values and restrictions change over time. We have already mentioned improvement potential at existing locations that influences assumptions about the productivity of these facilities. Even with new factories, this makes it possible to take start-up losses and learning curves at least implicitly[14] into account.

The additional **planning effort** for optimization and the associated delay in implementation therefore appear to be justified particularly in cases where the return on a relocation investment is not overwhelmingly attractive or the boundary conditions for relocation change over time. Using a computer-aided model to assist migration planning means that secondary conditions important to the company can be explicitly considered. These include the availability of financial resources for investments and one-off expenses and the impact of the migration on the profit and loss account, e.g., including the write-off of assets that become redundant.

4.1.7 Implementation and Management

Migration planning puts the strategic location concept on a concrete footing. Times for the implementation of individual steps are defined at the level of individual locations, which means that implementation can be effectively tracked. Investment budgets, long-term capacity planning, and savings objectives define large areas of the specifications and target agreements for works managers and project teams.

In defining and allocating the tasks for the individual locations or product areas, critical examination should be made of whether the human resources and operational capabilities of the companies and the individual teams and managers are sufficient to implement the location concept and manage the desired structure (also see section 6.2.1). There are numerous examples in which the allocation of resources, implementation plan, or even target structure have been adapted, ensuring faster and easier implementation as a result. However, there are also a considerable number of cases in which only hindsight revealed that insufficient thought had been given to the difficulty of the task and organizational skills. Decision makers should prevent a gap between the strategic objectives and the resources for their operational implementation from emerging right at the beginning of implementation.[15]

Besides implementation at the level of individual locations, a successful strategic location concept also demands further activities relating to the production network as a whole:

- Continuing the selection process and implementation at a local level (analyzing an individual location with its suppliers). Here, three different implementation modes can be chosen:

  □ Building or expanding a company's own **production capacities** or acquiring an existing company in the target country.

---

[14] *Explicit consideration of learning curves has its limits, at least with the application of hybrid integer programming methods, since linking the productivity of a location to the volume produced there can lead to a quadratic relationship that dramatically increases the complexity of the model and, thus, the computing time.*

[15] *Cf. Meijboom (1997), p. 790: gap between the strategic location choice and the operational management of locations (interviews: Dutch companies with a presence in Thailand).*

❑ Setting up a **joint venture** with an equity stake held by a local partner and developing or expanding a production site together.

❑ Relocating by **outsourcing** processes previously carried out in-house to suppliers in the target country.

■ Creating the **organizational skills** to implement the location concept and manage the global production network (analyzing the entire production network, including interfaces).

The implementation measures must be in harmony with the company's strategic direction, particularly the definition of the core competences. There is a significant dependence between the motivation for relocating production and the choice of implementation format. Having production processes close to the customer, for example, proves harder to reconcile with outsourcing or a joint venture than with the set-up of a site belonging to the company.[16] The industry, market, and corporate structures in the target countries are also vitally important, as well as other boundary conditions such as legal regulations. International partnerships between companies are difficult and sometimes have high failure rates. Partnerships therefore require a particularly high level of top management attention and a systematic approach, both during their development and later supervision.

When a company is planning its **own new plant**, the choice at the local level will be taken at least partially on the basis of criteria different from those applied at the country or continent level. Location factors such as real estate prices, development costs, or attractiveness to expatriates are subject to major variations at the local level and influence the ROI analysis significantly. The relevance of these factors is correspondingly high. After the local site has been chosen, the actual planning of the factory is performed, i.e., the plant and machinery, floor space and buildings, and human resources. The financial planning for the new location has to be firmed up in accordance with planning progress. New suppliers are selected and developed based on the specifications in the strategic location concept in parallel with planning and construction of the facilities. If the strategic objectives cannot be realized in the target country, or not under the assumed conditions, alternatives have to be reviewed in an iterative planning process.

The **organizational skills** needed to manage a global production network can be too much of a strain on the existing organizational structure. The globalization of business activities can demand both changes in the organizational structure and a redesign of internal and external interfaces. Agreements on targets with the works managers have to be revised in accordance with the new scope of functions at their factories. Corporate functions (e.g., quality management, personnel management of executive staff, or product and process development), management processes, and information systems need to be measured against the requirements of the future business system and modified if necessary. Logistics processes should be designed to ensure reliable and efficient operations even in the key phases of migration, e.g., during production ramp-up, which is particularly critical as innovation cycles become ever shorter.[17] Delays can easily outweigh other advantages of the location.

Implementation issues are part of the reason for the relatively limited success global production has had in reducing costs.[18] It is therefore all the more surprising that few companies institutionalize the learning process and gather and evaluate best practices for building and managing global production locations.[19]

---

[16] *Cf. Abele (2005), p. 105 ff.*

[17] *Cf. Abele (2003).*

[18] *Cf. Abele (2005), pp. 22-24.*

[19] *Results of the ProNet survey: Only one company collected best practices that had proven effective in setting up new locations abroad.*

4.2 Network Phenotypes

Production network phenotypes can be used to illustrate the main principles of a new configuration. Comparing these with the actual or planned production network setup for a company can help to reach a better understanding of relationships and their implications. The use of such patterns also allows better discussion of different setups and their specific advantages and disadvantages, as certain types of production network come with characteristics that are quite broadly applicable. Clarity in communication is important, as the redesign of the production network is a cross-functional undertaking. It can help to better integrate the key features of strategic production planning and the setup of other corporate functions – from HR management through to purchasing – into a consistent concept.

Specific network types are favorable if various factors are similar: supply chain requirements, the value den-sity of the product and its parts, the sales footprint, and the cost structure of the underlying production processes (the case studies in the last section of this chapter illustrate this). In turn, network types have specific characteristics, such as allowing final assembly close to the market, or maximum capture of economies of scale. They also highlight other network characteristics and requirements outside the production department, such as the ideal functional scope of the locations (collocating R&D and procurement resources is one example), or centralizing decisions and know-how. It therefore makes sense to consider these idealized structures in the discussion process and use them to illustrate the characteristics of individual location configurations and the differences between them. However, this should not be a substitute for quantitative analysis and the structured selection of locations for the specific case being considered.

The five network types shown in Figure 4.14 can be distinguished based on quantitative analyses of the

The "world factory" model captures economies of scale, while "local for local" minimizes transport costs

Fig. 4.14: Global production – network phenotypes

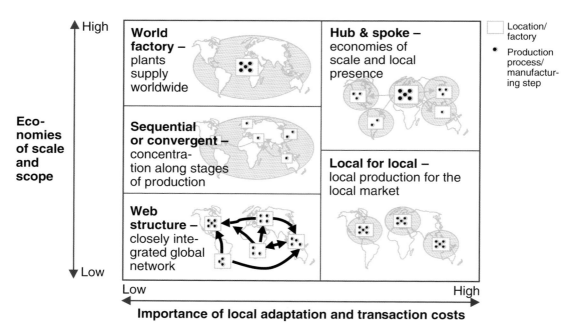

Source: McKinsey/PTW (ProNet analysis)

ROI of production networks, as well as examples of successful companies. In the 1980s and 1990s, the focus was only on two to three basic types of global production.[20] The number has expanded in line with spiraling global connectivity and the general decline in transport costs.

"World Factory"

Alongside "local for local," the "world factory" is a classic production network. Both types require only limited exchange of goods and information and therefore predominated in the early phase of the globalization of industrial production (see Chapter 1). Although global factories have lost their importance in many traditional industry segments, this model is very important in the high-tech industry. Factories that supply globally can realize maximum economies of scale in production and economies of scope with R&D and support functions.

> **The use of the "world factory" model only makes sense in industries with major economies of scale, economies of scope, high product value density, and reasonably long delivery lead times**

Manufacturing at only one location for the entire world market was also widespread in the past in the automotive and mechanical engineering industries. Economies of scale are no longer so dominant in these industries for two reasons:

- Companies have grown beyond the optimum size of operation and can no longer achieve significant economies of scale by expanding existing locations.

- Manufacturing technology has become more flexible in many areas. Short setup times are just one example.

Because of the diminished influence of economies of scale, the new factories opened by General Motors in the last few years are much smaller than tradi-

tional sites, whether you look at Eisenach (Germany), Gliwice (Poland), or Rosario (Argentina).

Economies of scale and scope are still highly relevant in high-tech industries, from semiconductor manufacturing to the assembly of large aircraft. The advantages of centralized production in these sectors go beyond the improved utilization of capital-intensive machinery and plant. Establishing manufacturing steps at only one or very few locations improves the availability of critical personnel and know-how, allows greater specialization, more intensive knowledge exchange, and shortens delivery times between the processing stages. The Korean electronics group Samsung, for example, has concentrated all its front-end factories for semiconductor chips in South Korea, achieving highly beneficial synergies from pooling its manufacturing capacity and staff.

"Local for Local"

The "local for local" model achieves the high level of market proximity critical for success in many markets. The reduced influence of economies of scale and greater importance of flexibility and short delivery times have led many companies to supply foreign markets via local factories that have relatively little interaction. This pattern has proven valuable especially for companies that make products with low value density, highly market-specific characteristics or short delivery times, and a large number of variants. One automotive supplier producing large-volume systems, such as fuel tanks, opens a new factory close to its customers in every national market.

> **The "local for local" model is suitable for market-specific products with low value density or very strict delivery requirements**

However, companies basing their location structure on the "local for local" model should be clear about the structural cost positions of their key competitors. The possible cost disadvantage relates here not only

---

[20] Henzler (1985), p. 169: The "world-scale factory" largely corresponds, for example, to the "global factory" shown in Figure 4.14.

to their own value added and the direct labor costs, which as a rule only amount to between 2 and 20 percent of the cost of goods manufactured. Competitors who use factor cost advantages along the entire value chain can often supply markets cost-effectively with short delivery times and a high number of variants even if they only base a small share of manufacturing and customization in the market.

Modern concepts for harmonizing product design and the logistics supply chain mean the "local for local" model is now the optimum solution only for a dwindling share of products, despite rising demands on delivery capabilities and market proximity. In an era where fashion products for the American market are made in China and the towels in Berlin hotels are laundered in Poland, the concept of entirely local production has become at least partially obsolete.

"Hub & Spoke"

In the "hub & spoke" pattern, manufacturing steps that are knowledge-intensive or demonstrate economies of scale are concentrated in one or just a few locations, while others (like assembly) are based at a larger number of close-to-the-market locations. This structure is particularly attractive to companies that want to deliver products with many variants and short delivery times to customers but are dependent on economies of scale in the production of parts and components. Close-to-the-market assembly often reduces expenditure on logistics and customs duties (as well as helping companies to respond flexibly and promptly to customer needs). The reason is that complex components and parts frequently have a higher value density and are subject to lower customs duties than the functioning end product.

> **The successful "hub & spoke" location model taps economies of scale and ensures market proximity**

In addition, in a typical supply chain most of the variants are not generated until the end product is assembled, so the safety stock required in the supply chain is much lower with local assembly. When developing new markets in low-wage countries, early relocation of labor-intensive assembly work is also advantageous from the perspective of the cost of goods manufactured.

The "hub & spoke" model is suitable for a large number of companies in different industries, and is used successfully by leading companies. Schmitz Cargobull, Europe's leading manufacturer of truck trailers, operates central production facilities for manufacturing components at its home base in Germany. These are then assembled at relatively small locations abroad. This greatly reduces both transport costs and direct labor costs in assembly.

A manufacturer of cellular phones produces a sufficiently large number of units to be able to operate assembly facilities in all three Triad regions, while critical components with a very high value density are produced centrally at one location. All the larger German automotive manufacturers use "completely knocked down" (CKD) assembly facilities abroad. These procure components or entire construction kits from central factories. In this case too, the "hub & spoke" network structure minimizes customs expenditure, since duties for the preliminary products are well below those for the end products. This is important particularly for supplying markets in countries such as India, Brazil, China, and the entire ASEAN region.

"Sequential or Convergent"

The "sequential or convergent" network has the strongest focus on the specific advantages of individual locations: Every manufacturing step is concentrated at a different location. However, the large number of international transport runs required in this structure limits its usefulness to products with high value density, such as electronics components. This location pattern means economies of scale and scope can be tapped optimally along individual manufacturing steps. For example, the assembly of most PCBs and production of LCDs are concentrated in Taiwan. Upstream manufacture of silicon wafers and their processing (front-end fabs) are focused in countries such as the US, Europe, Japan, Taiwan, and Korea, while a considerable portion of the further processing and wiring of the chips is carried out in Malaysia, Indonesia, and other low-wage countries. The extremely high value density of these products makes the share of transport costs virtually negligible.

The "sequential or convergent" network type maximizes economies of scale and scope in each manufacturing step

---

Much of the production of plastic parts and assembly of electronic components is also carried out in low-cost countries. However, the last manufacturing stage of products required in a large number of variants can also be based close to the market. PC manufacturers, such as Dell, Medion, and Gericom, assemble their products for the European market in Ireland, Thuringia (Germany), and Austria from components that are mostly imported from Asia. The "sequential or convergent" network has become the dominant structure in electronics manufacturing. It is also found in other industries, such as the production of food additives and fine chemicals.

"Web Structure"

The fifth network type is particularly relevant for companies that have high vertical integration, need to balance production capacity across factories, and make products with a relatively high value density.

A "web structure" helps companies smooth capacity utilization despite volatile demand in individual markets

---

One manufacturer of pistons for internal combustion engines, for example, uses closely interlinked production locations worldwide, achieving high utilization of the capital-intensive machinery via flexible order allocation. It is quite possible for a factory in North America to manufacture a product solely intended for the European market while a European factory is simultaneously supplying the North American market. Products need to have a moderately high value density for this network type to be worthwhile, and it requires a sophisticated logistics structures for effective distribution.

This model can be used to unleash internal competition for orders, since all production facilities are basically able to manufacture all products offered. This, in turn, drives efficient structures and high productivity.

Another automotive supplier combines this characteristic with having a lead function adopted by one specific facility in the network for each product. Even if the baseload of demand is produced in this factory (which accumulates product competence as a result), other factories in the network are also used for close-to-the-market production and to cover peak demand. This combination allows the company to tap benefits from centralization, while at the same time achieving high utilization of its production facilities and cost advantages from close-to-the-market production.

* * *

Idealized network structures are not a solution to location issues but a tool in the discussion and decision-making process. They are also valuable for designing and fine-tuning the interfaces between corporate functions (such as R&D and production). The following case studies illustrate how to combine these phenotypes with the systematic analysis described earlier in this chapter.

4.3 Production Footprint Redesign: Case Studies

The following case studies illustrate how specific companies applied the approach and optimization model described in this chapter to devise their location strategies. They are based on real examples, but the descriptions and figures are disguised.

Production network reconfiguration saved between 7 and 41 percent of total landed costs for the companies in these studies. Payback times were between three and eight years. These large bandwidths show how difficult it is to draw general conclusions about the impact of globalized production: Decisions always need to be made on a case-by-case basis.

Each case is from a different sector, and they all have diverging change needs. The first is elaborated in detail, modulating through the entire methodology step by step. The other three are explained more briefly, but the approach is clearly analogous.

First we consider the footprint of an **automotive supplier** that manufactures gearboxes, and whose product line for mid-sized automobiles is coming under increasing price pressure. Most of the factories in this legacy network are in high-wage countries. The restructuring backlog is considerable, because they have not adjusted in tune with market changes over the past 15 years.

The second case study examines the situation of a **consumer electronics** manufacturer. This company already moved large areas of its production to lower-cost locations years ago, but it is now under pressure from low-wage competitors in Asia.

The third case looks at relocation potential for structural **aircraft** components. The relatively small volumes mean economies of scale play an important role in production. As a result, every manufacturing step for a product will ideally be based at just one site, and production should be concentrated at only one or two plants in total. The company is particularly interested in gaining an understanding of the cost differences between different locations.

The fourth example describes an **appliances** manufacturer whose production has grown almost entirely through expansion of its home-based factories. Its domestic market has been stagnating for years, however, and only its foreign business is growing. The company is not just interested in reducing costs: It wants to produce closer to the market to shorten its delivery times. It is also highly vulnerable to the effects of exchange rate fluctuations, which cannot be hedged by financial instruments because of the volatile market situation and uncertain price trends.

Relocation of production to low-wage locations changes the entire cost structure

Fig. 4.15: Total landed costs*
Unit costs, EUR

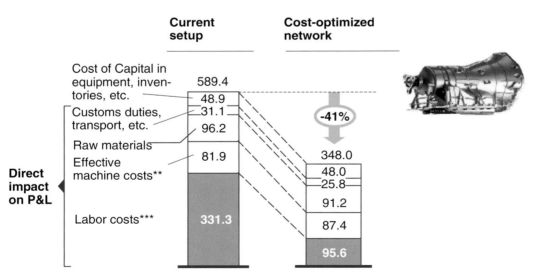

Current setup

Cost-optimized network

Cost of Capital in equipment, inventories, etc. — 589.4

48.9
31.1
96.2

Customs duties, transport, etc.

Raw materials

Effective machine costs** — 81.9

-41%

348.0

48.0
25.8
91.2

87.4

Direct impact on P&L

Labor costs*** — **331.3**

95.6

* Cost of materials, production, and logistics
** Incl. maintenance and fixed costs for machinery, plant, etc.
*** Incl. fixed costs for buildings, administration, etc.

Source: McKinsey

4.3.1 Case Study 1: Passenger Car Gearboxes

This case study from the **automotive supply industry** shows how optimizing the global production network can save around 40 percent of the total landed costs of standard gearboxes for mid-range automobiles (Figure 4.15). In the status quo, most of the company's manufacturing is based in high-wage countries. The production process chain extends across the automotive clusters in Germany, the United States, Japan, and Portugal. The manufacturing processes for the product segment being analyzed are mature, with a slow rate of innovation. Improvement potential at existing locations was not evaluated in this particular case, as existing sites had already gone through several rounds of operational improvements over previous years. The manufacturer's markets have shifted significantly in the last 15 years, while its own factories and those of its suppliers have largely remained in the traditional markets.

The **target structure** eventually chosen largely corresponds to the "hub & spoke" network type. It consists of two main locations in low-wage countries – the Czech Republic and Mexico – close to the manufacturer's large markets. All manufacturing steps are based at these main locations, from the casting of parts to final assembly of the gearboxes. Less capital-intensive machining and assembly processes with more limited economies of scale are carried out at branches in Brazil, China, and the Philippines.

Migration from the existing structure to the target network will take around 10 years and will involve the closure of several production facilities at high-wage locations unless they can be used to manufacture other products. The investment project has a payback time of around four years, taking into account restructuring costs. Maximum capital requirements during the migration phase are moderate: only around 11 percent of operating costs prior to redesign, i.e., around EUR 30 million for a business unit with expenditure of EUR 280 million per annum. Redesign of the production footprint will affect all five companies in the network.

Our model revealed that integrated optimization of the production network would lead to much more efficient structures than an isolated choice of location for individual manufacturing steps. This stand-alone approach would lead to total costs around 20 percent higher than with integrated optimization of the entire production network.

This case demonstrates the impact of two specific factors on choice of location: reduction of exchange rate instability and alternative manufacturing processes. We examined the costs of network configurations with low exchange rate imbalances. The approach we describe here achieved the greatest transparency on the conflict between costs and risk. The effects of alternative manufacturing technology are clearly perceptible but not overly high at 5 percent of total manufacturing costs. However, they do ultimately lead to a somewhat different network topology, because smaller, close-to-the-market production facilities depend on more manual processes with low fixed costs.

4.3.1.1 Identifying Strategic Objectives and Urgency

This gearbox manufacturer analyzed the competitive environment and its implications for the design of its production network. It recognized that action was needed in the segment that produces standard gearboxes for mid-sized automobiles.

The company is under pressure in this segment because its existing product platforms do not satisfy the requirements of premium manufacturers, yet the manufacturing costs of its current lines are too high for vehicles in lower market segments. However, the technical characteristics of the product line match OEM requirements very well. Redesign of the production network for this product segment would be an alternative to pursuing a divestment strategy and focusing exclusively on the premium sector.

Standard gearbox manufacturers mostly have their facilities in countries with medium wage levels, such as South Korea and Spain. They have a factor cost advantage corresponding to around 10 percent of total manufacturing costs. Large competitors' expansion of production via gradual relocation to low-wage sites in China, Poland, and Slovakia has accentuated this gap in the past few years. The relatively fast development of production capacity shows that sufficiently qualified staff are available at the low-wage locations, with a support structure of suppliers for materials and operating resources.

The company's costs and revenues are regionally very unbalanced. Its profits come under particular strain if the euro rises against the US dollar. This is especially problematic because competitors have local production facilities, so they do not need to make price adjustments due to exchange rate fluctuations. The share of imports from low-wage countries into the main customer markets has increased by several percentage points in the past three years, while

prices have fallen by 12 percent. This development in the share of imports is likely to continue, suggesting further price decline ahead.

The mass segment is strategically important to the company but has few production synergies with other business units. The company therefore wishes to redesign its own footprint and supplier base so it can achieve the cost position needed to remain competitive in this segment long term.

4.3.1.2 Modeling Existing Production

The **production process model** maps gearbox manufacture using 13 aggregated production processes (Figure 4.16). The characteristics of the production processes are captured in detail, especially the input factor volumes. For example, for chip-removing

The production steps along the entire value chain are recorded systematically

Fig. 4.16: Production chain for a passenger car gearbox

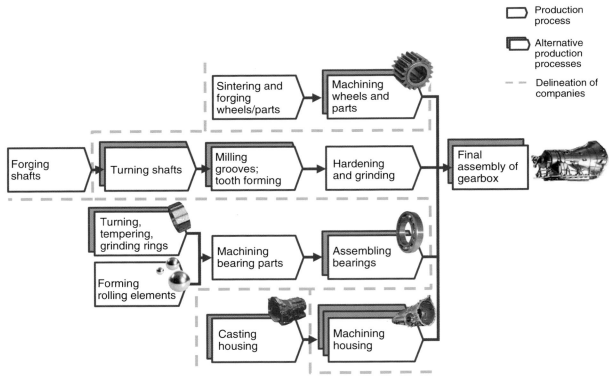

Source: McKinsey/PTW (ProNet analysis)

machining of the shafts, two minutes of untrained labor are estimated for materials transport, handling, disposal of the chips, etc., as well as 10 minutes of skilled labor for machining and inspection. Processing is assumed to be carried out on a machine with an investment volume of around EUR 150,000.

Staff skills for different manufacturing process requirements are also factored in. Low labor and capital productivity are assumed for countries with a poor supply of skilled automotive staff. Others are ruled out entirely if their level of education is inadequate and the manufacturer has no presence there.

The process model takes into account multiple regional factors, including the **location-specific prices** of steel and industrial buildings. Both these factors are relevant since they have considerable influence on total costs and vary greatly between potential production locations. Customs rates are also sometimes defined by production process or product and can have a significant impact on the network structure. In China and Malaysia, for example, duties for parts and components are significantly lower than for fully assembled gearboxes.

> **The large share of production at high-cost locations represents a structural disadvantage that is hard to compensate for with price-sensitive standard components**

The market demand is assumed to be 50 percent in North America, 25 percent in the Far East (Japan and South Korea), and 20 percent in Western Europe. The remainder is distributed across Brazil, Russia, and other countries. The total is estimated at 500,000

At the outset, the company manufactures predominantly in locations with high factor costs

Fig. 4.17: Production networks in the status quo (case study)

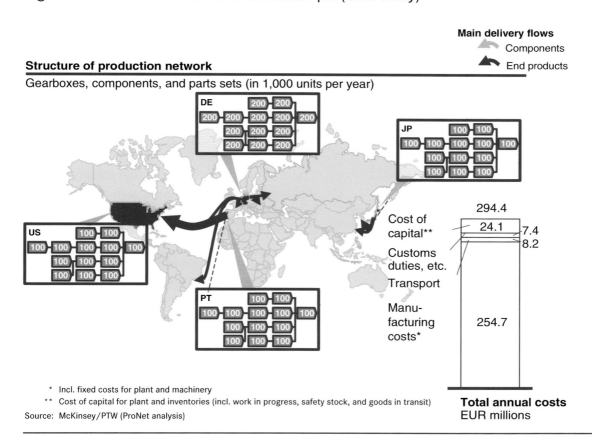

Main delivery flows
↗ Components
◤ End products

Structure of production network
Gearboxes, components, and parts sets (in 1,000 units per year)

Cost of capital**
Customs duties, etc.
Transport
Manufacturing costs*

294.4
24.1 — 7.4
8.2
254.7

* Incl. fixed costs for plant and machinery
** Cost of capital for plant and inventories (incl. work in progress, safety stock, and goods in transit)
Source: McKinsey/PTW (ProNet analysis)

Total annual costs
EUR millions

units per annum. The company supplies one hundred variants of the end product. Figure 4.17 shows the initial position of the production network before optimization. At this point, the average total landed costs are around EUR 590 per gearbox (or EUR 294 million for the total network).

4.3.1.3 Adapting Production Technology

Alternative manufacturing processes and shift models are defined for eight out of the total of 13 production processes. This means production can be mapped with two alternative levels of automation in two- or three-shift operation (in addition to the largely manual, standard two-shift manufacturing). In three-shift operation, the cost of capital and depreciation per work piece fall compared to working just

> **Incorporating the input factors step by step ensures precise decision-making support and transparency**

Fig. 4.18: Optimization stages and input factors

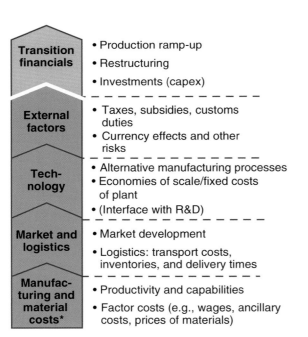

* Ex works; excluding fixed costs (included in "technology" category) and logistics costs (for parts and end products)

Source: McKinsey

two shifts per day. Capital productivity is higher. Higher labor costs have to be taken into account, however, resulting from night allowances and other additional requirements.

Modeling alternative production processes by process type also opens up the potential of **non-linear cost curves**. Alternative manufacturing processes with greater initial investment and high capacity lead to economies of scale that go beyond a simple linear decline in fixed costs accounting cf. Figure 2.22. This accounts for around 5 percent of total manufacturing costs of the gearbox for the entire production network.

4.3.1.4 Developing the Target Structure

The factors relevant to the choice of locations are structured into five categories: manufacturing and material costs, market and logistics, technology, external factors, and transition financials (Figure 4.18). The first four categories determine the long-term total landed costs in the markets. The fifth category covers the cost of relocation, i.e., the "price tag" for achieving the target network. This section demonstrates how the calculation gradually builds, revealing the exact difference that each category makes to the overall constellation (cf. the progression from Figures 4.17 to 4.22).

Step 1: Achieving Minimum Manufacturing and Material Costs

The main factors influencing alignment of the production network to **minimum manufacturing and material costs** are factor costs and productivity. Based on the assumptions outlined, the optimization

> **Based purely on manufacturing and material costs, Eastern Europe is attractive for parts manufacturing and Southeast Asia for gearbox assembly**

model shows that the best location for most of the manufacturing steps, such as turning, milling, grinding, and tempering of the shafts, is the Czech Republic. The shafts would best be forged and the bearings assembled in Russia.

Machining of the gear wheels and rod components should ideally be carried out in the Philippines. However, a focus on manufacturing and material costs does not take transport costs, economies of scale, and other factors into account.

The location factors incorporated in the optimization model correspond to current forecasts for 2007. Labor costs of EUR 6.50 per hour for an unskilled worker are assumed for the Czech Republic and EUR 2.00 for manufacturing in China (the coastal region, but not Shanghai). These values include the premium normally paid by international companies, which tend to attract and retain better skilled and motivated staff. This premium is particularly significant for China (see Chapter 2: Factor Costs) and is also necessary to achieve the level of training and experience in the company required for manufacturing high-quality products. Very high attrition rates are therefore unacceptable. Country-specific interest rates in line with the investment risk are assumed in evaluating the cost of capital.

The productivities and skills used as input factors are defined in relation to the benchmark factory in the United States. For Germany, for instance, labor productivity for machining (milling, turning) is pegged at 109 percent, 101 percent for the Czech Republic, and 63 percent for China (at the same level of automation).

> **Taking logistics costs into account – including the costs of inventory and delivery time restrictions – brings production much closer to the market**

Similar assumptions are made about the productivity of the other input factors (including capital). Estimates have to be made for countries in which no productivity values are available for comparable manufacturing processes. The value added of the industry, number of staff, level of education, and other factors can be based on data from statistical offices. These can be compared with corresponding indicators for existing manufacturing locations to obtain a reliable estimate of the productivity that can be achieved at a given location.

The ideal reconfiguration basely solely on manufacturing and material costs is shown in Figure 4.19.

Step 2: Including Market Trends and Logistics Costs

Market development and logistics costs have a considerable impact on the design of a cost-minimized production network. Taking logistics costs into account brings the last manufacturing steps, which generate a large number of product variants, closer to the markets (Figure 4.20). Market demand is assumed to be constant in this example because no increase in total demand or change in regional distribution is forecast.

The logistics costs and delivery time restrictions have a significant influence on the locations used and total network structure. To transport a fully assembled gearbox from China to Germany costs around EUR 45 per unit, for example. Only around half of this is accounted for by the actual freight rate for marine transport by container. The remaining costs are for packaging, loading, fees for customs formalities and container loading, unloading of the container, insurance, and land transport to and from the port. Long transportation times in intercontinental supply chains cause additional costs because inventories are tied up in transport, and higher safety stock has to be held in the network due to the longer lead times.

In addition to the tied-up capital, it is important to consider the depreciation in product value over time and the opportunity costs due to lack of flexibility. In the gearbox example, these costs could add up to as much as EUR 30 per unit.

Logistics costs and particularly delivery time requirements are a powerful driver for having production on the same continent as the customer market. This proximity to the market is necessary at least for final assembly of the system, which is delivered to the OEM on a just-in-time basis. The Czech Republic fits the bill for the European market and Mexico to supply customers in NAFTA states.

Step 3: Including Technology Options and Restrictions

Factoring in the **technology characteristics** of the production processes, i.e., alternative manufacturing processes as well as economies of scale and scope, indicates that production should be concentrated at three locations.

The use of **location-specific manufacturing processes** and product design can significantly reinforce the positive impact of global production (Chapter 5). In this example, three manufacturing processes with different levels of automation were mapped for the machining of the cast gearbox housings:

■ First, sequential processing on **conventional milling and drilling machines** is possible, with material flow, handling, control and tool changeover being performed manually. Machine control is only partially automatic, resulting in substantial waiting time during machining.

■ The second option available is production on **machining centers**. This frees up the worker during the machining time on the center, but material flow, handling, and control are still manual.

■ With the third option, work piece handling, tool changeover, and material flow are largely **automated**, so manual work on this manufacturing step is extremely limited and is largely restricted to monitoring, machine setting, retooling, and maintenance.

The optimization model automatically selects the manufacturing process geared best to a specific lo-

Comparatively high productivity with relatively low labor costs makes the Czech Republic attractive

Fig. 4.19: Locations chosen exclusively by manufacturing and material costs

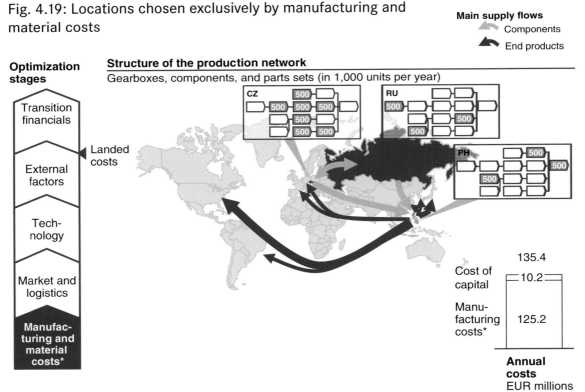

* Ex works; excluding fixed costs (included in "technology" category) and logistics costs (for parts and end products)

Source: McKinsey/PTW (ProNet analysis)

cation from the options described. The ROI is also evaluated against the backdrop of the unit volume to be produced, possible shift models, and economies of scale.

Smaller satellite manufacturing bases are unattractive if economies of scale and scope are factored in

The fixed costs per location and manufacturing process also need to be estimated. **Economies of scope** can be tapped in this example because production is close to the R&D function based in southern Germany. If manufacturing at the R&D location itself is not cost-efficient for this product segment even when economies of scope are taken into account, production in the Czech Republic still offers some proximity to the R&D location (Figure 4.21).

Step 4: Including external factors

External factors also have a high impact on the network structure. Customs duties and currency hedging are particularly important in this case.

Customs duties and **non-tariff trade barriers** are highly relevant in the automotive industry, particularly in some Asian countries, such as China, India, Vietnam, and Malaysia, and also for other locations such as Brazil. Most high-wage countries, in contrast, have much lower customs duties. Japan, for example, no longer levies any customs duties on the import of many industrial products, making imports from low-wage countries particularly attractive. The case study also considered the effect of direct and indirect subsidies. Direct subsidies were incorporated by a corresponding reduction of the acquisition costs of

Market development and logistics costs lead to the relocation of production to close-to-market, low-wage locations

Fig. 4.20: Locations chosen when market and logistics are included

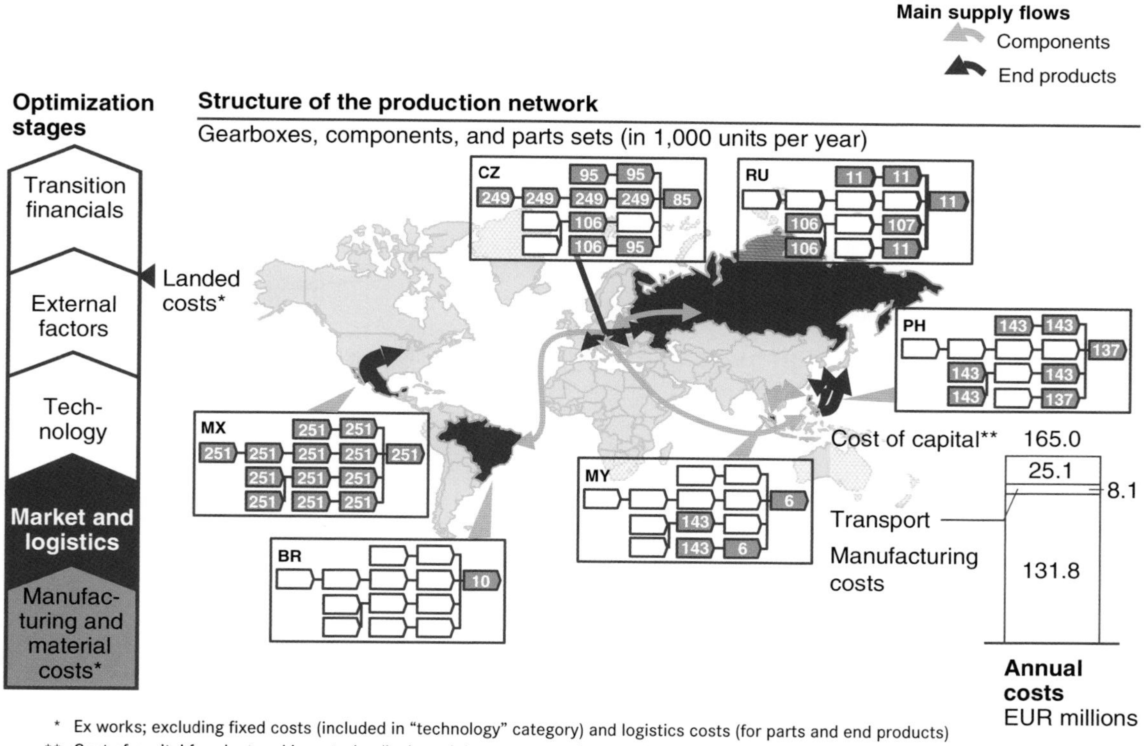

Main supply flows
- Components
- End products

Optimization stages

- Transition financials
- External factors
- Technology
- **Market and logistics**
- Manufacturing and material costs*

◀ Landed costs*

Structure of the production network

Gearboxes, components, and parts sets (in 1,000 units per year)

CZ: 95 95 / 249 249 249 249 85 / 106 / 106 95

RU: 11 11 / 106 107 / 106 11 11

PH: 143 143 / 143 143 137 / 143 137

MX: 251 251 / 251 251 251 251 251 / 251 251 251 / 251 251 251

MY: 6 / 143 / 143 6

BR: 10

Cost of capital**   165.0

25.1 — 8.1

Transport

Manufacturing costs   131.8

Annual costs
EUR millions

* Ex works; excluding fixed costs (included in "technology" category) and logistics costs (for parts and end products)
** Cost of capital for plant and inventories (incl. work in progress, safety stock, and goods in transit)

Source: McKinsey/PTW (ProNet analysis)

machinery, plant, and buildings, while indirect subsidies reduce expenditure on (for example) training and customs duties. For this case, tax motives had no influence on the choice of locations.

The desire to minimize **currency imbalances** and the risks associated with exchange rate fluctuations played an important role. The company's initial position is likely to seem familiar to companies from Western Europe. While a large share of the costs of in-house production and materials procurement is incurred in euros, much of its revenues are earned in the US dollar area. In the euro zone, costs exceed revenues by around 30 percentage points. The situation is roughly reversed in the United States. A substantial share of input factor prices is determined locally in these two large economic areas, so that even the adjusted currency imbalance is considerable. During a risk analysis, it was decided that the nom-

inal currency imbalance for each currency should be no more than 20 percent of the total network costs for each currency. The exception was Japan. Higher revenues than costs in yen were allowed since no significant devaluation of the yen was expected.

In the status quo, the maximum nominal currency imbalance is around 30 percent. The cost-minimized target structure of the network already has a much lower imbalance: a nominal figure of around 20 percent of total costs. Analysis of the sensitivity of materials prices to changes in exchange rates and the possibility of passing on part of a currency-related cost increase to customers shows that the real currency imbalance is around 10 percent. In concrete terms, the euro prices of some raw materials are actually somewhat aligned to world market prices in US dollars, so that if the value of the US dollar decreases, the prices of these goods also fall in local cur-

Economies of scale favor locations with large volumes

Fig. 4.21: Locations chosen when economies of scale, automation, and local skills are taken into account

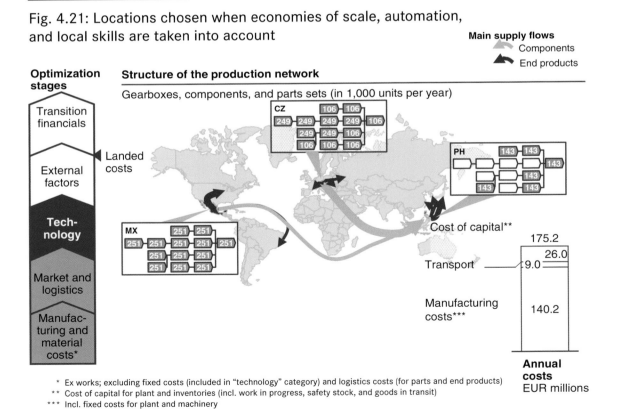

* Ex works; excluding fixed costs (included in "technology" category) and logistics costs (for parts and end products)
** Cost of capital for plant and inventories (incl. work in progress, safety stock, and goods in transit)
*** Incl. fixed costs for plant and machinery

Source: McKinsey/PTW (ProNet analysis)

rencies. Depending on the flexibility available for shifting costs between currency zones and the opportunities for hedging using financial instruments, an upper limit should be drawn for the maximum nominal currency imbalance.

Tighter restriction of the maximum nominal currency imbalance (this excludes the positive disequilibrium versus the Japanese yen) further reduces the risk of exchange rate fluctuations but leads to higher costs. A profit risk due to depreciation of the Japanese yen and the US dollar remains, as well as moderate risks from an appreciation of the Czech koruna, the Mexican peso, and the Philippine peso (see Table 4.1).

Restricting the maximum currency imbalance would lead to a shift in capacity from Mexico to the Philippines. This reduces the difference between expenditure and income in the different currencies. The correlation between exchange rates is integrated (approximately) based on historical data. The end result is that total network costs are slightly higher, but the profit risk from exchange rate fluctuations is reduced.

Figure 4.22 shows the production volumes per location and process step for the **lowest-cost network configuration**, taking into account all relevant factors influencing the long-term costs of the network and restriction of the nominal currency imbalance to 20 percent. Seven locations are used worldwide in total. Three of the locations have over 1,000 employees. Only 50 to 210 people are employed in each of the other countries. The configuration can therefore be described as "hub & spoke", with complete control of the entire process chain concentrated in two places worldwide.

The target structure uses two main locations and three smaller assembly plants

The sensitivity analysis of increases in labor costs shows that a location in Russia or Ukraine should be considered for toll manufacturing in view of the relatively rapid wage increases expected in the Czech Republic. Labor-intensive preliminary products could be delivered from there to the main location in the Czech Republic. The network would then develop into a hybrid of "hub & spoke" and "sequential or convergent."

Table 4.1: Net currency imbalances arising from different shares of revenues and costs in the currency zones. The figures apply to the production network in the case example.

Scenario	Currency imbalance (net, nominal)								Cost increase***
	USD	EUR	JPY*	CZK	MXN	PHP	CNY	Other**	
Initial position (base case)	0.314	−0.302	−0.134	0.002	0.011	0.003	0.031	0.074	n/a
Cost-minimized; no restrictions	0.149	0.070	0.183	−0.095	−0.192	−0.190	0.017	0.058	0%
Max. imbalance approx. 19%	0.114	0.050	0.183	−0.114	−0.111	−0.183	0.007	0.054	4.3%
Max. imbalance approx. 14%	0.100	0.051	0.183	−0.102	−0.081	−0.134	0.003	−0.022	7.4%
Max imbalance approx. 10%	0.046	0.055	0.183	−0.073	−0.058	−0.096	0.005	−0.063	12.1%

```
*    No restriction of the positive disequilibrium vis-à-vis the Japanese yen
**   Netted off
***  Increase in total landed costs assuming annual average exchange rates and 2003 prices
```

Source: McKinsey/PTW (ProNet analysis)

The remaining locations serve as close-to-the-market manufacturing and assembly sites. This is advantageous because of the high demands on delivery capabilities and the growing number of product variants along the value chain. A sensitivity analysis shows that a further small location in the US would be needed for final assembly if delivery time requirements became more stringent or the number of variants increased. Although this process is relatively labor-intensive and does not lend itself much to automation, locating it close to the market has benefits such as improving delivery capability and reducing inventory. An assembly location in the US would also further reduce the currency imbalance, but cause a moderate increase in total network costs.

In the configuration shown in Figure 4.22, the total costs of the network are 41 percent lower than the costs of the original network if the same unit volumes are produced. The delivery lead times meet the defined requirements, and all logistics costs are included, including inventory costs. The basic structure of this configuration is robust: It only changes incrementally if input parameters are altered within small margins. Incremental changes that would barely have any impact on total costs include relocating some parts manufacturing processes to Russia or Ukraine. Such flexibility is important, as alternative locations are helpful in implementing a location strategy. One example: Negotiating with multiple governments in parallel about setting up a production site can improve the odds of receiving investment subsidies or other support (also see Chapter 6).

However, the high savings – calculated by comparing total network costs of the target setup with the status quo – only reflect the long-term perspective.

Mexico and the Czech Republic are the main locations in the optimized network

Fig. 4.22: Locations chosen when external factors are included

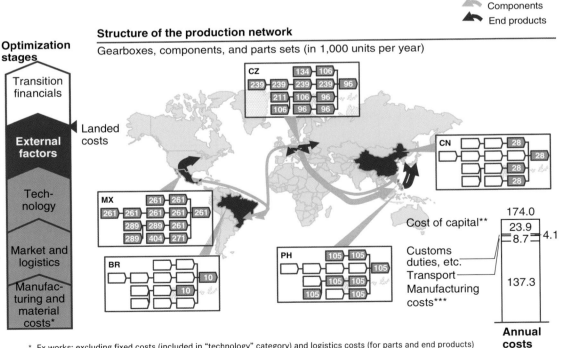

* Ex works; excluding fixed costs (included in "technology" category) and logistics costs (for parts and end products)
** Cost of capital for plant and inventories (incl. work in progress, safety stock, and goods in transit)
*** Incl. fixed costs for plant and machinery

Source: McKinsey/PTW (ProNet analysis)

Transition from the existing to the optimized network will initially create a capital requirement that needs precise calculation during migration planning.

4.3.1.5 Planning Migration

The ratio of annual savings that can be achieved in the long term (approx. EUR 115 million) to capital required within five years for rapid implementation (some EUR 160 million in restructuring costs, EUR 90 million in start-up expenses, and EUR 120 million in additional investments) is around 1:3.2. **Pragmatic migration** from the existing production network to the target structure would therefore not be optimal as it needs very high one-off expenditure (Figure 4.23). Together with the additional invest-

ments needed, this would generate a high capital requirement in the early years of implementation that would only be balanced out by savings in subsequent years. Slower migration would lead to a higher overlap between one-off expenditure and savings already achieved, reducing the capital requirement and, thus, the risks of the operation. The **speed** and **sequence** of relocating the manufacturing steps were therefore analyzed in detail and aligned with corporate objectives.

In planning the migration, three sets of issues are particularly relevant:

■ Expenditure for/constraints on production ramp-up

The model shows a positive impact after 3 years, but the entire relocation takes almost a decade

Fig. 4.23: Net cash flow in relation to the initial scenario
EUR millions p.a.

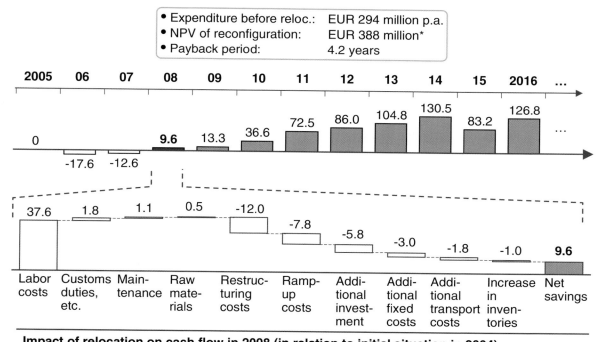

- Expenditure before reloc.: EUR 294 million p.a.
- NPV of reconfiguration: EUR 388 million*
- Payback period: 4.2 years

Impact of relocation on cash flow in 2008 (in relation to initial situation in 2004)

* Adjusted NPV of project contains country-specific risk premiums and is therefore lower than NPV of the unadjusted cash flow

Source: McKinsey/PTW (ProNet analysis)

■ Restructuring expenses

■ Investments in machinery and plant at the new locations.

The proceeds from the sale of the machinery, buildings, and land freed up at the locations to be closed should also be included when assessing the financial implications. The **end products of migration planning** are a business case for the reconfiguration of the production network and an action plan for the ramp-up and ramp-down of manufacturing capacity per location and manufacturing step.

Adjusted migration extends implementation over 10 years, reducing initial capital requirements

The **restructuring expenditure** is dominated by compensation payments. You need to enter into the model the country-specific values for severance payments (adjusted to the staff structure in each factory), natural attrition rates, and share of staff retiring over time. In Portugal, for example, the severance payment for a worker with a tenure of 10 years is estimated at around EUR 8,000. Non-cash-flow-relevant cost (such as write-offs) is not included in the business case since the NPV is calculated solely from expected cash flow effects.

However, this cost should be taken into account when calculating the influence of network redesign on the profit and net margin. The restructuring costs for all factories over a period of 10 years would amount to around EUR 80 million.

The spend on **production ramp-up** in the new factories results mostly from the training of new employees, additional expenses for expatriates, and low productivity during the start-up phase. The transfer of products previously made at other locations can lead to considerable costs, particularly if customers demand complex auditing and sign-off procedures. These items total expenditure of around EUR 50 million in this case. A further EUR 10 million in one-off costs arises from the physical transportation of ma-

chinery and plant, overtime allowances for the production of a buffer inventory, and reassembly and commissioning of the plant at the new location. This expenditure is distributed over 10 years, since the relocation of production and ramp-up of manufacturing steps takes place successively at the new location.

The additional **investment requirement** during the migration phase is moderate. Some of the equipment needed for manufacturing at the new sites can be transferred from existing locations. Even if the planned investment is considerably higher in the first few years of migration than if operations were continued within the existing structure, a large share of the investments will be covered by funds that would anyway have been needed for upkeep and modernization. The migration phase from planning until full ramp-up in the new factories extends over 10 years, with the most important changes in capacity distribution taking more than six years. The relatively long migration phase therefore also leads to a fairly low need for liquid funds to finance the relocation. Extending the time keeps restructuring costs low, makes training new staff less expensive, allows better continued use of existing machines and systems, and enables efficient development of suppliers. The additional expenditure from changing the location configuration is therefore outweighed by savings from an early stage.

More consideration should also be given in the migration phase to which companies in the network should take over which manufacturing steps at which locations. Constraints may also emerge here that should explicitly be included in the migration adjustment. The manufacturing steps "Housing: casting" and "Housing: machining" are not set up right at the start of production at the new location, since these processes involve significant economies of scale, and a supplier would want commitment to a relatively high unit volume to invest in new manufacturing capacity and molds. Supplying the finished housing also reduces complexity in the start-up phase.

In this particular case, no proceeds were estimated for the sale of land and buildings that were not required any longer at the status quo locations. It was assumed that all space would be rented both in the base case and target structure to improve comparability.

Strategic capacity planning is a vital element in the location concept and implementation planning

The adjusted migration path proposes gradual wind-down of production at the high-wage locations. The location in Japan will be closed five years after approval of the location concept and Germany three years later. Production capacity will be reduced by natural attrition and the transfer of employees to other functional areas, so layoffs for business reasons will not be necessary in this first phase.

The new locations in the Czech Republic and Mexico will be opened two years apart and powered up step by step with capacity increases of approx. 40,000 units per annum. Not all manufacturing steps will be carried out at the new sites at the outset. This approach minimizes production ramp-up costs and risks.

Figure 4.24 shows the strategic capacity planning of a new production location in the network. The schedule serves as a cornerstone for cross-functional planning. It is accompanied by a high-level action plan for each function specifying (for example) expatriate assignments and hiring needs. Strategic capacity planning and the action plan, thus, serve as an interface between central cross-network planning and local detailed planning and implementation. This gives the local project teams the key data for building their own locations, while Purchasing can see from it the

The migration plan provides a detailed breakdown of the requirements for each location and production step

Fig. 4.24: Capacity plan for each manufacturing step – Mexico factory
Thousands of units p. a.

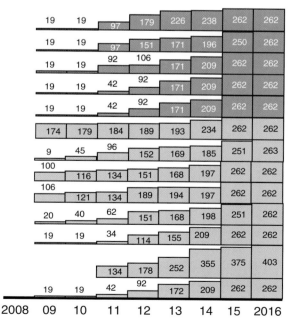

19	19	97	179	226	238	262	262	Housing: machining	Unit volume (in-house value added) / Unit volume (from suppliers)
19	19	97	151	171	196	250	262	Final assembly
19	19	92	106 171	209	262	262		Shafts: grinding and hardening
19	19	42	92 171	209	262	262		Shafts: turning
19	19	42	92 171	209	262	262		Shafts: milling
174	179	184	189	193	234	262	262	Bearings: turning, hardening, and grinding outer and inner rings
9	45	96	152	169	185	251	263	Bearings: assembly
100	116	134	151	168	197	262	262	Bearings: forming rolling elements
106	121	134	189	194	197	262	262	Bearings: grinding and hardening rolling elements
20	40	62	151	168	198	251	262	Gear wheels: grinding
19	19	34	114	155	209	262	262	Gear wheels: sintering
		134	178	252	355	375	403	Housing: casting
19	19	42	92 172	209	262	262		Shafts: forging

2008 09 10 11 12 13 14 15 2016

Source: McKinsey/PTW (ProNet analysis)

targets for selecting and developing the local supplier base.

The business case for the entire transition has an NPV of around EUR 390 million, i.e., some 130 percent of total network costs before optimization.

4.3.2 Case Study 2: Consumer Electronics

A consumer electronics case (Figure 4.25) demonstrates how savings of approximately 7 percent can be achieved through partial footprint redesign. The existing location structure had already been optimized to reduce costs around 15 years previously, so much of the manufacturing is based in the Iberian Peninsula. Now falling prices throughout the industry are creating additional need for action. A review

of the configuration has become essential, especially for products in the lower market segment.

4.3.2.1 Identifying Strategic Objectives and Urgency

The company is under severe competitive pressure in the consumer electronics sector. Most of the products in this market segment are sold through specialist retailers but do not earn the price premium of Apple or Sony high-end products. Manufacturers from low-wage countries are increasingly offering simple products at much lower prices through other sales channels such as discounters. Asian manufacturers are even offering units with identical functionality through the same distribution channels at prices around 20 percent cheaper. Competitors are achieving lower costs of goods manufactured via lower-cost

The savings potential is less when relocating from an area where wage costs are already relatively low

Fig. 4.25: Total landed costs*
Unit costs, EUR (for one representative product)

CONSUMER ELECTRONICS EXAMPLE

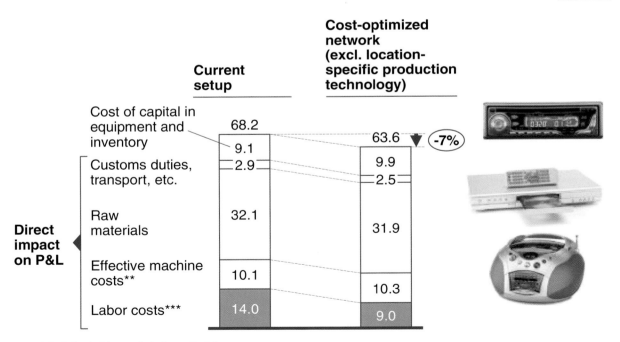

* Cost of materials, manufacturing, and logistics
** Incl. maintenance and fixed costs for machinery, plant, etc.
*** Incl. fixed costs for buildings, administration, etc.

Source: McKinsey/PTW (ProNet analysis)

procurement of simpler intermediate products, better leverage of Asian supply markets, very simple, non-capital-intensive manufacturing processes, and manufacturing at locations with wages of approx. EUR 0.6 per hour. Despite the company's existing brand advantage, the substantial price difference is leading to increasing erosion of market shares in its key markets. It has very little opportunity to expand distribution to countries with growing market volume because of its unfavorable cost position.

4.3.2.2 Modeling Existing Production

A representative product with few variants is used to define and model a production process structure that can represent the production network of the company and its supplies as a whole. Production was captured in 13 manufacturing steps, ranging from the etching of PCBs, insertion of components, assembly, and quality control through to packaging and shipping (please refer back to Figure 4.11). The production of semiconductor components and passive electronics parts was not explicitly modeled since these parts are procured globally and can be made available at every location worldwide at comparable costs because of their high value density. However, closer examination shows that local manufacturing often leads to much better access to the local supply market. This can lead to substantial price differences, even for standardized components. Some manufacturing steps at upstream suppliers (second- and third-tier suppliers) cannot be influenced in the mid-term because the company is tied to these providers by long-term contracts.

4.3.2.3 Developing the Target Structure

The target structure largely corresponds to the "sequential or convergent" network type. The cost-minimized production network uses the existing production sites in Hungary and Malaysia and a new factory in Romania. Suppliers in Malaysia are used to make most of the parts that are manufactured in-house. The assembly of a relatively small number of end products for the Asian market can also be assigned to these suppliers to minimize customs du-

ties, transport costs, and inventories. The more capital-intensive and complex manufacturing steps will then be located in Hungary, which already has an existing base of qualified staff, and where a key component is also assembled by a joint venture partner. Romania is also of interest for parts of the labor-intensive final assembly and manufacturing of simple parts due to its membership in the EU.[21] In contrast to many Asian countries, no customs duties are levied on imports from Romania into other EU countries – the company's main market. For the existing product, if the manufacturing process remains unchanged, the customs duty of 14 percent on electronics imports from Asia is higher than the additional labor cost advantage Asia has over Eastern Europe.

The relatively low savings of 7 percent of total landed costs to be gained from production network redesign are partly due to the fact that, in the status quo, manufacturing is already based in a region with relatively low labor costs of around EUR 6 per effective working hour. The small difference between the labor costs in the existing factory and potential new locations makes early payback of the one-off expenses and relocation investment difficult.

The cost effectiveness of relocation with a payback time of around seven years appears marginal. The restructuring and start-up expenditures and additional investments required would have to be further reduced to make the location concept sustainably attractive. Relocation alone does not solve the strategic issue of low-cost competition from Asia. Redesigning the product and production processes should be investigated as an alternative to a "copy & paste" relocation of the existing setup. Options could include a fundamental overhaul of the product design so that simpler, cheaper components can be used, or the complete outsourcing of production.

---

[21] *Before January 1, 2007 Romania already had associate EU status, enjoying comparable customs status to that of the EU member states.*

4.3.3 Case Study 3: Structural Component for the Aviation Industry

A case from the aircraft construction industry shows how global production can save over 20 percent of total landed costs using the example of one structural component (Figures 4.26 - 4.27). A reduction in the currency imbalance can also be achieved, as well as an improvement in local market presence, which can stimulate sales.

In the **status quo**, manufacturing is based in high-wage countries: Here, Western Germany was assumed in this example. Labor represents a relatively high share of costs. How much varies considerably along the manufacturing steps. While machining processes are highly capital-intensive and account for a large share of raw material needs, assembly processes are very labor-intensive. Automation is not cost-efficient for many manufacturing steps because of the small unit volume.

Cutting manufacturing costs was not the only concern when redesigning the production network. An-other focus was to reduce the exchange rate imbalance. The manufacture of aircraft components is relatively complex. The first step was therefore to identify a number of locations with attractive costs. A more detailed appraisal of the technological capabilities and likely productivity at these locations was to follow.

The analysis of only one fairly large component is revealing in terms of potentially attractive locations and the impact of location choice on cost structure. However, the product scope considered is too limited to determine the location configuration of the total production network. When only one component is analyzed, economies of scale dominate and indicate the need for a "global factory" network. Instead, the case study should be seen as a valuable snapshot of the approach. The model is more transparent and the information aggregated within it is easier to document because of the simple structure of the production network considered. The relatively compact model for the production of one individual part can then be integrated into a larger model with around 100 manufacturing steps for the entire aircraft.

The sequence of production steps describes the supplier relationships in the network

---

Fig. 4.26: Manufacture of a fuselage structural component – simplified process model

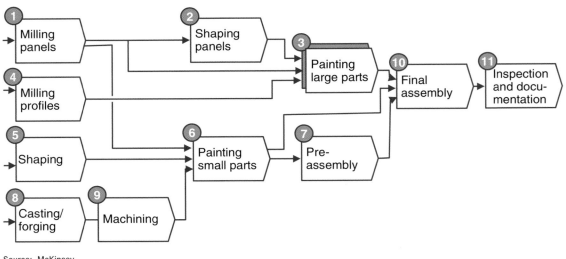

Source: McKinsey

4.3.3.1 Modeling Existing Production

The case example assumed that Germany and the US would each account for 25 percent of demand, and the remaining units would be delivered to France, Japan, Brazil, and China. Around 250 large-volume structural parts of these variants are manufactured per annum, with a large number of other components produced along a very similar process chain.

Since the model only considers one representative part with 11 production processes, the situation is comparatively easy to map. Process groups, i.e., process steps sharing the use of resources, were not defined. The individual process steps all use different resources, with corresponding fixed costs. Fixed costs are also estimated for locations that cover, for example, the fixed share of expenses for a works manager, administrative support, and facilities like the canteen.

The assumptions on the **input factors** per process step vary widely. While machining for panels and sections is highly capital-intensive, relatively little capital needs tying up in assembly, the paint shop, and inspection processes. The fixed capital employed per process step and process type ranges from EUR 50,000 for preassembly of parts (semi-automatic riveting tongs, simple mounts) to EUR 1.5 million for machining on large panels. The economies of scale are correspondingly high in some manufacturing steps.

Factor prices vary from location to location. It is very important to factor in the staff skills required for constructing aircraft components. The assumed labor costs vary significantly from the general average

The cost reduction comes solely from the labor factor if identical design is assumed

Fig. 4.27: Total landed costs*
Unit costs, EUR thousands

AIRCRAFT COMPONENTS
EXAMPLE

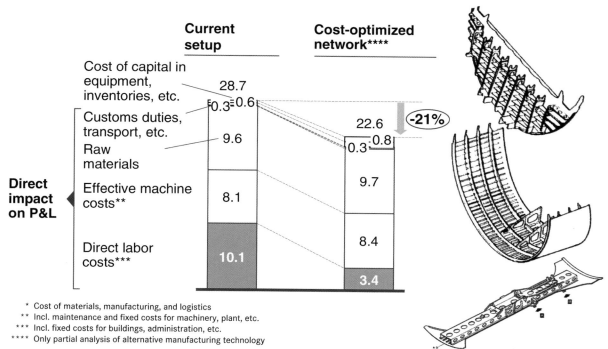

* Cost of materials, manufacturing, and logistics
** Incl. maintenance and fixed costs for machinery, plant, etc.
*** Incl. fixed costs for buildings, administration, etc.
**** Only partial analysis of alternative manufacturing technology

Source: McKinsey/PTW (ProNet analysis)

as a result. They range from EUR 3.60 per hour for a worker in Russia and EUR 5.53 for China to EUR 34.60 for West Germany. The relative labor cost differential for foremen and managers is somewhat lower. The physical productivity that could be achieved in the mid-term was estimated based on current aircraft construction activities in each country. For simplicity, eight years was taken as the depreciation period for all process steps and process types in all countries.

4.3.3.2 Developing the Target Structure

The structure of the optimum production network is characterized by economies of scale. The lowest-cost solution only uses one location in Eastern Europe. The Czech Republic appears to be the prime choice because labor costs are low and (at least some)

trained staff are available with a relevant background and experience in aircraft manufacturing. Fairly high mid-term productivity is therefore assumed and factored into the model.

If the company is forced to spread production capacity across different locations because of a limit on the maximum permissible currency imbalance, for example, all of one particular process step should be established at a given location. This would prevent the need to keep process-specific tools and machinery at several locations (which would duplicate fixed costs).

Limiting the nominal **currency imbalance** to a maximum of 35 percent means a second production location is required. This reduces the risk from appreciation of the Czech koruna. The total costs of the

Factories in the Czech Republic, Russia, India, and the Ukraine would be able to supply parts at very similar costs

Fig. 4.28: Ranking of locations for manufacturing parts
Manufacturing costs, EUR thousands; savings, in percent

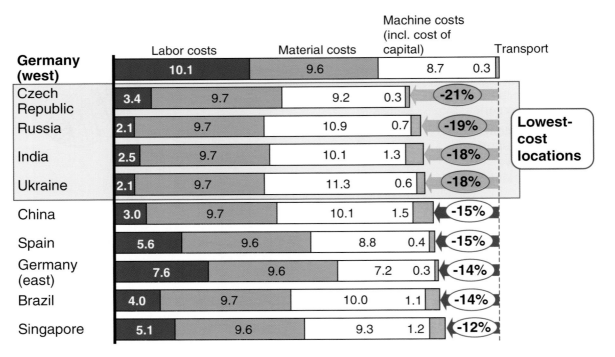

Source: McKinsey/PTW (ProNet analysis)

production network would rise by around 7 percent as a result. The lowest-cost configuration without a limit of this kind has a nominal currency imbalance of around 50 percent.

The model takes into account the fact that the trend in the exchange rate of the Czech koruna is compared against the euro rather than against other currencies, such as the US dollar. When the nominal currency imbalance is adjusted for the influence of price dependencies, the estimated imbalance drops significantly to around 10 percent. This is because both the prices of most raw materials, e.g., aluminum, and the prices of the machines used are generally determined on a global basis.

The robustness of the optimum location configuration is analyzed by successively eliminating the process/location combinations used in the lowest-cost structure. Figure 4.28 shows the results of this process.

The use of a production location in India, for example, results in 2.7 percent higher total landed costs than the lowest-cost location in Eastern Europe (the Czech Republic). In this case example, only assumptions on the productivities the company could achieve in the mid-term were applied. Specific minimum requirements on the currently available supplier structure and the supply of qualified staff at the location were not mapped. These factors and the concrete implications for the companies involved should be discussed in the decision-making process. If this discussion results in additional or modified assumptions on the costs or constraints for the individual locations, they can be integrated into the model.

The case example provides interesting insights into the development of the cost structure of a structural component if the production location is changed and indicates a number of potentially attractive locations. However, it also makes clear that restricting the scope of analysis to one component (even a representative one with costs of around EUR 5 million p.a.), does not deliver a suitable basis for an all-round evaluation of the production network. In practice,

the fixed costs would be distributed across a larger number of production processes by using the machines and locations for a large number of products. The target network type would then change from the "global factory" to the "sequential or convergent" network. The use of different locations makes sense due to the very different cost structures of the individual manufacturing steps. But economies of scale and transport costs make it uneconomical to split up production too much. It would therefore be advisable to extend the analysis to more and larger parts with a more superordinate role in the overall aircraft.

4.3.4 Case Study 4: Appliances Manufacturer

The case study of an appliances manufacturer (Figure 4.29) reveals optimization potential of around 18 percent of total landed costs. The savings potential is even higher for individual product areas, at around 24 percent. However, overall the company wishes to expand its strategy of keeping and developing core competencies in its home factories in the mid-term. Redesign of the location structure will still make the company more competitive by reducing the total landed costs and laying the groundwork for further expansion in foreign markets, since delivery times will be shorter and the supply chain can be better integrated with suppliers and customers.

The target structure includes aspects of the "hub & spoke" and "sequential or convergent" network types. The manufacture of parts such as valve components is relatively capital-intensive and exhibits significant economies of scale. These components are also quality-critical. All these factors favor centralization of the manufacturing steps. The large number of variants to be produced with short lead times demands close-to-the-market assembly and packaging. However, the hub & spoke network type is not fully realized because simple preliminary products do not have to be produced in the market, and their manufacture is not knowledge-intensive. These manufacturing steps should therefore be based at locations with low factor costs and be delivered from there – a feature of the "sequential or convergent" network.

Almost all of the company's production function was analyzed and modeled. The manufacture of its seven product lines (six are shown in Figure 4.30) is mapped in 41 production processes. The products surveyed represent around 85 percent of total revenues (excluding purchased finished goods). The mapping also included some suppliers' production processes. Technological alternatives were only developed and mapped for two production processes (steps 5 and 6).

The savings in this case example are distributed very unevenly across the manufacturing steps. For labor-intensive steps such as manual polishing, assembly, packaging, and shipping, significant savings can be captured from relocation to low-wage countries. This is not the case with production processes such as sintering and automated casting. The cost impact of

relocating steps with a high share of raw materials – the casting of brass parts, for example – is considerably reduced by the fact that these raw materials mostly have a standard price tag all over the world and therefore have no impact on the choice of location.

In this case example, migration to an optimized production network requires significant consolidation of the current location structure. It involves the closure of smaller factories in Germany. Production can then be carried out close to the R&D and design functions at the remaining lead locations. This scenario also sees a share of premium products continuing to be made at the lead factories whose manufacturing accounts for only a low share of total production costs and that have to meet high requirements on delivery time and delivery reliability. A new location in Eastern Europe can take on a large share of the production of mass and intermediate products for the Eu-

---

Relocating production to low-wage locations alters the entire cost structure

Fig. 4.29: Total landed costs*
Unit costs, EUR

APPLIANCES

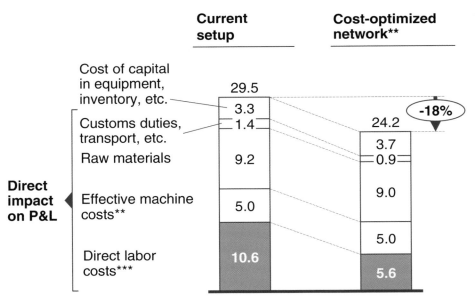

* Cost of materials, manufacturing, and logistics
** Incl. maintenance and fixed costs for machinery, plant, etc.
*** Incl. fixed costs for buildings, administration, etc.

Source: McKinsey/PTW (ProNet analysis)

ropean and Asian markets. An additional production facility in the NAFTA region will lead to slightly higher total costs but will support cost-effective supply to the North American market with short delivery times and also reduce the foreign exchange risk.

Above and beyond the concrete implications for the company, this case is also of interest because of the way optimization was executed. The level of detail employed clearly went beyond the bare essentials. By using a fairly small number of process steps, it is possible to focus the process model and concentrate on those manufacturing steps that will really make a difference. This would also greatly reduce com-

puting time for the modeling (by a factor of approximately 10), creating space to expand the scope of analysis.

In the optimum location configuration, the production process volumes are distributed such that a considerable number of processes can be aggregated without impairing the result. In the present example, the processes of "shaping sand core" and "casting" (now process 8, Figure 4.30) were combined when the process model was created. Even though the processes do not have a similar cost structure, separate modeling does not make much sense because the intermediate sand core product cannot be trans-

Fig. 4.30: Production process model (several product lines)

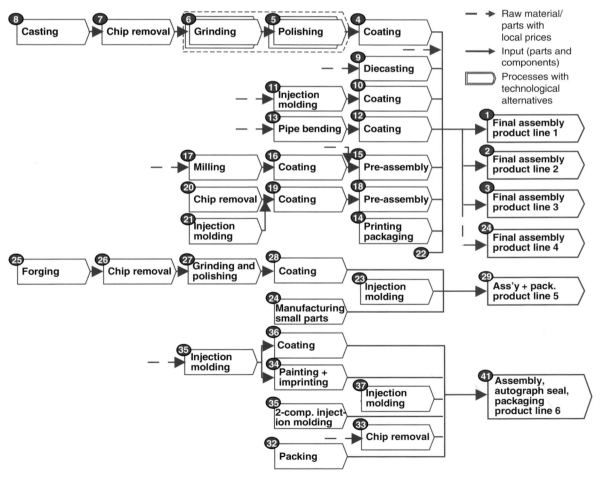

Source: McKinsey/PTW (ProNet analysis)

ported between locations without a disproportionately high effort. The final assembly for product lines 1 to 3 can also be aggregated into one process. This applies to the sequential steps 10 and 11, too. In these steps, the same volumes are made at the same locations in the current structure. Aggregation by simple addition of the input factor volumes therefore has no influence on the optimum solution. This means the number of process steps can be reduced by 16. Further aggregation across product line borders also appears feasible, so the model could be pared down to around 20 process steps (compared to the current 41).

* * *

Optimizing a global production network is a highly sophisticated operation. It has to be embedded in a company's strategic objectives, ensuring appropriate selection of the scope of analysis, input factors, and market assumptions. Its scenarios must be fact-based and transparent to ensure unbiased dialog in the decision-making process and buy-in once the course is set.

The methodology outlined in this chapter is an entirely new quantitative approach. In contrast to scenario-based analysis, its greenfield target structure delivers the optimal solution given the pre-established input parameters. This forces you to explicitly specify any restrictions: Why shouldn't a process be located at the site with the most compelling economics? At the same time, its comprehensive value chain perspective pushes the envelope: How could you align procurement to optimize along this model? Could you modify your technology, or even your products?

The following chapters detail certain elements in the design that were too comprehensive to cover here: procurement, product/process adaptation, or R&D – and implementation itself.

Further reading

Abele, E. and J. Kluge (eds.): *How to Go Global – Designing and Implementing Global Production Networks. Projektbericht "ProNet"*. Düsseldorf: McKinsey & Company, Inc., 2005.

Madura, J.: *International Financial Management*. 7th Edition. Mason: Thomson South-Western, 2003.

Paquet, M., A. Martel, and B. Montreuil: *A Manufacturing Network Design Model Based on Processor and Worker Capabilities*. Technical Report, Quebec: CENTOR Research Center, Université Laval, 2003.

Ritter, R. C. and R. A. Sternfels: "When offshoring does not make sense" in *The McKinsey Quarterly*, No. 4, p. 124-127. New York: McKinsey & Company, Inc., 2004.

TOBIAS LIEBECK, TOBIAS MEYER, EBERHARD ABELE

5 Production Technology: Adapting to Maximize Local Advantage

Summary

Selecting production technology for global manufacturing sites means balancing the advantages of standardization against those of local adaptation. Tailoring production technology to different locations can be highly beneficial. It enables manufacturers to leverage factor cost differentials more effectively, adjust production complexity to employee skill levels, and match plant and machinery to the unit volumes required at that site. Production technology may also need adapting to allow changes to the workpiece design, such as to suit local taste or customs regulations.

The downside is that local adaptation broadens the portfolio of production technologies and product variants that a company has to maintain. Variability creates complexity. And complexity drives costs. Companies therefore need to carefully control the number of design and production technology variants they have.

Companies basically have three options. Manufacturing technology that is complex and involves high development costs should remain largely unchanged across locations. If there is greater scope for reconfiguration, companies should first consider adapting a proven process technology without altering the workpiece design. Modifications to workpiece handling and transport are possible levers. Process changes that directly affect product characteristics such as automating metal cutting or using a different welding technology should also be considered, but with greater caution. The next step may include a change to the workpiece design, which often allows the use of completely different manufacturing methods and machinery.

When considering adaptation, alternatives need examining within the context of the overall production network. It is vital to consider the one-off expenses required to change the product design and manufacturing method in addition to the operations costs. Technology that cannot be leveraged globally is often too costly to develop and maintain.

5.1 Reasons for Adaptation

Setting up a new production location abroad should not automatically mean one-to-one transfer of a company's manufacturing technology.

Companies planning the move for **cost reasons** should consider adaptation to take advantage of the different factor costs and production volumes. They can make considerable savings by adjusting to local conditions (Figure 5.1). Similarly, it can be unwise for companies wanting to develop a **new market** using a new production location to transfer existing products and manufacturing methods without first testing them. The local market will often require modifications in functionality, design, and price positioning – which means production costs need to change, too. Altering the product design usually also makes it necessary and economically viable to adapt the manufacturing technology.

> Companies – with few exceptions – should consider at least two different sets of production technology: one for capital-intensive production in high-cost countries and one for flexible, simple, labor-intensive production in LCCs

Experienced companies take a global perspective to maximize the use of local adaptations at other manufacturing sites across the globe. While it may be tempting to use the same manufacturing methods and product design as at home to try to ensure fast, risk-free production ramp-up of high-volume commodity parts, this choice can be a false economy. Instead, replacing capital-intensive production methods with processes that make greater use of manual labor often makes it possible to set up more cost-efficient and flexible production at low-wage locations. Using the same production technology and product design in high- and low-cost countries often results in production that is too highly automated and thus too complicated, capital-intensive, and inflexible in a low-cost location.

Key questions, Chapter 5

- Why adapt production technology to local requirements?

- How do differences in factor costs, employee qualifications, planned production volume, customer requirements, and regulations impact the selection of production technology?

- How can production technology and product design be adapted to better meet local requirements and make production more profitable?

- What adjustments can be made to production technology without changing the product design?

- Under what circumstances should companies refrain from adapting to local conditions?

- What criteria should be used when choosing the optimum solution for a specific location? And for the company as a whole?

The framework outlined in this chapter should help companies to integrate the development and selection of site-specific manufacturing technology into their globalization strategy. We describe the basic options and demonstrate a procedure for evaluating them from the perspective of both the individual location and the company as a whole.

The extent to which product design and manufacturing technology need to be adapted is determined by conditions at the production location:

■ **Factor costs:** The fact that labor costs are lower and the cost of capital usually higher in developing or industrializing economies than in already highly developed nations makes the use of labor-intensive manufacturing technology more attractive than capital-intensive methods.

> **The main reasons for adapting manufacturing technology are differences in labor costs, cost of capital, unit volumes, qualifications, and product specifications**

■ **Skills/qualifications:** A low general standard of education and corresponding lack of qualified staff and experience can necessitate the use of less complex manufacturing technology. It is much easier to manufacture a threaded than a welded or glued joint.

■ **Unit volume[1] and flexibility:** The capacity of machine types and tools may make different manu-

Significant savings possible from adjusting production technology

Fig. 5.1: Savings resulting from choice of location and changes to production technology
Percent of total costs

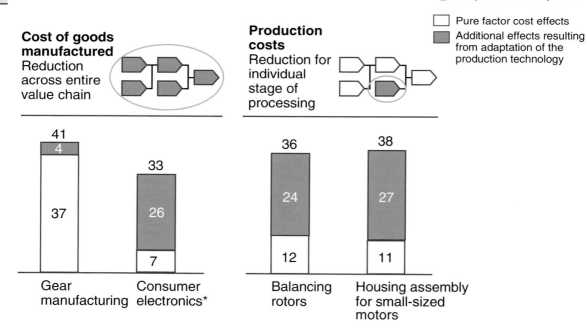

○ Scope of cost comparison

☐ Pure factor cost effects

▨ Additional effects resulting from adaptation of the production technology

Cost of goods manufactured
Reduction across entire value chain

Production costs
Reduction for individual stage of processing

Gear manufacturing: 41 (4 / 37)
Consumer electronics*: 33 (26 / 7)
Balancing rotors: 36 (24 / 12)
Housing assembly for small-sized motors: 38 (27 / 11)

* Incl. effects resulting from changing the product design and sourcing simpler components in competitive markets; manufacturing prior to optimization at locations where average wages are around USD 6 per man-hour

Source: McKinsey/PTW (ProNet analysis)

---

[1] *Cf. Lindemann (2005), pp. 14 - 19.*

facturing technologies attractive for the differing production volumes of each plant. Drop forging a part, for instance, may only be economical for over 200,000 units per year. Metal cutting is more economical for smaller volumes.

- **Customer requirements and local supplies:** Local customer preferences frequently lead to alternative materials, new tolerances, and different product features. Customers in developing countries, for example, geared much more to price than quality, often prefer simpler products with broader tolerances, a shorter guarantee, or more limited functionality – provided they cost less. Local sourcing is critical to achieve low product costs, particularly for products with low value density. Companies may also have to adapt production technology so they can process locally sourced raw materials and semi-finished products.

- **External conditions and risks:** Externally imposed regulations (such as customs requirements) can require the separation of product parts and a change in assembly sequence. Wishing to prevent a know-how drain and product piracy can also be important reasons for a certain choice of manufacturing technology and product design. Dividing up production steps over several locations or suppliers is one lever for minimizing the risks of losing know-how.

5.1.1 Factor Costs

Production technology alternatives allow companies to tailor plants to local conditions. Differences in labor costs are often the single most important factor in this decision.

To evaluate manufacturing methods, the costs of each option should be calculated based on local factor costs. The example in Figure 5.2 shows an aggregated calculation for several manufacturing steps. This reveals that automated manufacturing in India is around 16 percent more cost-efficient than in Germany (as an example of a high-cost location). Costs can be lowered by an additional 7 percent using manual production methods.

In practice, however, the advantage gained by switching to manual labor goes far beyond the additional savings in variable production costs. A manual process is significantly less capital-intensive and therefore has a shorter payback time. It is more suitable for supporting rapid expansion in a less stable market environment. Minimum production capacity is also lower: the machinery used for an automated process requires higher unit volumes to be economically worthwhile.

Differences in labor costs between locations affect more than just a company's own production processes. They also have an impact on the costs of supplied parts and semi-finished goods. More labor-intensive production can be favorable particularly for production in small unit volumes and batch sizes. For this reason, an increasing share of automotive spare parts, for instance, is produced in Romania and India.

Machine tools, jigs, and simple machines can be procured locally too – provided appropriately qualified suppliers are available. However, experience shows that more cost-efficient procurement of machinery and tools is often associated with substantial added expenditure for coordination or lower productivity and a shorter service life. The procurement of such equipment and tools from new sources should therefore be considered but not pushed too hard when setting up shop in a new country or region. Too many new elements in the production process can endanger production ramp-up and product quality.

5.1.2 Skills/Qualifications

The **qualification level** of **local employees** is a crucial factor when selecting production technology. The oversupply of personnel with only limited skills in low-cost countries keeps wage levels down. Well-educated, experienced staff are harder to find and expect substantially higher compensation.

To realize the potential of global production, companies need to take into account the supply and demand of workers with different qualifications. The type of production technology plays an important role here, as it determines the number of employees needed at different skill levels. Companies will find it particularly hard to use state-of-the-art manufacturing equipment effectively in developing countries such as Indonesia, Cambodia, Vietnam, Kenya, or Nigeria. Local staff usually lack the qualifications to operate and maintain such machinery. The machine manufacturers themselves also face similar constraints, and cannot provide the same level of support to their customers as in more developed countries. The economics of using expatriates to operate complex production plants are often so poor that any use of them beyond the initial startup phase makes production relocation economically unfeasible. If local staff cannot operate and maintain the chosen production facilities, a new plant will be unable to achieve a viable cost position no matter how low local pay-scales are.

Another critical issue is the ability of companies to retain tenured staff. A high churn rate entails a constant drain of knowledge and experience. Addressing this issue via training can become prohibitively expensive if employees leave the company before payback of the cost of their training. Lack of collective experience also means the manufacturing location cannot become self-sufficient or contribute to the build-up of global best practice in the production network. High attrition in the workforce is typical in rising and rapidly growing economies. In some countries the issue is structural, as individuals may choose to change location or employer frequently, or switch in and out of employment to fulfill family obligations.

Where general educational standards are low and staff churn is structurally high, it is hard to avoid or-

Low labor costs make manual production in low-cost countries economically attractive

Fig. 5.2: Comparison of cost of goods manufactured, high- and low-cost countries with varying levels of automation

	Labor content Hours/component			Labor costs EUR/hour	
	Manual	Automated		Germany	India*
Head of department/plant manager	0.01	0.01		60	6.1
Production planner/work scheduler/group manager	0.03	0.02		44	4.2
Foreman/engineer/systems supervisor	0.04	0.06	X	34	3.1
Skilled workers	0.20	0.01		24	2.5
Semi-skilled workers	0.20	0.05		21	1.1

Costs per component			Manual	Automated
Labor	Germany		14.2	5.2
	India*		1.5	0.9
Machinery **	Germany		0.9	2.0
	India*		1.0	2.6
Materials	Germany		10.4	10.4
	India*		11.2	11.2
Production	Germany		25.5	17.6
	India*		13.7	14.7

-16%

-7%

* Adjusted by estimated productivity differentials; at average 2004 exchange rates
** Depreciation, costs of maintenance and capital

Source: McKinsey/PTW (ProNet analysis)

ganizing work according to Taylorist principles.[2] Decomposing production processes into granular steps is the only way to achieve short training periods, a low reject rate, and sufficiently high productivity overall. If, however, better qualified employees are available or attrition is low, it is sensible to extend the scope of activities to make the operating system more cost-efficient and robust. This will also help increase job satisfaction and motivate workers to stay on.

Companies are able to respond specifically to the labor supply by gearing their manufacturing technology to local conditions. Companies should map alternative production technology against staff requirements and can then determine the typical qualifications and experience required (Figure 5.3). The local labor market should match this profile or at least offer the conditions to build a workforce that matches

requirements within a reasonable time frame (typically one to two years).

5.1.3 Unit Volume and Flexibility

The **unit volumes** planned and number of **variants** to be produced influence the choice of production method.

The design of manufacturing technology for one-off and small-lot production will always mean improvising to some extent. In these cases, assembly is predominantly manual. Rigid automation is not cost-efficient for very small quantities due to its low level of flexibility for variants and high setup costs. Flexible, computer-controlled machine tools and processing centers are, however, increasingly important for processing parts in small batch sizes. Capital-

Demanding activities call for high qualifications and extensive professional experience

Fig. 5.3: Level of qualifications and professional experience required

Qualification level	Examples of typical skills	Examples of typical activities	Professional experience normally required for full effectiveness
Head of dept./ plant manager	• Ability to manage and develop independently	• Managing production	Approx. 5 years
Production planner/work scheduler/group manager	• Comprehensive specialist knowledge and innovativeness	• Designing products and tools	Approx. 3 years
	• Organizational skills	• Deciding work plans and machine utilization	Approx. 3 years
Foreman/ engineer/ systems supervisor	• Enhanced specialist knowledge	• Installing and starting up machinery	Approx. 2 years
	• Cooperative skills	• Planning own work within the team	Approx. 1 year
Skilled workers	• Planning skills	• Planning supply of parts at work station	Approx. 0.5 years
	• Sophisticated manual skills	• Retooling	Approx. 1 month
Semi-skilled workers	• Specialist process- and product-related knowledge	• Testing product function • Overall assembly	1 week - 1 month
	• Simple manual skills	• Assembling parts • Inserting parts	0 - 1 week

Source: McKinsey/PTW (ProNet analysis)

---

[2] Cf., e.g., Eversheim (1996), pp. 12 - 50.

intensive equipment can boost labor productivity to a level where production in high-cost countries is attractive provided the qualification of workers in low-wage locations is insufficient to operate such complex machinery. Materials flow and assembly of products in small lot sizes is relatively similar in high- and low-cost countries.

As output increases, however, it is useful to further differentiate between manufacturing methods depending on the location and prevailing factor cost structure. In a low-wage country, manual operating systems are generally retained as volumes increase. The system is simply duplicated: new units are created that work on the same principle. With increasing production volumes, activities can even be fur-

ther decomposed into very simple, granular process steps. This allows the use of unskilled workers and minimizes the need for training. In electronics assembly, for instance, many manufacturers use assembly and packaging lines with some 20 to 40 workers each. Smaller volumes are mostly produced in alternating batches with other products. With increasing volumes, the setup of these lines remains unaltered, but more lines produce the same product using an identical, largely manual process. Further work-step decomposition does not lead to higher productivity, and the use of identical production lines allows standardization and flexibility of total output volume. The picture is quite different in high-cost countries, where the high costs of manual production mean it may be cost-efficient to use automation

High- and low-cost countries follow different strategies for increasing production output

Fig. 5.4: Operating systems with different levels of automation

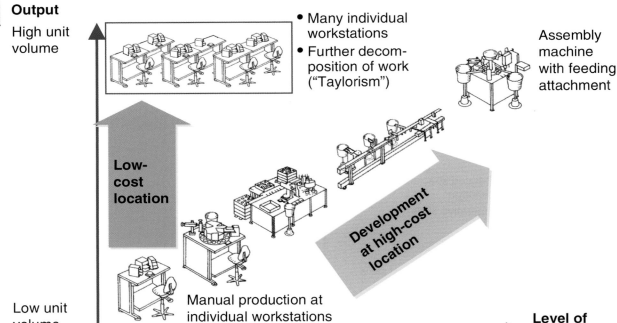

- Many individual workstations
- Further decomposition of work ("Taylorism")

Assembly machine with feeding attachment

Output

High unit volume

Low-cost location

Development at high-cost location

Low unit volume

Manual production at individual workstations

Level of automation

Low degree of automation
- High degree of flexibility
- Low fixed machine costs

High degree of automation
- High economies of scale
- Smaller share of labor costs

Source: Dubbel (1994); McKinsey/PTW (ProNet analysis)

(Figure 5.4). Automation often results in higher economies of scale for the use of larger machinery. As a result, automated production lines often have such high capacity that only one line per product is required.

Nonetheless, automation is not always the right decision – even at high-wage locations. The scalability of the machinery, demand fluctuations, and the number of variants all need to be weighed up. Automated assembly today is still often less scalable and less flexible than manual assembly. It is essential to factor in demand fluctuations and number of variants when determining the level of automation appropriate to the location to avoid low average utilization and frequent retooling. **Automated plant and machinery** are usually designed for comparatively **high production volumes** and can capture substantial economies of scale (Figure 5.5). However, volume forecasts and the number of variants need to be accurate. The benefits of automation often remain unrealized because actual market demand deviates from forecasts.

Centralized, highly automated production proves economically superior if there is constant, high-volume demand. It is often more suitable for parts and components, as these have a smaller number of variants than finished products.

With manual production, fixed costs for the individual work stations are low, meaning that only minimum economies of scale can be achieved – mainly learning-curve effects.

5.1.4 Customer Requirements and Local Supply

Depending on the industry and product, markets in different regions and cultural groups place value on **specific functionalities** or **designs**. Country-specific and possibly even location-dependent product design and manufacturing methods are therefore necessary.

Gearing production locally to the needs of the relevant market is especially advisable in the consumer

Adjusting the automation as production volumes increase ensures much greater economies of scale

Fig. 5.5: Economies of scale at ten times the volume
Percent of cost of goods manufactured

	Replication of production units (same level of automation)	Automation level adjusted to production volume
Hydraulic cylinder Basis: 100/day	-8%	-18%
White goods Basis: 1,000/day	-3%	-10%
Automobile gears Basis: 250/day	-9%	-15%

Source: McKinsey/PTW (ProNet analysis)

goods industry. Products might not only differ in their specifications to cater for local taste or customs. Often ingredients need to be varied, or local supplies have characteristics that require different processing. Even in industries where components are largely standardized, lower customer expectations can require local adjustment. Companies should consider adapting quality, accuracy, and usable life of a product to **local expectations** if this can help to significantly reduce costs. Many Western companies have excessively stringent product specifications, making their products relatively expensive. Introducing higher tolerances in metal-cutting processes, for instance, can make the use of less capital-intensive machinery feasible.

Make products more affordable for lower-income consumers by offering smaller package sizes

While the right balance between cost, product functionality, and quality depends on the product and specific context, companies that make the trade-off in favor of lower costs seem to be more successful.

The cost aspect goes beyond pure production costs. Companies need to make their products more affordable to successfully compete in developing countries. Consumers often have only limited cash available or are unwilling to spend it on one product only. The entry price point therefore needs to be very low. Hindustan Lever Limited, for instance, introduced very small package sizes when entering the Indian consumer goods market. The lower price per package was one crucial element that made the company India's leading player in the fast-moving consumer goods sector.

Companies are often not aggressive enough in defining their product features based on target unit cost. This is important, though, to compete successfully in markets, where most consumers are low on the income pyramid but make up a very broad base. Management should ask production engineers what they can produce for a target cost using the simplest manufacturing technology possible. They can then decide whether the outcome will satisfy customer preferences in these markets, and whether it is compatible with the company's brand. It is also useful to consider the lowest possible cost position a competitor could potentially achieve.

Companies in low-cost countries that offer products for the large lower-income segment should design them with a specific price point in mind

This also applies to raw materials and semi-finished parts. These might not be available in the same form or quality as in other markets, and thus require adaptation of the production process. Food products in Asia, for example, are much more often rice-based than in Western Europe or North America, where the focus is more on wheat or corn. Basic materials such as ores or coal also have different characteristics in different parts of the world. It is important for companies to realize the impact of such differences on the production technology required and build a technology portfolio that suits most occasions, while using as few variants as possible.

5.1.5 External Conditions and Risks

Regulations relating to foreign trade policy, such as different customs duty rates for individual parts, subassemblies, and finished products, can be the key factor in adjusting product design and production technology. By placing higher customs duties on nearly completed modules and end products, low-cost countries often try to force companies to manufacture locally. This poses a challenge if local demand does not make an independent production facility economically viable.

For manufacturers, this primarily means having to replace capital-intensive manufacturing steps with alternative production solutions that still allow competitive manufacturing, despite the small quantities produced. Automotive OEMs, for example, have developed the concept of SKD or CKD assembly[3] (see

---

[3] SKD: semi knocked down, CKD: completely knocked down.

Chapter 1), where vehicles are shipped in individual modules as assembly kits and put together locally. This approach makes it possible to continue performing capital-intensive manufacturing steps centrally, while avoiding the high customs duties for finished products (see case example in section 2.6.2).

The risk of **product piracy** and **loss of proprietary know-how** can be another reason for choosing a certain production alternative. Some developing and newly industrialized countries do not have the appropriate regulations in place, nor the will and skill to enforce them. Pressure from industrialized nations aims to improve the situation, but compliance is still poor. Companies therefore risk infringement of their trademark and other industrial property rights, as well as the loss of company-specific know-how.

When setting up local production sites, companies initiate **extensive transfer of knowledge** to local employees, partners, and suppliers. This increases the danger of know-how being passed on unchecked. Examples that would be considered outrageous in Western Europe and the United States are quite commonplace in other parts of the world – poaching of entire or large proportions of the workforce, for instance. Local entrepreneurs – sometimes even the partners in a joint venture – set up their own production facilities alongside the international company and acquire critical know-how by deliberately siphoning off the international player's staff after they have received training and some experience. This endangers startup of production, plus the international company has to deal with a local rival. The economic loss caused by such product piracy is often accompanied by damage to the company's image if the imitations are of inferior quality.

In view of such risks, an increasing number of manufacturers are avoiding full-scale local production and the use of certain technologies to prevent the exodus of proprietary knowledge. Instead they distribute individual manufacturing steps among different locations. This ensures that any copying of their production process does not lead to a marketable product. The trade-off, however, is higher production costs.

5.2 The Options for Adaptation

Decision makers have three main options. They can leave production technology and product designs unchanged, building a new plant by (in effect) copying an existing one. Alternatively, they can change the production technology but leave the product design unchanged. The third option is to adapt both production technology and product design. Obviously, they can also vary the degree to which production technology and product design are modified.

Each of these options has some aspects in common and some distinct pitfalls. To make informed choices, decision makers need to examine the relevant alternatives and concepts very carefully.

5.2.1 Basic Adaptation Models

There are a thousand ways to make a product. The number of efficient manufacturing methods for delivering a product that meets specification may be more limited, but there are still numerous alternatives. Each differs in its cost structure, product characteristics, and requirements on staff. This makes it important to maintain an overview of the options and encourage cross-functional collaboration in order to select those best suited to the specific situation.

While this applies to manufacturing across industries, the following sections exclude the chemical and other process industries for the sake of space and simplicity.

5.2.1.1 Manufacturing Process Alternatives

There seems to be insufficient awareness among management regarding potential manufacturing processes and the trade-offs involved. There are countless possibilities, particularly when you consider the combination of different manufacturing processes, tools, process parameters, degrees of automation, and related selection of machinery.

Figure 5.6 provides an overview of basic manufacturing processes. The exhibit is structured along the typical sequence of product manufacture, starting with primary shaping on the left. While each of these processes can be automated to a greater or lesser degree, some methods are more suited for automation and high-volume production than others. Forming under conditions of tensile and compressive pressure, for instance, requires large-scale presses, and thus expensive tooling. This method is highly effective for producing sheet-metal parts in high volumes, ranging from cans for drinks to car body parts.

There are multiple alternatives for each basic manufacturing process. Figure 5.7 reviews the options for metal casting. Ceramic molding, for instance, is a potential alternative to chill casting. It achieves similar tolerances, and both methods are suitable for parts weighing up to 100 kg. Ceramic molding requires lower tooling investments, as a costly permanent mold is not required. However, the process is more labor-intensive and thus typically only used for small to medium runs. While unsuitable for mass production in high-cost countries, ceramic molding can be an alternative for the production of parts in a high number of variants in a low-cost country.

5.2.1.2 Ways to Adapt an Existing Process

When expanding or relocating production for an existing product, it is important to understand the implications a change in production technology and product design will have. This goes both ways. Changes can enable a reduction in manufacturing costs and the investments required. Changes may, however, also require additional time to plan and implement and cause greater costs and complexity.

> **Changes in manufacturing technology and product design are interdependent**

To assess the relevance and impact of these effects, managers should analyze the implications of small and more radical changes. The spectrum is broad, as

> **A thousand ways to make a product – but usually little transparency on the trade-offs**

Fig. 5.6: Classification of manufacturing processes

■ Detailed in next figure

Principle	Create cohesion	Preserve cohesion	Reduce cohesion	Increase cohesion	
Pro-cesses	(1) **Primary shaping** • Metal casting • Forming of plastics • Powder metallurgy • Electroforming • Autocatalytic plating	(2) **Forming** • Compressive conditions • Combination of tensile and compressive conditions • Tensile conditions • Bending • Shearing conditions • Sheet/plate vs. massive forming	(3) **Cutting** • Severing* • Machining with geometrically defined tools** • Machining with geometrically undefined tools*** • Chipless machining • Disassembly • Cleaning and evacuation	(4) **Joining** • Assembling • Filling • Pressing against/into • Joining by primary shaping • Joining by metal forming • Joining by welding • Joining by soldering • Gluing	(5) **Coating**

(6) **Changing of material properties**

     * Includes blanking, wedge-action cutting, tearing, and breaking
   ** Turning, drilling, milling, planing and shaping, broaching, and sawing
 *** Includes grinding with rotating tool, belt grinding, reciprocating, grinding, honing, lapping, barrel polishing, and machining by abrasive blasting

Source: Based on DIN 8550

There are multiple alternatives for each basic manufacturing process

Fig. 5.7: Metal casting processes – overview

EXAMPLE

Type of mold	Expendable molds						Permanent molds			
Type of pattern	Permanent patterns				Expendable patterns		No pattern			
Process	Hand molding	Mechanical molding	Shell molding	Ceramic molding	Precision casting (lost-wax process)	Full mold casting	Pressure die-casting	Chill casting	Centrifugal casting	Continuous casting
Materials	All metals	All metals	All metals	All metals	All metals	All metals	Al-, Mg-, Zn-, Cu-, Sn- or Pb-based die-casting alloys (iron-based materials developed	Light metals, special copper alloys, high-grade zinc, lamellar and nodular graphite cast iron	Lamellar and nodular graphite cast iron, cast steel, light metals, copper alloys	Lamellar and nodular graphite cast iron, cast steel, copper and copper alloys, aluminum, and aluminum alloys
Weight range (approximate values)	No limit, available transport facilities and melting capacity determine maximum weight	Up to several tons, restricted by size of machines	≤ 150 kg	≤ 1,000 kg	1 g ≤ several kg (≤ 100 kg in special cases)	No limit (maximum transportable weight); particularly suitable for heavy components	Al alloys: ≤ 45 kg Zn alloys: ≤ 20 kg Mg alloys: ≤ 15 kg Cu alloys: ≤ 5 kg (limited by size of pressure die-casting machine)	≤ 100 kg (more in special cases)	≤ 5,000 kg	Up to several tons
Quantity range (approximate values)	Single items, small production runs	Small to large production runs	Medium and large runs	Single items, small to medium runs	Single items, small runs. Series production of suitable components	Series production. Life of mold (in thousand castings): Zn ~ 500 Mg ~ 100 Al ~ 80 Cu ~ 10	Series production. Life of mold: Al~ 100,000 castings	Series production. Life of mold: 5,000-100,000 castings depending on size of workpiece, casting material and type of mold	Length of billet depends on machine	n/a
Tolerance range* Percent	2.5 - 5	1.5 - 3	1 - 2	0.3 - 0.8	0.3 - 0.7	3 - 5	0.1 - 0.4	0.3 - 0.6	1	0.8

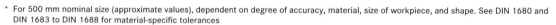

* For 500 mm nominal size (approximate values), dependent on degree of accuracy, material, size of workpiece, and shape. See DIN 1680 and DIN 1683 to DIN 1688 for material-specific tolerances

Source: Dubbel (1994), p. K5.

the following comparison shows (see Table 5.1). They range from adjusting the flow of materials to fundamentally altering the production technology and product design. Some measures, such as modifying workpiece handling technology and subordinate components, can be implemented at reasonable expense. Others – such as changes to the manufacturing methods of central product components – call for extensive new development work.

Figure 5.8 illustrates the options for adapting manufacturing methods and product design. The graphic shows that one cannot fundamentally change the production technology without revisiting product design and vice versa. Production technology and product design are interdependent. There is some flexibility, but it is limited. Workpiece handling and materials flow can be changed with relatively little impact on the product characteristics. Altering the tools, jigs, or processing technology, however, will inevitably call for changes to the product design or impact product features.

A dramatic change in production technology will inevitably alter a product's characteristics

Managers need to be aware of this. If pushing for a radical reduction in capital expenditure for a new manufacturing method in a low-wage country, you may end up creating a new product variant. This variant need not be better or worse, but it will be different. Companies should therefore take this opportunity to also review product functionality and quality requirements. Once a new variant is created,

Table 5.1: Ways to adapt production technology and product design

Level of adaptation	Adaptation of ...	
	Production technology	**Product design**
1. Low	**Adaptation of materials flow** ■ Automation of materials flow and handling ■ Storage technology (e.g., automated placing/ releasing from stock in high-stack racks) ■ Linking of work stations **Adaptation of quality control** ■ Scope and automation of quality control	**Adaptation of auxiliary parts** ■ Mounting elements ■ Covering elements ■ Fasteners
2. Moderate	**Adaptation of workpiece handling** ■ Loading and unloading of machines ■ Automation of workpiece mounting and assembly **Adaptation of jigs and tools** ■ Adaptation to local supply (e.g., change in quality) ■ Adaptation to local requirements (unit volumes and tolerances) **Adaptation of process control and parameters** ■ Computer-controlled vs. manual processing ■ Processing parameters, e.g., cutting tools and speed	**Adaptation of product design** (core components) ■ Material ■ Shape ■ Tolerances ■ Surface quality
3. Comprehensive	**Adaptation of the production technology and process chain** ■ Altered machinery and plant ■ Alternative manufacturing process, e.g., forging vs. metal-cutting ■ Alternative process sequence, e.g., hardening and grinding	**Adaptation of product functionality and value to the customer** ■ Modified functional principle ■ Change in size and capacity ■ Modified area of application

you are already halfway there and may as well develop a solution that fits local requirements best. Otherwise, it could be more effective to maintain the standard production setup and not incur the one-time costs and greater complexity from additional variants of product and equipment.

5.2.2 Adapting to Local Requirements – Concepts and Case Examples

Three strategies can be distinguished for configuring production technology at a (low-cost) location. They are presented in the following subsections, with examples.

Adapting production technology and product design can help to eliminate production complexity

■ **Adapting production technology without changes to product design:** Companies can use this strategy to vary the capital intensity or level of automation. An example of this is the use of simple, manual workstations for assembly. These allow supply of the local market at competitive costs, even with low production volumes per variant.

■ **Adapting production technology using a modified product design:** With this strategy, adaptations go beyond changes to materials flow, handling, and assembly technology. Design changes can apply to individual product components or to the structure of the entire product. Changes can be as radical as an entirely new design or as simple as a change in tolerances. The switch from a glued to a threaded joint allows you to simplify both the manufacturing process and quality control. Lower hydraulic cylinder tolerances (i.e., a change of the product specification) mean you can dispense with capital-intensive precision machining.

■ **Standardizing production and products globally:** If global standardization is called for, a new location's production will not differ from that of others in terms of either production technology or product design. An example is the manufacture of semiconductors. It is not economically viable to adapt the expensive and complex manufacturing process to each location.

> **Engineers in product development and production should systematically select and modify product design and production technology to make use of the specific strengths of different locations**

The different strategies for aligning site, production technology and product design also highlight the opportunities for different types of location. Low-cost countries are suitable locations for many but by far not all products and production steps. Engineers in development and production should instead aim to systematically leverage the specific strengths of different locations.

Production in a different environment can require a change in both production technology and product design

Fig. 5.8: Options for adjusting production technology and product design   SCHEMATIC

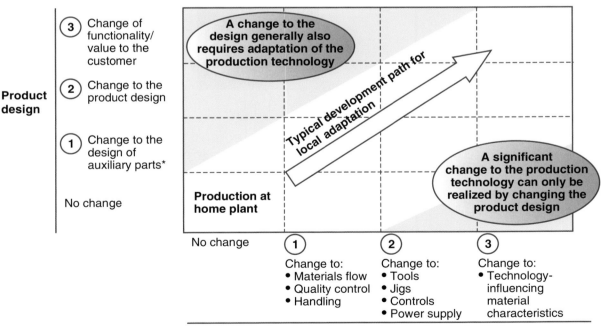

* E.g., connecting parts; key functional parts and functional surfaces remain unchanged

Source: McKinsey/PTW (ProNet analysis)

Aligning Technology with Local Conditions: Weighing Equipment Manufacturing at Sartorius in China, Germany, Malaysia, and the US

Sartorius AG – a EUR 500 million business headquartered in Göttingen, Germany – offers its customers a wide range of mechatronic products, from low-cost standard scales and industrial weighing machines to high-precision measuring instruments for laboratories. In product development, the company focuses on its core competence of mechatronics, i.e., the combination of electronics, mechanics, and IT, which makes a crucial contribution to the precision of its products. In the mid-1990s, the production of electronic elements at its home factory was under considerable cost pressure. As a result, it planned to purchase electronics parts from a cheaper supplier in Malaysia. The decision had already been made when a conversation with the Asian supplier gave Sartorius' executives a new perspective.

Electronic metrology was part of the company's core technology, enabling competitive differentiation of Sartorius' products. Yet it had become clear that the Malaysian supplier was unable to produce these electronic elements to a quality-assured standard, and would be unable to continuously improve the manufacturing processes going forward. Production of these components in house was superior, backed by many years of experience and

Components are divided up according to their complexity and importance, enabling location-friendly production

Fig. 5.9: Technology differentiation: example – measuring device

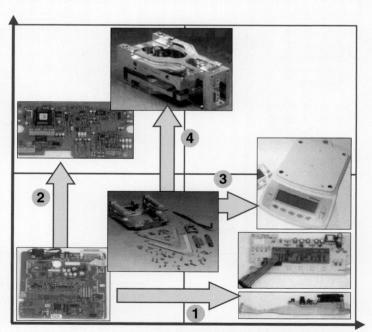

Innovation and competence advantage of high-cost locations Göttingen (GER) and Denver (US)

Cost advantage of low-cost locations Penang (MY) and Beijing (CN)

Source: Sartorius AG, McKinsey/PTW (ProNet analysis)

specific know-how in developing and producing the electronic measuring elements and integrating them into the end product. At the same time, however, the printed circuit board (PCB) also contained simple electronics such as a display unit of the type commonly used in consumer electronics. This component, found in any modern scale, is not a distinguishing feature of Sartorius' product. Sartorius' management realized that the company was unable to produce such components cost-efficiently in high-cost countries. Even if automated, the capital-intensive process in Germany would still be far more costly than manual assembly in a low-cost country, such as Malaysia. In addition, the supplier also bought parts, equipment, and raw materials in Asia itself, leveraging its better access to the market for electronic parts and purchasing in larger quantities. Outsourcing production would therefore provide additional advantages that Sartorius alone could not achieve.

These insights led Sartorius to the concept of technology differentiation: the overall product design was altered, such that knowledge-intensive components could easily be separated from simpler parts. These critical components could then be manufactured within the knowledge cluster in and around the company's central locations in Göttingen and Denver, close to critical suppliers and their own R&D department. Design and integra-

tion of the simple components could be organized for production in a low-cost country abroad.

Sartorius now designs the overall product to allow separation of the components for manufacture in different locations. It also selects the PCB layout according to conditions at the factory location (Figure 5.9). In Asian low-cost countries this means manual assembly of the PCBs for the display (1). In Germany, the PCBs of the electronic measuring elements (2) allow automated assembly of the SMD[4] components. Sartorius also subdivides the mechanical components: the simple, labor-intensive production of plastic add-on pieces (3) is based at the low-wage location in Asia. The monolithic base unit, which directly affects the instrument's functional quality, is manufactured at the home plant with a computer-controlled high-speed machining center optimally adjusted to the components' requirements.

Bottom line: Sartorius managed to achieve significantly lower production costs without compromising product quality by fundamentally redesigning its product and production processes. The company accomplished this by matching the different manufacturing requirements for the product components to the specific strengths of each location.

---

[4] *SMD: Surface-mounted device.*

5.2.2.1 Adapting Production Technology Without Changes in Product Design

Adapting production technology without changing product design typically means varying the level of automation. Little more can be done without affecting product design and characteristics. But even altering the level of automation can have quite an impact, particularly when a new production site is being selected. Automation allows the substitution of labor with capital. Changing the level of automation is therefore an important lever if production can

be shifted between locations with a different factor cost structure.

Different **levels of automation** can be used for controlling the main manufacturing process as well as materials flow and workpiece handling. Companies establishing production capacity in low-cost countries should significantly lower the degree of automation and thus the capital intensity of their production processes. Otherwise the economics of shifting production could be doubtful given that productivity in emerging markets is often lower than in developed countries. However, companies based in low-cost countries will need to replace manual labor with machines when establishing production in higher-wage countries, a move they will find necessary to supply these markets with short lead times and a wide variety of products.

Automate handling and workpiece positioning in a high-cost country; this is also vital to increase machine throughput

The example of a car manufacturer with alternative processes for the installation of spare wheel pans illustrates the concept of adapting production technology without changing product design (Figure 5.10). For its production in high-cost countries, this OEM opts for fully automated installation of the spare wheel pan. The industrial robot grabs the part from the removal area as it is delivered, carries the pan past an adhesive dispenser, and then presses the component into the body of the vehicle. The use of manpower is limited to maintenance, repair, and control of the system, enabling the company to minimize unit labor costs. The pan is also installed in countries with relatively low labor costs, such as South Africa and China. In these locations, the assembly process for the spare wheel pan is organized differently. Because the use of a robot such as the one in the lead plant is impractical and too expensive due to the lower labor costs and lower staff qualifications, the OEM adopted simple, manual installation of the spare wheel pan. The worker takes the component from the delivery point and applies the adhesive to the join patches, then inserts the pan and presses it in place.

The OEM has to make additional checks to achieve high process reliability comparable to automation. Before start-up of production, the production and

Adjusting the automation level without changing the product is easiest in assembly

Fig. 5.10: Economic viability of alternative manufacturing methods in assembly

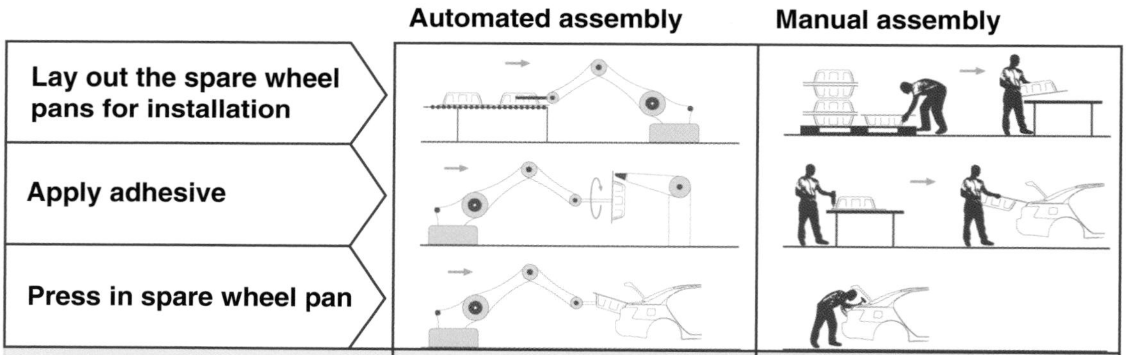

	Automated assembly	Manual assembly
Lay out the spare wheel pans for installation		
Apply adhesive		
Press in spare wheel pan		
Investment volume EUR	200,000	10,000
Direct labor	–	1 employee
Takt time Seconds	60	60
Unit costs (using the example of dual-shift operation in a country where the labor costs are EUR 7/hour)	• EUR 0.18 machine costs*	• EUR 0.01 machine costs* • EUR 0.12 labor costs

* Depreciation, maintenance, and cost of capital

Source: DS Engineering, McKinsey/PTW (ProNet analysis)

quality planning departments develop and supply additional guidelines, process descriptions, and quality control cards to make the process reliable. Ramp-up planning therefore needs to take into account changes in production technology and budget in time to implement support initiatives. Additional material is needed, such as manufacturing notes and examples for training.

Hero Honda as an Example of Automation in Low-Cost Countries: as Little as Possible, as Much as Necessary

Hero Honda is the market leader for motorbikes in India. Operations at the company's main factory in Gurgaon outside Delhi provide a good example of the effective use of automation.

In 2004, Hero Honda sold 2.6 million motorbikes in India.[5] Labor costs, including ancillary wage costs, amounted to approximately EUR 2.5 for every hour worked at Hero Honda in 2004. Employees received higher-than-average wages that year because of high utilization and the fact that wages were linked to total plant production volumes. The use of less capital-intensive manufacturing methods at Hero Honda is cost-efficient, given a ratio of around one to ten for labor costs in India compared with labor costs in high-cost countries such as Japan or Germany. Consequently, the core components – the gearbox and engine – are assembled manually (Figure 5.11). The final assembly to complete bikes is also mainly manual. However, even Hero Honda has processes with a fairly high degree of automation. Automation is primarily used to meet **quality requirements**. Core manufacturing processes that are difficult to master manually are automated. However, even at Indian wage levels automation is sometimes used entirely based on **economic considerations**.

The welding of fuel tanks is one example of the use of automation for **quality reasons**. Figure 5.12 shows a robotic arm and tungsten inert gas welding plant (TIG). The robotic arm turns through 90 degrees once a weld seam is completed and welds a tank in a second clamp. During the time it takes

Fig. 5.11: Manual assembly of motorbike gearboxes at Hero Honda in Gurgaon, India

Source: T. Meyer

Fig. 5.12: Automated welding of fuel tanks by a robot with five axes

Source: T. Meyer

---

[5] *The company reported unit sales of over 3.3 million in FY 2006/07.*

to weld the second tank, a worker removes the first welded workpiece from the clamp and clamps the next two semi-finished parts in place. The system was purchased for the plant in Gurgaon; the plant configuration and process design were taken from another Honda plant. As a result, the process development costs were low.

Economic considerations were the decisive factor in automating the way parts were fed to a punching machine. Automated feed, taken in isolation, is not cost-efficient because depreciation, repair and maintenance, and the capital costs of the automated equipment exceed the savings in labor costs. However, factoring in total systems costs does make automation economically viable: the automated feed allows shorter cycle times and more effective use of both machine and tools. This meant the company was able to avoid the purchase of an additional tool set. Automating the subprocess is therefore feasible even though an isolated comparison with manual execution could lead to a different conclusion. The increased efficiency of the overall manufacturing systems makes automation of the subprocess the better choice.

Bottom line: Quality criteria play an important role in determining the optimum degree of automation for a manufacturing process in a low-wage location and can require the automation of core manufacturing processes. Automation of other processes such as materials handling and machine loading can be economically viable, but generally only if this significantly improves efficiency of the manufacturing system overall.

The degree of automation in handling and manufacturing steps that are not quality-critical can be adjusted relatively easily

When expanding or shifting production for an existing product to a location with lower labor costs, companies often find an "older" **manufacturing method** most suitable. This offers a practicable way to adapt production technology with relatively little effort and risk, as the following example shows.

When establishing a small plant in a low-wage country, a manufacturer of worm gears had to decide whether to use the same process setup as at home or reconfigure production. An analysis quickly revealed that it would be uneconomical to replicate the home factory's setup. The leading-edge, high-volume thread rolling machines and grinding machines would have required too high an investment, and plant capacity would have exceeded planned sales volumes. Operating and maintaining the machines would also have required technological competence that was not available at the new location, so there would be no guarantee of high productivity.

Instead of using the highly productive but capital-intensive rolling method, the company decided to use an older production technique: a turning process common 15 years earlier. The old process and plants provided the basic concept and this was further refined. Newer lathes with modern high-speed cutting (HSC) technology were installed, reducing the machining time by a factor of three. This enabled the manufacturer to achieve production costs at the new location comparable to those in the home plant despite the relatively small unit volumes (Figure 5.13).

"Simplicity = productivity," particularly in low-wage locations

However, such a change in the core manufacturing method also entails changes to the product characteristics. The implications are far greater than from changing the degree of automation in materials flow or workpiece handling. This means quality concerns are legitimate, but they are hurdles that can be overcome. Admittedly, automated operations usually achieve a higher level of precision and repeatability than manual labor, and fluctuations in the manufacturing process are less frequent. "Robots don't sweat" is how a supervisor in the welding department of a plant in India put it (small drops of perspiration on the weld metal can cause severe quality issues). Many companies refrain from changing the core manufacturing method as a result, citing quality problems as the reason.

However, the experience of companies that have a long record of success in manufacturing in low-cost countries – or are headquartered in these locations – demonstrates that it is also possible to achieve and guarantee high quality standards. These companies use manual processes in combination with supporting measures and principles such as *Poka-Yoke* and *Jidoka*.[6] In contrast to competitors who refrain from changing their production for quality-related reasons, these companies adopt targeted measures to avoid manufacturing errors and product defects wherever possible.

For example, they implement **simple control mechanisms** in the core process that ensure work steps are carried out correctly. An automotive manufacturer at a low-wage location, for example, uses an inductive sensor for a manual welding process that monitors the number of weld points per body,

Highly automated production can be attractive for large volumes and simpler methods when unit volumes are low

Fig. 5.13: Concepts for manufacturing worm gears

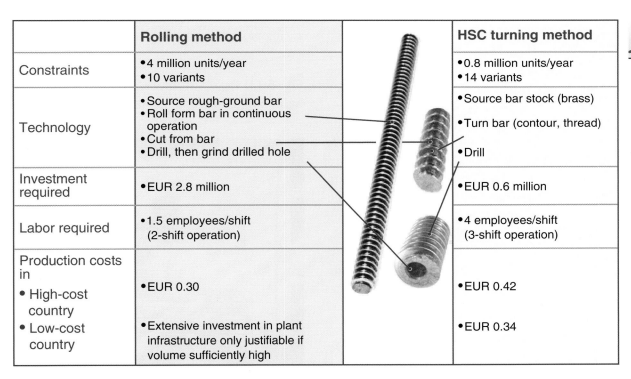

	Rolling method		**HSC turning method**
Constraints	• 4 million units/year • 10 variants		• 0.8 million units/year • 14 variants
Technology	• Source rough-ground bar • Roll form bar in continuous operation • Cut from bar • Drill, then grind drilled hole		• Source bar stock (brass) • Turn bar (contour, thread) • Drill
Investment required	• EUR 2.8 million		• EUR 0.6 million
Labor required	• 1.5 employees/shift (2-shift operation)		• 4 employees/shift (3-shift operation)
Production costs in • High-cost country • Low-cost country	• EUR 0.30 • Extensive investment in plant infrastructure only justifiable if volume sufficiently high		• EUR 0.42 • EUR 0.34

Source: Company data, McKinsey/PTW (ProNet analysis)

[6] *The Japanese expression* Poka-Yoke *(English equivalent: "error-proofing") describes a principle that consists of several elements, including technical precautions or facilities for the immediate detection and avoidance of errors. The Japanese expression* Jidoka *(English equivalent: autonomation, which stands for autonomous automation) describes the operation of a machine without human supervision. The machine includes components and functions that allow it to detect any deviations from normal operation.*

Alternative Production Technology: Industry- and Location-Specific Solutions for Balancing Rotors

A dentist's drill, a crankshaft, or a power station turbine, whether customized or mass-produced, have a key common denominator: all rotating and oscillating parts have to be balanced to ensure high quality, convenience, and long life of the machinery and plant. Schenk RoTec, a subsidiary in the Measuring and Process Systems division of the Germany-based Dürr Group, offers its customers tailored concepts for every aspect of balancing. In addition to the location, this customization depends heavily on the field of application, as illustrated by its electric motors.

Coiled electronic armatures are used in motors for many different purposes – from servomotors in vehicles or handheld tools and domestic appliances to industrial plants such as power stations. Balancing constitutes a key process for the quality of the component. Smooth operation, low vibration, high energy efficiency, and long life are critically determined by the balancing technology. Particularly in the automotive industry, where large numbers of small and medium armatures are used, the equipment has to meet high standards in terms of precision and balancing quality, while also maintaining high productivity levels. The utmost accuracy is imperative at every step. An automated process is therefore essential.

The balancing process takes place on fully automated, integrated four- or six-station machines (Figure 5.14). The electronic armature is clamped, and the imbalance is measured automatically. It is then equalized at the next two stations by milling on two levels. Finally, a control measurement is made before the electronic armature is passed back to the conveyor system. Total automation and the close interaction of measuring and balancing ensure high productivity as well as high quality and process reliability. In less quality-critical indus-

tries, such as the use of electric motors in toy cars, such accuracy is not needed. Schenk RoTec offers a simpler, manual machine concept for these applications. In low-cost countries, this more labor-intensive concept is also superior to an automated process from a cost perspective.

In the less automated plant concept, the machine operator inserts the electronic armature manually and starts the measuring process. Next the weight is balanced by attaching extra weight, such as small metal plates. The precision of this method cannot be compared to that of the automated process; there is also the risk that the added weight(s) could come off in operation. But the machine requires much lower investment. The process is, of course, more labor-intensive. Takt times are higher by a factor of 4 to 5, so overall output of the manual process is much lower than at the automated plant. However, this manual balancing process is an economically viable alterna-

Fig. 5.14: Fully automated system for balancing electronic armatures

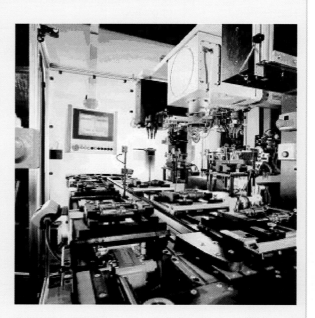

Source: Schenck RoTec

tive especially for products with lower quality demands, relatively short product life cycles, and production in low-cost countries.

The cost impact of the technology selection is significant. Relocating a fully automated plant from a typical high-wage to a low-wage location can lower the costs of armature balancing by some 12 percent (assuming the same productivity levels). Adaptation of the process technology can lead to a further 25 percent savings compared to production in a high-wage location. In this example, process technology selection actually has a greater impact than choice of production location.

Bottom line: Adaptation of process technology can make a huge difference. The change can improve the cost structure considerably. But there are constraints. The applicability of a production technology can be limited by the customer's industry and requirements (particularly with regard to accuracy and quality). The production process technology has to fit both the location and the customer's requirements.

guaranteeing that all activities are performed according to the work plan (Figure 5.15). Sources of error can also be greatly reduced in manual material flows by, for example, color coding the workpiece carrier.

5.2.2.2 Adapting Production Technology Using a Modified Product Design

Changing **workpiece design** can allow fundamental adaptation of the core manufacturing processes. It gives the production engineer much more flexibility to define an optimal manufacturing system for a specific location. There are drawbacks, though, that management has to consider. Fundamental redesign entails one-time costs. The number of product variants increases, and the standardization of processes and equipment becomes more difficult. Depending on the industry, design and process changes may require customer approval that is both expensive and time-consuming, particularly in the automotive supply sector. So when is redesign worthwhile?

The first step in finding an answer is to look at the process steps with the highest contribution to overall manufacturing costs. This will identify areas for improvement that can overcompensate for any one-off costs associated with a change of design and process. Allowing for higher average surface roughness and setting a requirement of 1 instead of 0.3

micrometers, for instance, typically means grinding processes can be used, and makes lapping or honing unnecessary. Compared with grinding, these technologies entail two to five times higher manufacturing costs and increase complexity. Savings can be substantial if requirements can be reduced. Given the mutual dependence of manufacturing technology and product design, a change to the product design may be required to allow broader tolerances. The interdependencies are often complex and not sufficiently transparent to decision makers.

Examples of alternative concepts in product design and manufacturing technology include:

- Avoiding complex cut components that require capital-intensive milling processes by joining together several individual components

- Avoiding expensive automatic test devices for quality control by using error-proof design (according to *Poka-Yoke* principles)

- Avoiding capital-intensive connecting systems by altering product design to support simpler joining methods (e.g., screwing rather than gluing or forming).

The economic impact achieved by changing the manufacturing method varies greatly both from industry to industry and even from part to part. But savings often extend significantly beyond the cost reduction achieved when only the automation of materials flow, workpiece handling, and quality control is adjusted. In the case of **differential design** (where the component is manufactured from different individual parts) and manual joining/assembly processes, for example, the costs of tools, jigs, and machinery are generally significantly lower than for **integral design** (where components are more monolithic, made via forging, forming, or milling processes). Integral design is typically far more capital-intensive as it requires both larger machines and large, complex casting molds, forging tools, or high-performance metal-working tools.

Changing product design greatly increases the flexibility to adjust manufacturing technology to local requirements

Figure 5.16 illustrates the three production technologies for manufacturing a specific aircraft component – a center wing box rib. The two basic design alternatives available are integral and differential design.[7] **Differential design** of the wing rib requires the cutting, forming, and machining of aluminum plates as well as the assembly of machined parts via riveting. The labor content of these processes is high, while only very simple machines are required to process the parts. **Integral design** of the component

---

Critical processes can be adapted with the help of appropriate quality assurance

Fig. 5.15: Approach to reducing the level of automation when shifting production to low-cost locations

① Reduce capital intensity of auxiliary processes	② Reduce capital intensity of core process	③ Ensure quality in core process	④ Develop quality assurance measures
• Reduction of automation – Monitoring – Testing – Cleaning of components – Materials flow	• Spot welding with manually placed clamps • Manual welding with guided electrode (e.g., WIG)	• Manual welding – Self-fixing jig – Power flow meter – Weld point counter • Simple automation for critical parts	• Labor-intensive test routines (e.g., visual, induction, and ultrasonic inspection) • Dynamic monitoring of random samples

Fully automated welding line

Low "design-to-low-cost" competence: Quality problems lead to a return to automation

Welding by hand

High "design-to-low-cost" competence: Quality problems are overcome via targeted improvements

Simple automation at critical points only

Source: McKinsey/PTW (ProNet analysis)

---

[7] Cf. Lindemann (2005), p. 39: discussion of the advantages and disadvantages of an integral/composite design compared with a differential design for a pinion cage.

essentially involves processing a large aluminum plate on an NC milling machine. Machining time accounts for some 80 to 90 percent of total manufacturing time. The actual labor content is low (excluding installation, retooling, and maintenance of the production machine).

Traditionally, integral design is still primarily seen as a way of reducing the weight of a part.[8] However, this view may be overly one-sided. The choice between integral and differential design of a part leads to an entirely different set of manufacturing processes and

cost structure (Figure 5.17). This provides an opportunity to substitute labor with capital or vice versa on a substantial scale, making the choice highly relevant to the design of a global manufacturing network.

Frequently, a change to the product design or at least to selected parts and components is triggered by lack of skills or the more limited production volumes required at a new production location. Low local demand often makes capital-intensive production non-viable for suppliers. Semi-finished products that

Alternative manufacturing methods are possible if the product design is changed

Fig. 5.16: Alternative manufacturing methods and product designs – example from aircraft construction   SCHEMATIC

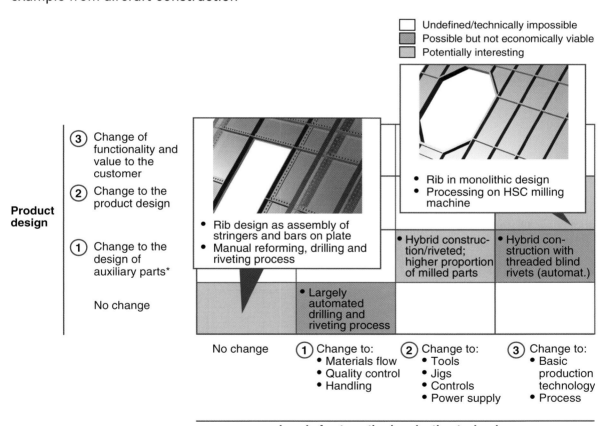

* E.g., connecting parts; key functional parts and functional surfaces remain unchanged

Source: McKinsey/PTW (ProNet analysis)

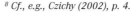

---

[8] Cf., e.g., Czichy (2002), p. 4.

require capital-intensive equipment such as presses and forming tools (e.g., extruded sections) may therefore not be available on the local market. A redesign of the part and change in specifications can allow the use of local parts and avoid the high cost of imports. Being able to adapt to (local) suppliers' lower level of technological and engineering skills in low-cost countries[9] can be a source of competitive advantage. The advantage is particularly high in industries that require local product features but less standardization across plants.

Even when localization is favorable, it may initially still be necessary to import selected parts and components from the company's high-cost locations. Phasing in locally supplied parts may only be grad-

ual to establish local production fast while still allowing time for changes in product design.

Adapting the product design and its production process to local conditions can open up new opportunities, particularly in a low-cost country. If done well, the adaptation should lower costs significantly. The product may also become more attractive for areas of application and groups of customers not previously considered. These decisions are therefore of strategic importance, potentially enabling the company to serve new markets or market segments. Pushing for adaptations can be crucial.

The competitive strategy and market segment also determines to what extent companies should adapt

Changing from a monolithic design to the assembly of intermediate parts can allow production in smaller volumes

Fig. 5.17: Alternative manufacturing methods with a change to the product design – examples

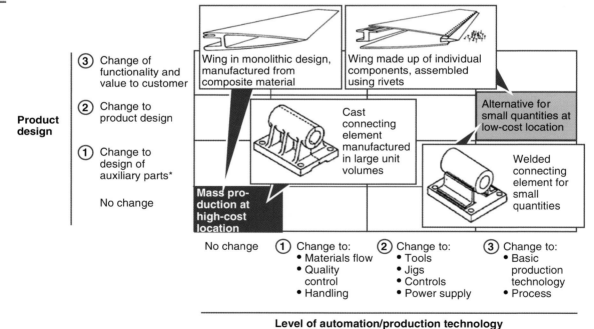

* E.g., connecting parts; key functional parts and functional surfaces remain unchanged

Source: McKinsey/PTW (ProNet analysis)

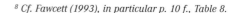

[8] Cf. Fawcett (1993), in particular p. 10 f., Table 8.

product design and production technology to specific locations. While globally renowned OEMs of high-end products usually sell products that are standardized across the world, niche- and cost-oriented suppliers are more likely to opt for products and processes adapted to suit local conditions (Figure 5.18). These smaller manufacturers can only compensate for their smaller scale compared to large competitors by using adaptations.

It must be stressed, however, that adapting product design and production technology is neither possible nor economically advisable in every industry. In some industries, it is better to only vary the degree of automation without changing the product design, or to use a uniform production process and product design globally.

5.2.2.3 Standardizing Production and Products Globally

Having varying production processes in locations with different environmental conditions is not ap-

propriate in all industries. Quality guidelines or strict customer specifications often prohibit changes to product design and process technology.

In some industries, customer requirements are fairly similar worldwide (Figure 5.19). As a result, customers do not appreciate changes in product or process design – quality concerns are too high. In other industries such as semiconductor manufacturing, the fixed cost of a redesign can be prohibitively high. Developing a new production process with acceptable levels of reliability and performance is very costly. Instead, products and processes are standardized across the globe, with identical or very similar factory design across locations.

Diversity may also prove the wrong path if a manufacturing method has special characteristics that cannot be achieved using another technique. Examples range from ordinary products to high-tech applications.

The freedom to make local adaptations can vary considerably depending on the industry

Fig. 5.18: Economically viable options for local adaptation in different industries

	Product design	**Production technology**	
Furniture	Adjustment to country-specific customer expectations as well as to specific availability of materials	Highly automated or manual production depending on location; no cross-locational production	
Machine tool	Adaptation of all peripheral components (e.g., chip box) to local customer requirements, and some changes to allow use of different manufacturing technology (e.g., more sheet-metal forming and assembly instead of milled parts)	Adaptation from automated to manual process depending on the relevance/function of the part and expected production volume/size of facility	**Increasing possibilities for local adaptation**
Valve for fuel system in the aviation industry	Adaptation not possible/cost-efficient in aviation and aerospace industries due to high quality and documentation requirements, and small total volume	Adaptation not possible/cost-efficient in aviation and aerospace industries due to high quality and documentation requirements, and small total volume	

Source: McKinsey/PTW (ProNet analysis)

Sophisticated manufacturing processes, such as clean-room production in the semiconductor or biotech industries, can make local adaptation of the manufacturing method uneconomical, and may even cast doubt on the whole idea of establishing production in emerging markets. Companies that focus on constant economic optimization and technical development of their manufacturing technology, evolving specific core competencies as a result, can gain a strategic competitive advantage that secures them worldwide superiority in terms of either product quality or costs of goods manufactured. When used at all locations worldwide, such **dominant production methods** are more cost-efficient than alternative methods, although they can make demands that only a few locations can fulfill.

However, dominant production technologies can also be relatively simple, which means they can be used in most locations. As an example, companies use the same process to manufacture cans for soft drinks worldwide (second-operation drawing of a metal sheet). This is the most effective way to make the cans independent of location. It makes little sense to try to vary the basic production technology from one site to another with this kind of product. The benefits of adaptation would be very limited.

In the case of highly complex, dominant manufacturing methods, it may be necessary to keep production at locations where there is a high level of competence, and to accept the factor cost disadvantages. With processes of this kind, being able to guarantee stable production (and therefore product quality) often has a much greater impact on economic viability than labor costs. Remaining in a specific knowledge cluster may also be important to allow a company to build a leadership role, with the right to set industry standards in that field.

The share of product designs that are standardized worldwide is considerably higher in global industries

Fig. 5.19: Share of alternative product designs in different industries*

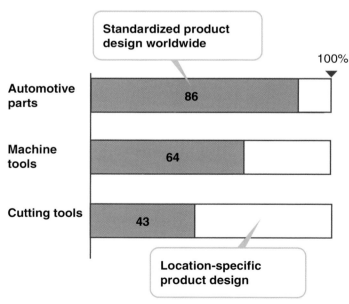

Automation in High-Cost Countries: Integrated Optimization of Product, Machinery and Sequence at SGF

Süddeutsche Gelenkscheibenfabrik (SGF) based in Waldkraiburg (Bavaria), Germany, is the leading producer of flexible couplings (Figure 5.20) for installation in the powertrain and steering systems of passenger cars and other vehicles. SGF is open to the opportunities of producing in countries with low labor costs. For example, it uses a location in the Czech Republic as an extended workbench. This location focuses mainly on simple and labor-intensive processes, e.g., deburring the couplings. Core production processes are, however, primarily carried out at German locations, even for high-volume products. These processes are highly automated and this makes production at a high-wage location viable. Automation of the core manufacturing process was a major challenge. The winding of the flexible couplings demonstrates the technical challenge of automating the complex manufacturing process intelligently (Figure 5.21).

Fig. 5.20: Flexible coupling disks

Source: Süddeutsche Gelenkscheibenfabrik

The thread design in the flexible coupling allows transmission of high torsion forces while also ensuring pliability and damping. This balances angular misalignment and longitudinal deformation in powertrains and the steering mechanism. Pro-

Fig. 5.21: Production of a flexible coupling disk: manufacturing steps

Cut and machine bushings ⟩ Paint bushings and disks ⟩ **Wind disks** ⟩ Vulcanize ⟩ Deburr and test

Manual winding
- Bushings clamped manually
- Thread guided via former
- End disks pressed on

Fully automated winding
- Automated material feed, winding, pressing on of end disks and delivery
- Capacity of around 250,000 units p.a.
- Improved quality

- Factor input (estimated)
  - Approx. 3 min. labor per disk
  - EUR 0.10 depreciation and cost of capital per disk

- Factor input (estimated)
  - 1 min. machine input (approx. EUR 0.45 depreciation and cost of capital per disk)
  - Approx. 0.5 min. work per disk

Source: McKinsey/PTW (ProNet analysis)

duction of the thread design does, however, require a complex winding process. Traditionally, the thread is wound over clamped bushings and controlled via a template (a process similar to copy turning, but here the template is used to steer the thread instead of the bit). This process is very labor-intensive. Simple productivity increases are hardly possible. Increasing the speed or number of parallel wound couplings would make the process more susceptible to faults.

SGF achieved cost-efficient and high-quality production of flexible couplings with a **rigorous automation program** (Figure 5.22). The winding machines, which are designed and largely manufactured in-house by SGF, can produce the drive-shaft coupling right through to depositing it in the shipping container. The process is entirely com-

puter-controlled. The preliminary product feed is highly efficient: the thread is on rolls, with the bushings, disks, and rubber solution as bulk material. The bushings and disks are automatically aligned and fed into the collets of the winding spindle. The coupling is then wound in the core of the machine. The equipment's design is thus optimally adapted to the product, with the entire know-how **in the hands of one company**.

Bottom line: This plant's economic feasibility is firmly centered on the location's specific strengths and employee skills. The integrated optimization of product, machinery and sequence has been key to SGF's success. This example also demonstrates that high levels of knowledge and innovation are prerequisites for the viability of a high-cost location.

Fig. 5.22: Automatic winding machine for flexible couplings

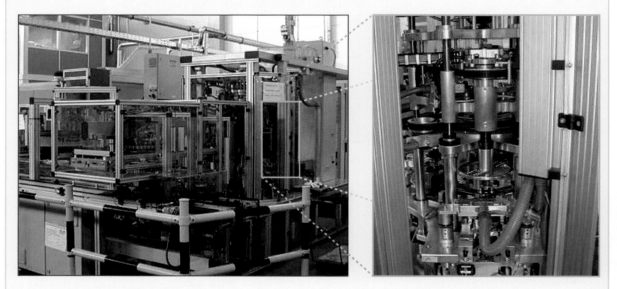

Source: Süddeutsche Gelenkscheibenfabrik

5.3 Evaluation and Selection

Companies setting up a new production location must consider their existing **production technology portfolio** and review the need for further adaptations. The number of adaptations overall should be reasonably small. Managers should take a global perspective and include the cost of complexity when evaluating a related proposal.

Managing this process is difficult. Too much central control kills continuous improvement ideas and innovativeness in the manufacturing locations. Too little, however, leads to a complex portfolio of various manufacturing technologies that makes quality control difficult, and prevents the company from reaping the rewards of standardization.

> **The optimum level of automation is determined by the costs of labor, the cost of capital, technical know-how, and the number of variants to be produced**

Companies' experiences with reconfiguring their production processes have not all been positive. Shifting from automated to manual processing is particularly tricky when it comes to the core processes that determine key product features. International companies have found that a greater share of manual activities often results in quality problems. Completely abandoning automation and capital-intensive equipment in materials transport and handling, for example, can have a negative effect on the productivity of core machines and processes – for instance in a sheet-metal press shop. Overall output does not meet expectations if manual feed leads to lower production frequency and greater downtime than with an automated materials handling system. These concerns regarding quality and overall productivity are legitimate. But they are often self-inflicted – a result of international manufacturers' focus on reducing labor content in production processes and increasing automation. Reducing labor content has been the top priority for most product and process design engineers at companies with their roots in Western

Europe, North America, or Japan for decades. Making simple, labor-intensive processes work is a capability that often needs to be relearned.

Selecting the right production technology is not an easy task and requires a sound understanding of the trade-offs involved, as well as staff capabilities and mindset. We therefore suggest the following steps to determine which production process technology and product design is best suited for a specific location:

- Create **transparency around the trade-offs** involved. Specify the advantages and disadvantages of adapting the production technology and product design to specific locations. Analyze customer expectations and the mindset and capabilities of the company's own personnel.

- Analyze the impact of adaptation for a **specific location**. This helps to clarify whether an existing production technology is suitable. Analysis of a particular location also provides insights into how large the opportunity of adaptation would be.

- Broaden the perspective to include the **entire production network**. The findings for an individual location need to be put into the broader context of the global manufacturing network. Management should assess whether a new production process could be leveraged for other locations and markets, or whether an existing technology is close enough to ideal for a specific location. Assessing the production technology may also require revisiting the suitability of a specific location for production. If adaptation is required to make production at a specific location viable but redesign is too costly, it may be better to expand production at existing locations.

Location-specific analysis reveals the trade-offs for a specific market. A holistic perspective on the technology portfolio and all production locations will allow the company to achieve the lowest costs for the production network overall.

5.3.1 Creating Transparency Around the Trade-Offs

Adapting the production process and product design to local conditions at the manufacturing site requires trade-off decisions. This means it is important to create transparency and consensus around the evaluation criteria. Similarly, the decision process needs to reflect that a redesign with all its benefits may conflict with other objectives. It may be particularly hard to combine fundamental redesign with a fast production ramp-up at the new factory.

5.3.1.1 Pros and Cons

Despite the potential opportunities, adaptation can incur costs and increase overall complexity. Complete transparency on the benefits and costs of redesign is therefore essential.

■ **Opportunities.** Typically, redesign opens up the following opportunities:

□ **Substitution of labor by capital** and vice versa: This lowers production costs if locations with significant differences in wages and cost of capital are involved.

□ **Adjustment of production complexity and requirements** to local conditions: This allows optimal leverage of local skill levels, and avoids overstretching staff. Adapting production technology can also open the door for local raw materials and intermediate products that are unsuitable for the original process.

□ **Flexibility in scaling capacity:** Production volumes at established locations are often significantly greater than required for new sites. The machinery/plant capacity of the existing production process is therefore often high. Adaptation can allow the efficient production of smaller unit volumes. It can also provide greater flexibility to produce different products and product variants.

□ **Adjustment of product specifications to local market requirements and preferences:** How customers value product features varies by market. Adaptation can make it easier to fulfill market-specific preferences. It can also eliminate process steps and costs for features that customers are unwilling to pay for.

■ **The downside.** Adapting production technology and product design also has negative implications. Local adaptation can increase costs, particularly from a more holistic perspective. There are four major cost drivers:

□ **Local adaptations hinder standardization:** Companies can lower costs by avoiding one-off costs, bundling volumes (to boost their purchasing power), and reducing inventories. Standardization of products and processes is the main lever to achieve these savings. At best, local adaptations make standardization more difficult. In combination with a low degree of centralization and organizational discipline, adaptation makes the standardization of parts, tools, machinery, raw materials, and intermediate products impossible.

□ **Process and product redesign entail one-off costs:** Product or process redesign requires multiple activities. Development engineers redesign the part. Production engineers develop and test the new process, producing and inspecting parts to ensure that quality requirements are met. Raw materials, intermediate parts, and tools are required to produce output for test runs that generally cannot be sold. All these and other related activities incur costs. Not all these costs may be directly visible or allocated to the product.

□ **Adaptations take time:** Redesign of a production process or product design requires time. Not just the actual design: you need to procure new equipment and input materials as well as test the process and product quality. Redesign may prevent early market entry, defeating the object of the exercise.

□ **Changes entail market risk:** The involvement of customers in redesign is necessary but tricky. On the one hand, involvement is crucial to comply with typical quality requirements. Customers often reserve the right to inspect new production setups and test product quality. On the other hand, involvement can trigger demands to lower the price of the product as the customer may insist on benefiting from lower production costs. This entails the risk that the manufacturer is left with the one-off cost of the redesign but loses part of the ongoing benefits required for payback of the initial expense.

Analyzing the pros and cons of adapting production technology and product design to location conditions provides insights into which approach is most suitable in a specific situation. Cost drivers suggest the spectrum is discontinuous when it comes to related costs. There are three discrete thresholds:

The first involves **any change whatsoever** to the process. Even if small, it requires documentation and some engineering work. Use of machinery from different suppliers makes it harder to leverage scale and negotiate better rates. The benefits from standardization are harder to realize.

The second involves **changes to the core production process**. This can be the use of different machinery, process parameters, tools, or control methods. Such changes potentially impact parts quality. This means inspecting and testing a sizable batch of parts produced with the new setup and the process itself. Such testing can be costly and make the batch of parts or finished goods unusable. In industries with high quality requirements, such changes to the production setup also require customer approval, which can trigger elaborate and costly consequences.

The third threshold involves **changes to the product design**. A new product variant has to be administered and changes documented. The new design must be tested and approved. Additional variants can cause future costs, as changes to other parts and components require alterations to multiple variants

of a part, not only one. Spare parts management becomes more complex. Overall, more variants create complexity. And complexity drives costs.

It is therefore important to maximize the benefits once a threshold is crossed, so the additional costs are justified. To achieve maximum benefits, the technological comparison of production methods should focus on alternatives that can fundamentally transform a product's cost structure. As we have seen, capital-intensive manufacturing steps have to be substituted by manual labor at locations in low-cost countries. Engineers should look for alternatives that allow the reduction of capital expenditure by at least half. Changes have to be significant to make an impact. This goes for more than just cost and capital expenditure – overall process design and plant layout should also be included in the scope. Companies rooted in high-cost locations should not underestimate the difficulty of managing complex production processes in developing countries. If you cannot reduce complexity and skill requirements, it may be better to increase capacity at home rather than venture abroad.

Particularly in global industries such as the automotive sector, product and process design changes are often undesirable or not financially worthwhile. The share of standard product designs worldwide is correspondingly high in these industries. This helps to keep the overall complexity of the production network within limits. Design changes are more commonplace in industries that are more local in character or produce simpler products.

Interestingly, companies come to very different results when selecting production location and process technology. These decisions usually become clear when seen in the context of the overall corporate strategy.

Renault, for instance, follows a "one platform, one plant" philosophy when it comes to high-volume, small and mid-sized passenger cars. In contrast to other automotive OEMs, Renault does not aim to produce multiple models on the same line in this market segment. Instead, it tailors its production technology and degree of automation to the plant's location. This is in line with the strategic objective: cost leadership in this market segment. A case in point is the Dacia Logan, Europe's most competitively priced mid-sized passenger car. Manufactured in Romania by Renault's subsidiary Dacia, its production is matched to the characteristics of the region – with correspondingly low automation and capital intensity.

5.3.1.2 Perceptions and Process

Companies struggle with "softer" factors, too, when trying to reap the benefits of adapting production technology and product design to local market conditions:

- **Buy-in and incentives:** Redesign of the production process or product design to local requirements is new to most manufacturing companies, especially those that have only recently started to expand their production footprint. Organizational skepticism can therefore be significant. Incentives are also often not in line with requirements to achieve fast and successful technical redesign.

- **Collaboration and timing:** Redesign needs the involvement of different functions and departments. But the task of technical redesign is often new to companies. This means formal processes are not in place to facilitate cross-functional collaboration. Yet this is critical to achieve a positive impact. When it comes to timing, it is important not to overload the production ramp-up phase with additional complexity. If necessary, the redesign should be postponed.

Companies need to consciously manage these issues, particularly in industries where adapting the production process and product design to local conditions is crucial. This involves changing incentives and perceptions where they are detrimental, defining a process that can absorb the remaining risks.

Perceptions that hinder redesign are often customer-related, and are therefore hard to challenge. Customers are often perceived as unwilling to **accept changes** to the product. Undoubtedly there are hurdles, such as the need for testing, inspection, auditing, and documentation costs for the manufacturer and also the customer. This particularly applies in industries with very high **quality requirements**, such as the automotive supply and aviation industries. But this does not mean customers would object to design changes that dramatically improve a product's cost position.

Another hurdle to successful redesign is the low priority that such efforts are often assigned, particularly in engineering departments, even if the potential cost impact is high. Revisiting an old design is often perceived as less interesting than developing something new from scratch. Engineers also often have little incentive to present alternative manufacturing technologies for a product to management. They are left to make the decision themselves – sometimes without sufficient context, as they are not involved in the location planning.

To ensure efficiency, companies should first examine their existing technology portfolio. This review may include processes that are no longer used at high-cost locations or are earmarked for replacement. This has proved effective, particularly for companies based in high-cost countries that have expanded into lower-cost locations. The organization should only look for additional alternatives if existing production technology does not suit the new location.

Collaboration between functions is critical when it comes to the process of reviewing and adapting production technology to local conditions. The lack of **cross-functional collaboration** between marketing, product development, and production engineering is often a hurdle to effective redesign. This is not surprising. Redesigning product and processes to the local conditions of a location is new to many companies and certainly to those globalizing their footprint for the first time. Most companies have followed industry best practice, establishing a formal process for ensuring cross-functional collaboration in the development and production ramp-up for new products. However, when it comes to redesign of the global production network and a related review of production technologies and product design, this best practice is rarely applied. Making interdependencies and trade-offs transparent is therefore an important task that management needs to facilitate. It helps to closely involve production engineers in strategic location planning, so they have greater incentives to provide a suitable portfolio of production technologies.

Finally, the timing needs to be realistic. If redesign of the product design and production process is to be conducted ahead of production ramp-up, sufficient time needs to be budgeted. The risks are substantial if the complexity of a new process is combined with a production ramp-up under time pressure. If sufficient time is not available, redesign should be limited to changes to the process without impact on the product design. To realize the full cost reduction potential of the low-wage location, more fundamental changes can be introduced later. This may mean compromise and writing off equipment and tools, but will allow a smooth ramp-up.

5.3.2 Analyzing the Impact for a Specific Location

What does it mean to adapt production technology to an individual location? How can a company achieve minimum costs?

There are multiple trade-offs and relevant interdependencies. Comprehensive automation requires both high capital intensity and highly qualified employees. Qualified employees, however, are comparatively expensive even in low-cost countries. This means the relative gap compared with labor costs in high-cost countries is much greater for low-skilled than highly skilled staff. Labor costs in China are roughly 20 to 30 times lower than in Germany, while for a (local) plant manager the difference is only around a factor of 5.

On the other hand, due to higher depreciation and higher maintenance, capital, and transaction costs, the use of complex machinery is often more expensive abroad than at home. Less complex production technologies frequently pay off.

Against this backdrop, established companies from high-cost countries commonly have to develop new manufacturing technologies that suit local conditions in emerging markets. Occasionally, they can also use older, tried-and-tested methods. This is because mature technologies are particularly suited for use in low-cost countries and at new locations as the standardization of the process is already advanced, making it relatively foolproof.

Figure 5.23 illustrates the production technology and economic viability of two connecting methods using the example of a gear motor. A manual screw fitting is to be employed at the low-cost location. Manual production is more cost-efficient than using an automated process due to the lower unit labor costs. In contrast, an automated rolling process is advisable in the high-cost country. Automation leads to a significant reduction of costs compared with manual production. To determine the cost of goods manufactured in relation to labor costs, the fixed costs for each machine and plant for each year are calculated and allocated to the individual component in accordance with the planned production volume. The variable costs, particularly labor costs, are then determined based on the standard times. The level of qualification needed for each specific manufacturing step should also be taken into account. This information can then be used to perform a simple calculation. Other cost drivers may also be relevant, such as shift allowances, variable operating costs, and machine utilization.

5.3.3 Broadening the Perspective to the Entire Production Network

The motor shaft case study shows that alternative manufacturing technology can greatly contribute to reducing production costs. But new manufacturing methods and product variants also result in extra costs. More variants complicate production management as they generate one-off development expenses for changes to production and product design.

Local factor costs determine which manufacturing technology is most efficient

---

Fig. 5.23: Alternative manufacturing technology and design for an electric motor-gear connection

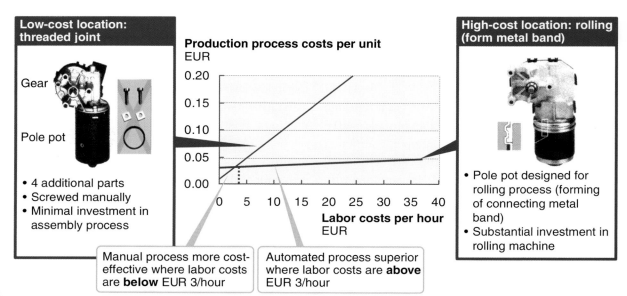

Low-cost location: threaded joint

Gear

Pole pot

- 4 additional parts
- Screwed manually
- Minimal investment in assembly process

Production process costs per unit
EUR

Labor costs per hour
EUR

Manual process more cost-effective where labor costs are **below** EUR 3/hour

Automated process superior where labor costs are **above** EUR 3/hour

High-cost location: rolling (form metal band)

- Pole pot designed for rolling process (forming of connecting metal band)
- Substantial investment in rolling machine

Source: McKinsey/PTW (ProNet analysis)

Using Alternative Manufacturing Methods: Armature Shafts for Small Motors

Every year, millions of small motors and micro-motors are produced for all kinds of applications. The armature shaft is a fundamental component of every electric motor to which the rotor is attached, so they are produced in large quantities (Figure 5.24).

In high-cost countries armature shafts are made using a fully automated production process. The material is uncoiled, straightened, and cut. Next a conveyor system feeds the shaft into a lathe, where it is clamped. The lathe machines the outside contour and ends. The shaft is then inductively hardened to protect against wear and tear to the surface. Finally the product is finished using centerless grinding, with automatic control of the diameter, roundness, and surface quality.

Fig. 5.24: Armature shaft

Source: McKinsey/PTW (ProNet analysis)

This process chain can be translated into a much less automated system concept better suited for production at a low-wage location. This requires stepwise adjustment of the manufacturing concept and affects the core manufacturing methods as well as auxiliary process steps.

Step 1: Adjust the Materials Flow, Work-piece Handling, and Control Processes

The initial changes affect the materials flow, workpiece handling, and control processes. These processes do not directly impact product features, so adjustments to them are comparatively easy to realize. In this example, the first step would be to cut out the automated material feed. In place of the capital-intensive, automatic feed via a coil, the parts are delivered in low-cost countries by semi-

skilled logistics staff. Bar stock is used instead of coiled material, allowing straightening to be omitted entirely. Manual feed is used for both hardening and grinding, saving on comparatively capital-intensive materials-handling technology. A visual sample check replaces the automatic monitoring of the diameter.

Step 2: Reduce Automation of the Core Processes

In Step 2: of the process adaptation, automation of the core production processes is reduced. The cutting process is now carried out manually using clippers, and the workpiece is clamped manually

Fig. 5.25: Alternative manufacturing methods – example of armature shafts for electric motors

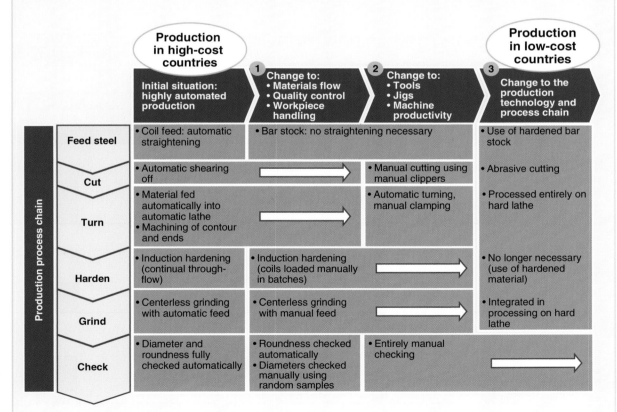

Source: McKinsey/PTW (ProNet analysis)

in place in the automatic lathe. In quality control, manual inspection can be used for the roundness test, which was carried out by machine in the previous step.

Step 3: Use Alternative Manufacturing Technologies

The final step in adapting production processes is to use manufacturing technologies more suited to low-cost countries due to their specific input factors. In the case example, the plant buys pre-hardened bar stock, allowing the hardening process to be left out. Next, abrasive cutting is used for the bar stock, and hard machining to size is carried out entirely in one machining center (Figure 5.25).

This example illustrates that adaptations outside the core technology are relatively easy to implement. But they also only have a comparatively minor impact on the reduction of production costs and investments. It takes a fundamental overhaul of production processes to achieve a more substantial cost reduction. Even for core process steps, the use of a new manufacturing technology better suited to the location needs to be considered. Additional quality assurance measures can ensure that the reduced degree of automation does not impair quality (Figure 5.26).

Bottom line: Several production process alternatives are available even for a simple part like the shaft of an electric motor. The impact can be significant, but may require fundamental changes to the manufacturing technology even for core process steps.

Fig. 5.26: Changes to the production costs, capital intensity and quality of components

FIGURES FOR INDICATIVE EXAMPLE

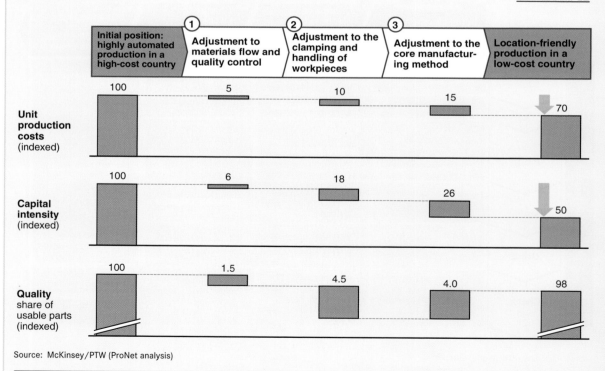

Source: McKinsey/PTW (ProNet analysis)

The fact that new product and process variants can be used in multiple locations across the network leads to interdependence between plants. A decision on the production technology in one location affects the others. This should be considered when drawing up the strategic location concept and production technology portfolio.

The following factors should be analyzed to decide on the **potential of a new production process** within this broader context:

■ One-off costs, the cost of increased complexity, and other opportunity costs related to developing a new production process or product variant (e.g., more limited bundling of purchasing volumes, linked to a decrease in purchasing power and discounts)

■ Savings in variable costs that could be achieved by adapting the production process at a specific location, compared with transplanting the most suitable existing production process

■ Additional savings if the new or adapted production technology is used in other locations as well.

When these factors are examined, many alternatives begin to look less attractive.

The consideration of one-off and recurrent expenses triggered by additional variants typically reduces the number of viable options substantially.

A more precise analysis of the one-off costs reveals the different influencing factors (Figure 5.27). First, there are costs directly associated with adapting the

Extensive adaptation causes significant one-off expenses

Fig. 5.27: Adapting the production technology – **one-off** expenses

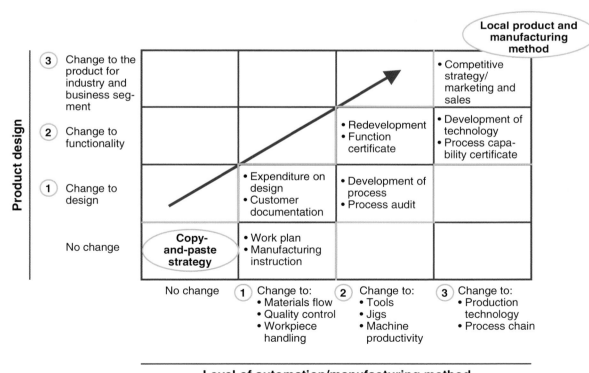

Source: McKinsey/PTW (ProNet analysis)

production technology. Changes to the manufacturing method and – if applicable – to the product design need to be agreed upon with the relevant customers. Appropriate function and quality certificates have to be furnished in accordance with the agreed testing, sampling, and auditing guidelines. These requirements are particularly stringent in the automotive supply and aviation industries.

The effort needed to meet these requirements can be considerable depending on the extent of the adaptation. Such adjustments are generally uneconomical for products that have already passed the growth phase in their life cycle. Involvement of the customer can destroy the expected cumulative savings effects in other ways, too. If customers base their price demands on the variable costs of production, they will insist on a price reduction when asked to approve a new, more cost-efficient process. OEMs often fail to anticipate the one-off expenses and investments involved in developing and implementing new manufacturing methods.

> **The one-off costs associated with modifying products and manufacturing methods can make adaptations uneconomical for mature products**

In the example of the motor manufacturer, renewed sample approval at the new location led to internal expenses of approximately EUR 1,000 per variant. The additional expenditure for customer acceptance would have been at least EUR 3,000. Because one of the motor variants being examined was a critical el-

Additional variant management and expenditure on technology development are important additional expenses for locally adapted production

Fig. 5.28: Adapting the production technology – **recurrent additional** expenses

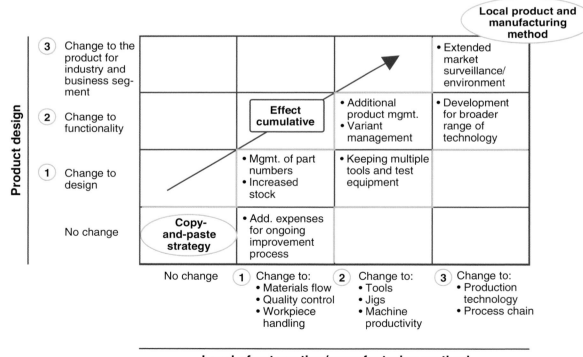

Source: McKinsey/PTW (ProNet analysis)

ement for the customer, a new process audit had to be performed in the presence of that customer. For the manufacturer this meant additional costs of EUR 6,000 - 10,000, and expenses of EUR 4,000 - 5,000 for the buyer.

Changes to the product or production process can trigger substantial costs even if these changes appear minor. When switching from a glued to a threaded joint, for instance, the manufacturer has to bear the costs for the change to the design, revision of the bill of materials, manufacturing notes, work plans, and inspection schedules, including the relevant customer documentation. If simple joining elements are changed, this leads to additional expenditure of around EUR 5,000 for just one variant. These costs can be substantially higher for more fundamental changes.

The level of such costs is very specific to the industry and product. It is therefore essential to analyze and consider each case individually – especially if changes are extensive.

Expanding the **manufacturing technology portfolio** also generates additional costs. It becomes harder to realize savings from standardization, for example. Companies can minimize expenditure considerably when buying machinery, plant, and installations if a standardized manufacturing concept applies to all their locations worldwide. **Economies of scale** in purchasing can be reflected in more favorable terms and conditions. Selecting a location-specific method, however, generally cancels these out. Additional costs are incurred when processes are structured differently at individual locations, as illustrated in Figure 5.28.

Variants trigger additional costs: economies of scale are lost, and the added complexity requires extra management attention

---

Companies should develop a portfolio of manufacturing processes and specifically highlight the locations for which a process is suitable. A portfolio approach to the selection of manufacturing technology helps

avoid fragmentation into too many product and process variants. At the same time, it also allows a manufacturer to reap the benefits of local adjustments.

To understand the different factors that may come into play when taking a global perspective rather than focusing purely on a specific site, let us revisit the manufacturer of small motors profiled in the box on page 226 ff. ("Using alternative manufacturing methods"). This automotive supplier was initially faced with a choice of four technological options for a plant in China (Figure 5.29).

- A highly automated rolling process (as used in the home plant)

- A manual rolling process

- A threaded joint

- A joint made of two sheet-metal shells connected by a riveted joint.

All options were technically feasible and fulfilled the product specifications. But which one would be the most economical choice within the context of the entire production network?

As a starting point, you need to calculate the net present value of the savings relating to variable costs for the product's remaining production period. But this is only one component. Based on comparison of the direct variable costs alone, adaptation of the production process usually appears very attractive. The expected one-off outlays for redesign then reduce this initial savings figure. In the next step, you should subtract the net present value of the expected recurrent additional costs. Fundamental adaptation of the production process or the product design is often no longer attractive for many products after factoring in these additional costs. The cost of the redesign and the additional effort to introduce and continuously improve an additional production process frequently wipes out the benefits. This approach identifies which production processes are most economical within the global footprint.

In the case of the electric motor manufacturer, analysis of the alternatives quickly revealed that neither the manual rolling process nor the new design of two sheet-metal shells with a riveted joint were cost-efficient for the network as a whole (Figure 5.30). Both cases predicted high expenditure for the initial development work and ongoing management. Low one-off costs were only expected from using the threaded joint, as older product variants were available for this design. The company could also leverage the experience it already had with this technology.

* * *

The ability to adapt value-added activities to local conditions can give companies a significant competitive advantage, particularly if demand characteristics are not global and the share of R&D costs is moderate or low. Successfully adapting to local conditions means assessing, developing, and implementing modified production processes and product variants. Companies can use a portfolio approach and staged analysis to determine their ideal set of manufacturing technologies. Smart management of this global technology portfolio and the entire range of variants is an essential competence.

Adjusting production technology to local conditions can be the basis for further improvements. It can lead to specialization of a location in a certain type of technology, enabling a company to broaden its capabilities and operate in more geographical markets and market segments.

A well-managed technology portfolio allows companies to leverage the specific advantages of global locations, from low wages to knowledge. It is an important enabler for reaping the benefits of a global production network.

Various more labor-intensive manufacturing methods are possible for low-cost locations, taking into account variable costs

Fig. 5.29: Economically viable production alternatives for low-cost locations

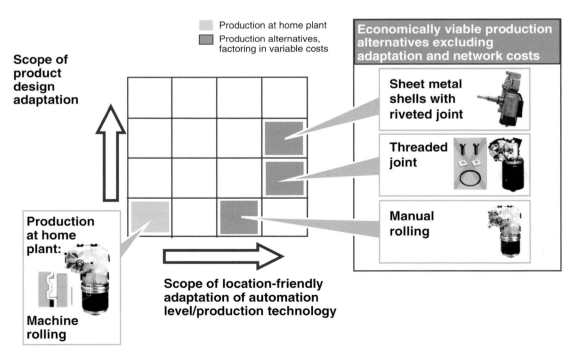

Source: McKinsey/PTW (ProNet analysis)

Both one-off expenses and recurrent additional expenditure reduce the number of economically viable alternatives from the viewpoint of the company overall

Fig. 5.30: Economically viable production alternatives – overall evaluation
Percent of the measure's initial NPV (i.e., without taking into account one-off and additional expenses)

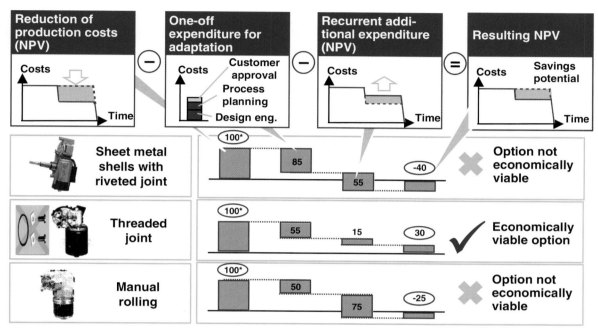

* Indexed

Source: McKinsey/PTW (ProNet analysis)

Minimizing One-Off Expenditure: An Automotive Supplier Uses Older Machinery and Plant Concepts

An automotive component supplier decided to locate the assembly of its products – including a fuel filter – in the immediate vicinity of a large customer's plant abroad (Figure 5.31). Factor costs at the new location were very different, so management had to decide whether or not to redesign products and manufacturing processes.

For the assembly of the fuel filter, a fully automated rotary system as used at the high-wage location did not appear to make sense. Unit volumes at the new production facility were too low to jus-

tify the investment. An older production concept was reviewed instead – an earlier configuration from the home factory that used two manual work stations. This choice reduced the initial investments for the new location from EUR 210,000 to EUR 14,000 (Figure 5.32). Investments could not be reduced any further due to the test devices needed.

With a maximum production volume of 400 units per shift and work station, this production concept allowed better scalability, making risks at the foreign site manageable. The process was already familiar; it did not need extensive development work, nor were high investments required for equipment.

Fig. 5.31: Fuel filter – structure and operation

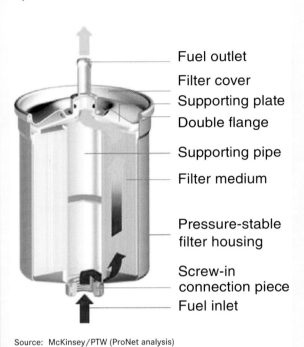

- Fuel outlet
- Filter cover
- Supporting plate
- Double flange
- Supporting pipe
- Filter medium
- Pressure-stable filter housing
- Screw-in connection piece
- Fuel inlet

Source: McKinsey/PTW (ProNet analysis)

When local market growth called for further investment after two years, the manufacturer had gathered enough local experience to venture into the next stage of development. Management decided not to duplicate the existing plant concept because the location now had more highly skilled employees, and wage costs had increased. Instead, the company slightly increased levels of automation in individual manufacturing steps. The entire production sequence was distributed across a total of six stations, with more highly specialized employees for the sophisticated operations. As with the first assembly layout, the company based this scheme on earlier concepts from the home plant, keeping the one-time costs for development and customer approval to a minimum.

Bottom line: One-off expenditure for adapting production facilities can often be minimized by resorting to previous techniques from the home factory. Investments can be reduced even further if machinery and plant are still available.

Loosely combined manual workstations

Fig. 5.32: Comparison of different assembly systems

	Fully automated rotation system	Manual workstations using partly automated manufacturing steps	Manual workstations using entirely manual manufacturing steps
Investment volume	EUR 210,000	EUR 60,000	EUR 14,000
Max. unit volume per shift	6,500 units	2,000 units	400 units
Max. no. of manufacturable variants	4 variants	10 variants	12 variants*

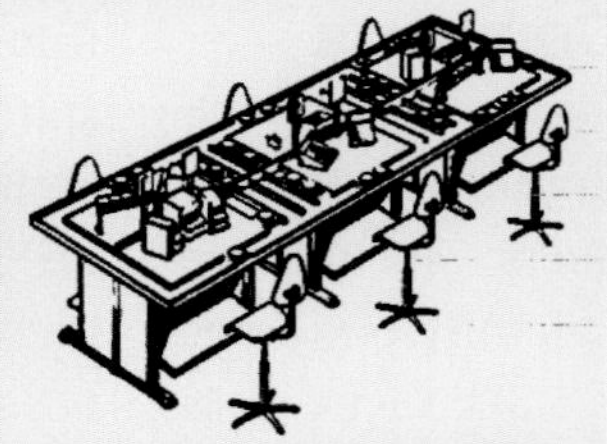

* Number of variants possible in principle; only 2 variants were being manufactured at the time the production site was set up

Source: Dubbel (1994), McKinsey/PTW (ProNet analysis)

Further reading

Dubbel, H., W. Beitz, and K.-H. Küttner (eds.): *Handbook of Mechanical Engineering.* London: Springer, 1994.

Bhattacharya, A. and S. Nandagaonkar: "Hidden competitive advantage" in *The Economic Times*, September 26, 2006.

Dorf, R. C. and A. Kusiak (eds.): *Handbook of Design, Manufacturing and Automation.* New York: Wiley, 1994.

Fritz, A. H. and G. Schulze (eds.). *Fertigungstechnik* [Manufacturing Technology]. 6th Edition, Association of German Engineers (VDI). Berlin: Springer, 2004.

Kalpakjian, S. and S. Schmid: *Manufacturing, Engineering & Technology.* 5th Edition. Upper Saddle River, NJ: Prentice Hall, 2005.

Sebastian Simon, Marina Dervisopoulos, Frank Jacob, Ulrich Näher

6 Implementation: Ramping Up New Facilities for Top Performance

Summary

Manufacturers with international experience manage to establish new sites abroad much faster and more cost-efficiently than companies with a relatively limited global footprint. Analysis of our interviewees' strategies and practices revealed a number of clear success factors as well as traps to avoid.

Many relocation attempts fail because firms overestimate their capabilities. Companies need to carefully align the prerequisites and complexity of the undertaking with their skills and resources. If a mismatch is apparent, they should either reduce the project complexity or provide additional experienced personnel.

The actual site chosen can heavily impact the project's economic viability, as location-related factors may differ significantly within regions. Local partners' contacts with public authorities, customers, and suppliers can also make a huge difference when setting up in developing countries.

Posting expatriates to a new location is usually much more cost-intensive than using local skilled workers and executives. Their know-how and companywide connections are indispensable, however, especially in the initial phase. The role of human resources management (HR) is vital in dovetailing foreign postings with swift skill building for locals to take over.

Companies can achieve rapid ramp-up of their targeted capacity and quality by applying best practices that are readily transferable from one industry sector to another. Global leaders plan relocation with painstaking foresight to ensure high delivery reliability and capacity utilization even during the move. A phased start-up with the sequential introduction of uncoupled manufacturing processes, new suppliers, and products helps to stabilize production. This staggered approach also limits the risk of downtime from technical faults.

6.1 Improvement Potential Revealed by the Survey

The key to a successful production network is perfect planning. Existing locations have to be downsized or expanded, and new ones opened. Although the company will have roughly defined the target countries or regions for new locations in the network design phase, details of what the location should be like and how it will be put into operation require further extensive planning. It is the planning and preparation phases that lay the foundations for later success (or failure). They determine the time and resources needed to set up the site and make it operational.

A comparison of start-ups at more than 30 new locations shows that the time and cost involved span a considerable bandwidth. One-time costs vary by up to 50 percent. Expenditure fluctuates particularly in the fields of quality and personnel. Differences are also similar for time spent, with the most striking variations apparent in the length of the ramp-up phase (Figure 6.1).

The winners save 50 percent of the one-time costs and set up their new locations in half the time required by average players

An analysis of the huge differences based on the sectors our survey targeted – automotive supply, mechanical engineering, and the electronics industry – unexpectedly reveals no significant disparities between the mean values for the three, but a wide gap between individual companies in the same business (Figure 6.2). A close examination of the reasons for such major discrepancies reveals several key factors closely linked to a company's specific situation:

■ The **complexity** of a project inevitably has a major influence on one-time costs and the time factor. The location, manufacturing processes, and product range dictate requirements. The acquisition of an existing facility or relatively simple manufacturing processes that require little training speed up project completion.

Key questions, Chapter 6

■ What factors determine the success or failure of a new site?

■ By how much can a company typically reduce the time and effort required to set up a new location?

■ What is the best way to integrate expatriates and local employees?

■ How can companies select the best start-up strategy, ramping up capacity with the maximum possible speed?

■ **Operational excellence:** Rigorous project management, including the swift recruitment of local staff, likewise reduces the cost, time, and effort involved in establishing a new location. Companies with a long history of globalization have a broader spectrum of experience and greater skills when it comes to operational implementation. Their new locations have a particularly high success rate due to their effective site selection, superior personnel policy, and high product quality even during the start-up phase.

Besides the differences in expenditure and duration, product quality variations are also apparent. In a re-

cent survey, over 40 percent of companies that had relocated production reported an unscheduled drop in quality. Some 12 percent even mentioned significant unscheduled downtime.

The ProNet survey's participants from global manufacturers emphasized three crucial success factors. First, it is important to plan every new location abroad carefully and systematically. Another crucial element is excellent HR management and skill building (from production workers through to executives) at the new location. The third is optimal design of the ramp-up phase.

Significant differences in the cost and time required to establish new sites abroad

Fig. 6.1: Cost and duration of new site setup

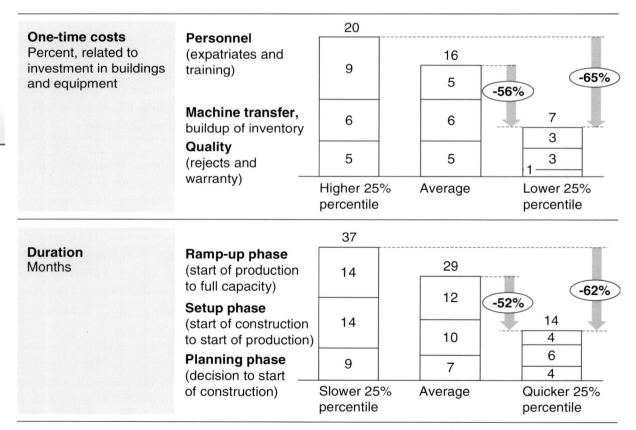

One-time costs
Percent, related to investment in buildings and equipment

Personnel (expatriates and training)

Machine transfer, buildup of inventory

Quality (rejects and warranty)

Duration
Months

Ramp-up phase (start of production to full capacity)

Setup phase (start of construction to start of production)

Planning phase (decision to start of construction)

Source: McKinsey/PTW (ProNet analysis)

6.2 Detailed Regional Planning

Setting up a new location without sufficient preparation can prove expensive, as the first case example shows. What could this European mechanical engineering firm have done to reduce risk? How could it have prepared better?

To set up any new location abroad, it is vital to conduct a precise analysis of local conditions, cross-referencing them with the company's own skills and plans. What degree of complexity can the new location handle? What skills are available locally, and which can be contributed by other locations? Are there any alternatives to going it alone? Would local partnerships be helpful?

6.2.1 Balancing Complexity Versus Capabilities

Successful models from a high-wage location cannot simply be transferred to a new plant abroad. A common mistake is excessive complexity – processes are too complicated, vertical integration is too high, and the product range too wide. But complexity alone is not responsible for success or failure. The company's skills and experience determine the extent to which it can handle complexity. Companies need to take a critical look at their own capabilities compared to the project requirements. What skills can existing employees contribute if they are posted to the new location? What experience has the company already accumulated with similar start-ups?

Major differences exist within industries

Fig. 6.2: Examples/factors impacting the setup of new sites

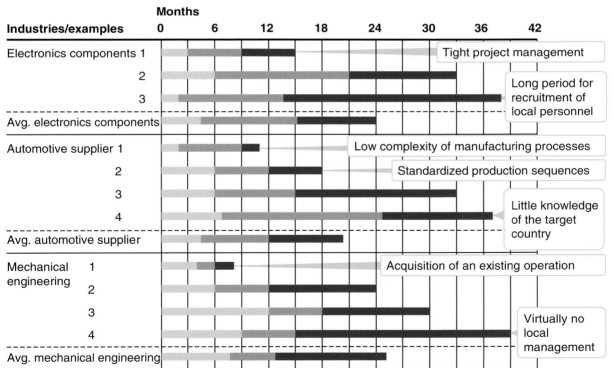

Source: McKinsey/PTW (ProNet analysis)

Pulling Out All the Stops – Too Far: A Mechanical Engineering Company Overreaches Itself in Asia

A few years ago, a European mechanical engineering group founded a production and sales facility in Asia. It planned to manufacture its entire range of expensive, high-quality machines for the Asian market with around 1,000 staff, and to generate annual revenues of around EUR 150 million. The company put up a production hall on a 20-hectare plot of land, with a floor space of eight hectares and a ten-story administrative building. Total investments ran to approximately EUR 200 million. The mechanical engineering manufacturer more or less copy-pasted its manufacturing processes and organization structure from Europe.

Management consisted predominantly of expatriates without any experience in Asia. In the first few months, it failed to stabilize manufacturing processes because local workers were not trained to European standards. A decision was therefore taken to fly some 30 skilled workers and foremen in from Europe for several months. The company set up a training center to provide appropriate skill-building. Around 15 interpreters were hired to liaise between the German experts and local staff. Once the first generation of local skilled workers had been trained, neighboring companies discovered their qualities and poached them.

In contrast to the attractiveness of the newly trained staff on the local labor market, the demand for the high-quality, expensive machines fell far short of expectations. The land and buildings were much too large, but the lease had been signed for 20 years. The factory halls remained mostly empty or were used for storage and three floors were unoccupied in the administrative building. Even ten years after it had been founded, the site was posting high losses and achieving barely 25 percent of its intended capacity.

Bottom line: Companies should not undertake too much at once when establishing new locations, building up more complexity than they can control. They need to adopt a step-by-step approach or make sure they have sufficient management capacity first.

It is difficult to assess whether complexity and skills are properly balanced using just a qualitative description of the complexity drivers and capabilities. A quantitative evaluation supplies a better basis, perhaps in the form of a scoring model (Figure 6.3).

The key is to align complexity and skills

For the first case example, this quantitative assessment reveals a clear discrepancy between complexity and company skills, even though there is a certain scope for subjectivity in the individual dimensions. The assessment system helps to both diagnose and solve problems. It systematically identifies weaknesses in the setup plan and approaches to reduce complexity.

One thing is very apparent in the case example: the new plant places high demands on both the location and product dimensions, with numerous complexity drivers. Could the company have initially started with a limited product range? Would it have helped to take along established suppliers? Would a different location nearby with a richer supply of qualified staff have been a wiser choice? The dimensions of complexity control can be examined similarly by asking targeted questions. Would partnership with a local company have been smarter? Could the foreign company have assembled a core team of competent local managers early on? The company would have been much better placed if it had considered these issues earlier.

Three different scenarios are conceivable if you consider both complexity and skills. Each requires a different approach:

1. Overcomplexity: The level of complexity does not appear manageable with the skills available. This suggests streamlining the complexity using a targeted approach. In these circumstances, companies should only use tried-and-tested manufacturing processes, retain familiar suppliers, and engage more external support. They also need to focus on appointing managers with local experience.

2. Balance: This is a scenario where the manufacturer has the skills to manage the complexity. Is it perhaps worth replacing low-value complexity drivers by some with higher value? Would a wider product portfolio with little local vertical integration be preferable?

3. Surplus skills: Skills exceed complexity requirements in this constellation. Could the company leverage its potential better, choosing a more demanding location further inland that has lower wage rates or using some of its management capacity for other projects?

6.2.2 Selecting the Right Site

The target country or region is decided when the production network design is finalized (see Chapter 4). However, to decide on the exact location, further tests are needed to see how to best leverage the great **differences in investment-relevant criteria between local sites**. These criteria can be divided up into three categories: human resources, costs, and logistics (cf. Table 6.1).

Failures can be prevented

Fig. 6.3: Balancing complexity and skills EXAMPLE

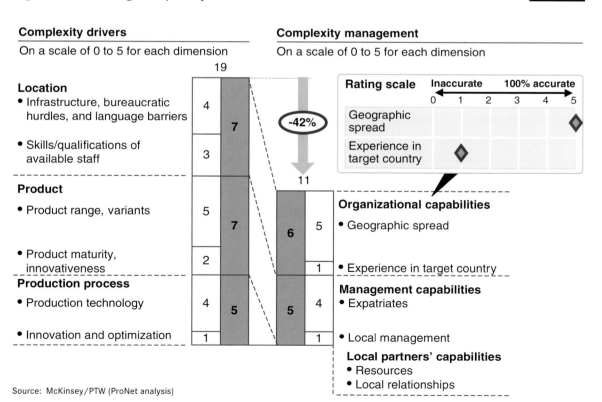

Source: McKinsey/PTW (ProNet analysis)

Tab. 6.1: Criteria for choice of location

Field	Subcategory	Selection criteria (examples)
Human resources	Local staff	■ Availability, education, wage costs, working hours
	Expatriates	■ Attractiveness of the new site (leisure facilities, comfort, children's education), distance from airports, accommodation/domestic staff
Costs	Land	■ Cost of land, development costs, duration of development, expansion options
	Neighboring facilities	■ Shared use of facilities (e.g., sewage treatment, power generation), local service companies (agencies for temporary staff, catering companies, etc.)
	State support	■ Investment grants, tax exemptions, bank guarantees (often negotiable)
	Customs duties, taxes, and other fiscal charges	■ Local customs duties, tax rates, other fiscal charges
	Utilities	■ Quality of local utilities (electricity, water, gas, etc.), regulations for waste disposal/recycling and emissions
Logistics	Outbound logistics	■ Proximity to customers (production facilities, R&D locations, purchasing offices)
	Inbound logistics	■ Proximity to suppliers (production facilities, R&D locations, sales offices)
	Infrastructure	■ Road network, rail network, waterways, distance from airports

In the event of major wage-level differences within the country, often amounting to a factor of up to two or three, personnel costs prove a major lever. Other cost levers are membership of special economic or customs areas and local government support, such as subsidies or local tax benefits.

Investment conditions vary within the target country or region

Negotiating with local public authorities is advisable especially for projects with significant investment volume, highly prestigious high-tech products, or a strong positive impact on the local labor market. This may enable companies to obtain financial incentives such as tax exemptions, investment and training subsidies, and cheaper land deals, which can be equivalent to between 10 and 40 percent of the investment. Companies should keep multiple options open for as long as possible and play local sites off against one another. A microchip manufacturer ne-

gotiating several location options in parallel in an Asian country secured substantial state funding by taking this approach right up until its final decision.

A systematic selection process should precede the final decision on location (Figure 6.4). The group of potential locations is drawn up using the criteria described and then shortlisted to three to five. These locations are given a detailed appraisal, including a precise analysis of economic viability and key qualitative factors.

In the next case example, an automotive supplier reaches a location decision within just four months using this systematic selection process. The company quickly reduced the list of potential locations in South Korea based on defined selection criteria.

An argument frequently cited against such a short selection process is lack of information about conditions in the target area. True. Accelerating the process

Bargaining Power: Raising the Stakes in Back-End Chip Production

A global chip manufacturer wished to increase its capacity with an additional site for back-end production, i.e., chip wiring and testing. Compared with front-end production (the manufacture of chips on silicon wafers), this stage of the process is usually less sophisticated and more labor-intensive.

Once the choice of location had been narrowed down to a country in Asia (because of its proximity to existing locations and favorable factor costs), three sites were shortlisted. The executive board negotiated with representatives of all three business parks in parallel, keeping the final decision open to the last possible moment. They took advantage of competition between the locations, achieving extensive concessions with direct subsidies and tax exemptions that amounted to around 30 percent of the investment total.

Bottom line: Companies – especially in industries highly attractive to the target country – should negotiate investment terms with several locations in parallel to optimize local support.

The supplier rapidly narrows down the number of possible alternatives

Fig. 6.4: Site selection of an automotive supplier in South Korea

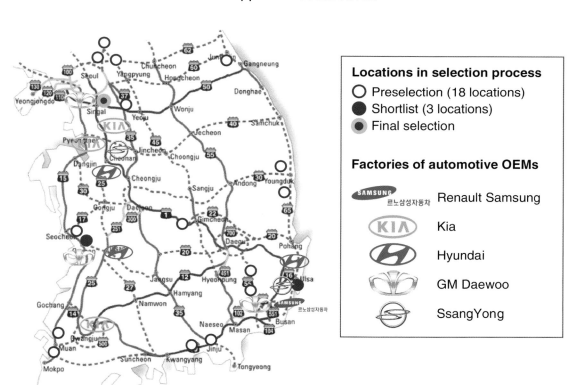

Accelerating Location Planning: Best-Practice Site Selection in South Korea

It usually takes six to eight months to decide on a specific location in South Korea. One of the world's largest automotive suppliers managed to reduce this to just four months by installing a tightly monitored system:

Preselection: A local agency was commissioned to compile a list of suitable plots. The knockout criteria simply required the location to be developed (with freeway access and utilities, in particular) and situated in an industrial zone. This process yielded a list of 18 sites throughout the country that were basically suitable. Even though this was still a very large figure, at least it was a manageable number of alternatives.

Short list: In the next step, attention focused on the distance from automotive manufacturers' factories and land prices. The list was reduced to three sites (Figure 6.5). Short distances were not just important for supply reasons – geographic proximity to as many plants as possible belonging to the five major Korean producers would also cut travel time and costs.

Detailed evaluation and final selection: Attractiveness to expatriates (international school, quality of housing) and international transport links (main roads, distance from Seoul airport) were of prime importance in the final selection. Senior executives visited all the shortlisted locations before the final decision, to gain their own personal impression of the sites.

Bottom line: In countries where companies are unfamiliar with local conditions, a systematic selection process with the support of local agencies can significantly speed up location planning.

A systematic selection process ensures the best location

Fig. 6.5: Approach for choosing a site in the target region

■ Focus of this chapter

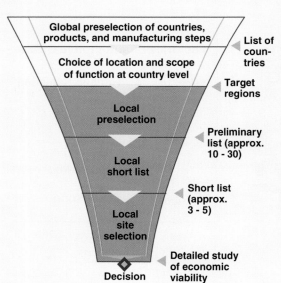

Success factors

- Ensure that requirements are adequate
- Decide on consistent targets and preferences

- Evaluate network efficiency quantitatively
- Consider interdependencies between locations, products, manufacturing steps

- Apply appropriate knockout criteria and minimum requirements
- Possibly use local agencies, lawyers, and auditors

- Contact government bodies if state support is possible
- Review actual conditions on site (e.g., possibility of expansion)

- Consider all relevant factors
- Perform separate monetary and qualitative evaluation
- Negotiate with several states/regions in parallel

Source: McKinsey

like this is only possible using the services of realtors or agencies. In many countries, other bodies are also available to advise companies on direct investments, whether government departments, embassies, industrial associations, or chambers of commerce. However, these sources of information are only sufficient for the preselection process.

For the short list and final selection procedures, it is essential for representatives of the company to perform a thorough analysis on site. This is the only way to ensure realistic assessment of a number of crucial questions, for example, in discussions with managers of local companies. Do the premises genuinely offer opportunities for later expansion? Are there leisure activities in the vicinity that will make the location attractive to expatriates staying for longer periods? Are there already multinational companies competing locally for the few highly qualified workers? How much progress has been made on freeway connections to the site?

6.2.3 Going it Alone Versus Partnering

A stand-alone operation tends to be the most common approach for new locations abroad. Nevertheless, it

may also be useful to involve **local partners** – in the form of a joint venture, an acquisition, or subcontracting. In some countries, it is actually a legal requirement to collaborate with a local partner in a joint venture. A partner in the locality can support and simplify the process in many different ways (see Table 6.2). This is particularly important for pioneering companies in newly industrialized countries. Local partners can contribute their knowledge of customers and supply markets, overcome bureaucratic hurdles, and help with procuring land and personnel.

According to the ProNet survey, companies entered into local partnerships or joint ventures for around 5 percent of new plants driven by cost motives and some 16 percent of new growth-driven sites (Figure 6.6). A similar number of companies opted for acquisition as their form of entry. However, joint ventures failed in 40 percent of cost-driven cases and 19 percent of those motivated by growth. Collaboration actually did more harm than good. Joint ventures appear to be better for opening up local markets than creating cost-efficient production facilities.

Partnerships have to be carefully selected and legally safeguarded to prevent any nasty surprises. Com-

Independent start-ups tend to be the most frequent and successful

Fig. 6.6: Frequency and success of foreign start-ups
Percent

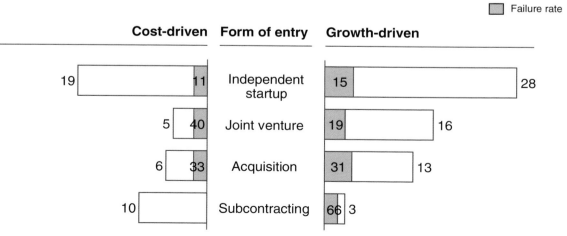

☐ Failure rate

	Cost-driven	Form of entry	Growth-driven
	19 — 11	Independent startup	15 — 28
	5 — 40	Joint venture	19 — 16
	6 — 33	Acquisition	31 — 13
	10	Subcontracting	66 — 3

Source: McKinsey/PTW (ProNet analysis)

panies should meticulously analyze their partner beforehand: What are their interests, relationships with local institutions, and how well is the company embedded in the local environment? It is essential to create a win-win situation if collaboration is to be successful. Otherwise, financial losses may be incurred, weakening the company's market position significantly. This is what happened to a mechanical engineering SME that lost most of its investment, while at the same time creating a new competitor (next case example).

There are two main reasons why joint ventures fail. The first is if the partners have divergent, unrealistic expectations. The local partner working closely with local government institutions may mainly be interested in seeing the industrial infrastructure grow and the transfer of technological know-how, with profits and revenues being only secondary. Local partners also often have far less contact with customers in the region than the foreign company expects. The second is when local partners do not keep to agreements. Breaches of contract by the partner,

such as the violation of patents or retention of profits, are a great danger in some countries. A lack of commitment is also a frequent issue: the partner may provide personnel with inadequate qualifications or little support for sales.

To minimize dependence on a local partner, especially where the legal system is not very reliable, companies should act with foresight when designing the **control** and **management** of the joint venture, and offer sustainable **collaboration incentives**:

Control: An equity share of over 50 percent secures complete transparency on all transactions and control over all business decisions. If a company is forced by law to have an equal share or less, it needs to at least ensure that there is an effective mechanism in place for the settlement of disputes, such as arbitration by a neutral third party or parent companies that can act as mediators. Even with a minority share, the joint venture can be controlled de facto via management interdependencies and veto rights.

Tab. 6.2: When to look for support from local partners

Field	Specific function	Indicators that support is needed	Countries where this especially applies
Markets	Sales	■ Small market share to date ■ Opening up market is top priority ■ Poor local payment patterns	■ Asia
	Sourcing	■ Little transparency in supply market	■ Asia, South America
	R&D	■ Needs of local customers are highly specific ■ Scope of own local R&D capacity is small	■ Asia, South America
Production	Operations	■ Frequent failure of utility supplies (e.g., gas, water, electricity) ■ Local partner has advantageous relationships with utilities companies	■ India, China
	Buildings	■ Bureaucratic hurdles ■ Difficulty finding reliable local construction firms	■ Eastern Europe, Asia
	Land	■ Restrictions on foreign companies buying land ■ Few plots available	■ Eastern Europe
	HR	■ No local recruitment agencies ■ Narrow labor market ■ Frequent labor disputes	■ China, South America

Management and leadership: Companies can ensure that control rights are exercised operationally by filling positions of influence with people who enjoy their trust and who are familiar with local conditions. The law of the host country may contain certain peculiarities that must be given special consideration. The CEO of a Russian company, for example, has unusually far-reaching powers (including sole power of representation) and should, if possible, be appointed by the company itself or at least monitored locally.

Collaboration incentives: Mutual benefits can be agreed that extend beyond the joint venture to safeguard the commitment of the partner in the joint venture. For example, a contract can be drawn up stipulating that each partner will incorporate the other party's (complementary) products in its domestic sales. A balance of interests is achieved if the services of each partner are difficult for the other to substitute.

Particularly in newly industrialized countries, regulations for joint ventures with foreign countries tend to change rapidly. A few of these countries, such as China, are relaxing their regulations as their markets open up. This can mean that local partnerships are no longer mandatory several years down the line, and the local partner can be bought out. Joint venture contracts should therefore include clear terms on terminating the deal and voluntary withdrawal of partners. (These agreements must, of course, conform to the law in the country of incorporation.)

6.3 HR Management

The primary task of human resources management when a new facility is set up is to provide suitable personnel at low cost. The aim must be to recruit as many local staff as possible with the requisite skills and know-how. However, people with the technical and management expertise so urgently needed are often not available at the new location. Developing and newly industrialized countries have a particular dearth of employees who can become acclimatized to the production and management processes of multinational companies fast and effectively.

Choose your Friends Carefully: How an SME was Exploited by its Joint Venture Partner

A mid-sized mechanical engineering company established a plant in China as a fifty-fifty JV with a local partner in the 1990s. The two parent companies each appointed a director. The foreign company's director had no experience whatsoever of business practices in China. The Chinese partner was very quick to supply staff, but they all needed training. The partner also helped plan the manufacturing processes without actually contributing any know-how worth mentioning.

Within a matter of weeks after production had started, all the local employees handed in their notice and switched to a neighboring location belonging to the Chinese partner in the joint venture.

They had procured exactly the same machinery and plant without the knowledge of the foreign company. Production was able to start immediately with these now semi-skilled employees – without involvement of the foreign firm.

The foreign company failed in its attempt to assert its claims against the joint venture partner through the courts. It paid dearly for its lack of local experience: it had placed too much trust in its partner and had no effective means of applying pressure. What is more, it had failed to look for a partner without any conflict of interest.

Bottom line: Joint venture failures could be prevented by deploying an assertive managing director with appropriate intercultural skills, and giving the local partner an incentive to collaborate over the long term (by only gradually disclosing technical details).

As a result, companies send a higher share of skilled workers and executives from home to the new site during the ramp-up phase. These are usually employees with relatively long tenure. They can be transferred to the new location abroad either from their home country (expatriates) or from somewhere else. How can a company achieve a favorable ratio of staff assigned from existing locations to spending on local personnel to ensure sound, rapid, and cost-efficient skill-building?

The results of the ProNet study show that successful companies spend much more on training local employees than less successful ones do (measured by defective units in the ramp-up phase) (Figure 6.7). Spending on expatriates and training is balanced equally for less successful new sites, while for successful ones more than twice as much money goes into training locals as into expatriates.

Training, or training on the job, is just as important for skilled workers and executives as it is for operational staff. Staff retention measures are required to ensure that expensively trained employees actually use their newly acquired knowledge and skills in the service of the company. The following sections outline the key aspects of HR management for a location abroad.

6.3.1 Filling Skilled and Executive Positions

Different options exist for companies to obtain the skilled staff and executives they need when setting up a new facility. They can send employees from the company's corporate center, from other parts of the business/other countries, or hire local staff. The right mix for a new location has to be decided on a case-by-case basis. A company needs to consider fundamental strategic staffing issues, as well as the suitability of available employees for specific tasks. If expatriates are used, it is particularly important to choose the right people and prepare them sufficiently for their jobs. Companies usually try (largely for cost reasons) to meet their requirements for skilled staff and executives at foreign plants with local employees, in the longer term at least. They are

increasingly developing innovative assignment and management concepts to achieve this.

6.3.1.1 Fundamental Staffing Strategies

The decision whether to use expatriates or local skilled workers and executives depends particularly on the location's role in the production network (e.g., whether it is supplying parts to other plants or to the local market) and the requirements of the individual positions. The position of sales manager is more suited to a local employee, for instance, since it demands a lot of interaction with local customers. An expatriate may well be more suitable for the position of controller if close interaction with head office is required. The cost of filling skilled and executive positions also needs taking into account, as does the balance of skills on the entire management team at the site abroad. There are three possible staffing strategies (Figure 6.8).

Successful companies prioritize the training of local staff

Fig. 6.7: Additional staff expenditure for a site start-up abroad
Percent

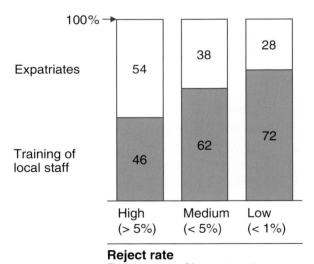

Reject rate
Percentage of investment

Source: McKinsey/PTW (ProNet analysis)

Expatriate-oriented strategy. Using a high proportion of expatriates allows transfer of the parent company's proven approaches and corporate culture and facilitates contact with the company's other locations. It is often vital to assign expatriates with a deep insight into company-specific products, equipment, and management processes to make up for any gaps in the skills of managers recruited locally.

The use of expatriates also has its drawbacks. One issue is high expatriate compensation, which is inevitably a burden to the new sites. Local skilled staff and executives tend to be more cost-efficient, especially at low-wage locations. This varies significantly, however, depending on the function and country concerned. It is therefore highly advisable to perform a detailed analysis of the local market for skilled staff and executives.

Difficulties can also arise when expatriates try to integrate into their new work context and cultural surroundings. They may find their efficiency declines or wish to return home early. Poor knowledge of the local language and national peculiarities can make it difficult for foreign managers to interact with local staff and their environment. Their effectiveness in making staffing decisions and negotiating contracts can also suffer.

Often it is not essential to send skilled staff and executives abroad for long periods of time. This is especially true if the new location is supposed to be run largely independently, being supplied by the local market. In this case, regular visits to the branch abroad by skilled workers and executives from the corporate center and other factories will generally be sufficient. Besides defining targets and review-

---

Companies can adopt three basic strategies for staffing skilled and executive positions

Fig. 6.8: Approaches to staffing new locations abroad

E Expatriate
L Local manager
O Other (manager from a different country)

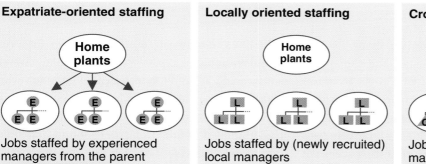

Expatriate-oriented staffing	Locally oriented staffing	Cross-country staffing
Jobs staffed by experienced managers from the parent company (expatriates)	Jobs staffed by (newly recruited) local managers	Jobs staffed by experienced managers from a branch in another country

Typical applications

• When initially ramping up production	• For simple or locally adapted manufacturing processes and products	• When taking over products and manufacturing processes from locations in other countries
• When taking over manufacture of products or manufacturing processes from home factories	• When production is already up and running	• When using the location to supply other countries
• When using the location to supply other markets and factories (parts manufacture)	• When using the location to supply the local market	• At multinationals
		• When suitable staff are available for posting internationally

Source: McKinsey/PTW (ProNet analysis)

ing performance, these visits also serve as a means to transfer company-wide best practices and introduce new manufacturing processes.

However, if the aim is to integrate the foreign site more closely into the international production network, it is usually necessary for skilled workers and executives from other plants to stay there long-term. This particularly applies if the site uses complex manufacturing processes and needs to coordinate closely with central R&D.

Locally oriented staffing. Employing managers with local roots makes it easier in the initial phase for the company to interact with public authorities, suppliers, and service providers on site and to communicate with staff. If the company is working together with local companies (in a JV, for example), its business partners can often contribute these abilities.

The use of local skilled staff and executives is preferable for more than just cost reasons. It often has a positive effect on the recruitment of local employees, not least because candidates expect a management style they are more familiar with and better career opportunities.

Cross-country placement. There are many good arguments for assigning skilled staff and executives with international experience who come from another country altogether, i.e., who are neither local nor from the company's home location. The transfer of staff between low-wage locations is relatively cost-efficient. Employees from other countries are often more suitably qualified than expatriates from the home factory, and in multinational companies these staff also tend to be more readily available.

Cross-country staffing with skilled workers and executives from an international pool of employees makes a particularly good human resource development tool, enabling companies to systematically build up an international executive team and promote a uniform corporate culture across all locations. Sometimes, however, problems can occur with the acceptance of managers from different countries –

cultural sensitivity should always be the prime consideration.

In general, managers from other international locations with a broad spectrum of experience are in very high demand. The concomitant risk is that these staff are all the more likely to be poached by competitors at the new site abroad.

6.3.1.2 Criteria for Filling Higher-Level Positions

Decisions on when to take expatriates, locals, or staff from other international locations for more highly qualified positions have to be tailored to each new move. It is useful to apply a set of criteria to systematically evaluate the suitability of the three groups (Figure 6.9).

1. Personnel costs – particularly the cost of skilled workers and executives – have a major impact on the economic viability of new locations. The costs involved in assigning an expatriate are often between two and ten times higher than those for employing a local worker. The number of expatriates and managers from other locations should therefore be as low as possible, and their stay as short as is feasible.

Guidelines cited by successful companies for a reasonable number of expats are approximately five executives and around ten technical experts for every 1,000 employees at the new location. As the size of the location increased, the proportion of expatriates needed tended to fall.[1] Companies found that sending out executives for between two and three years and technical experts for around three months produced the best results. In industrialized countries and regions, such as the United States, Western Europe, or Japan, the number of expatriates required is somewhat lower.

The high costs for expatriates and managers from other countries are due to differences in basic salaries, foreign service allowances as compensation for political risks and climate differences/social pres-

---

[1] *Cf. results of the ProNet survey.*

sures, possibly purchasing power compensation to cover additional costs, an education allowance (e.g., school fees), a relocation allowance, and a separation allowance (e.g., regular flights home).

2. The high demands on **general qualifications** set by international companies drastically limit the number of competent skilled workers and executives available, especially in developing and newly industrialized countries. To reap the benefits of better insight into local customer requirements and business practices, companies have to accept costly staff development initiatives (such as training in other factories) long before production starts.

Companies wishing to enjoy these advantages right from the start usually have only one option: to assign a large number of expatriates and managers from other locations to the new site. This start-up team must gradually be replaced by local skilled workers and executives. As the ProNet study showed, it makes little sense to send in a large team of managers and engineers from elsewhere for five or more years. The company will lose some of the cost benefits of its low-wage location, and the new site will never get up and running independently. The continuing presence of expatriates blocks career paths for local employees and prevents the location from building up its own technical competences.

3. How much **know-how** needs **transferring** to the new location is a major factor in the decision whether or not to deploy local managers. Extensive knowledge transfer is required particularly in the

---

Six criteria should be factored in when staffing executive positions

Fig. 6.9: Evaluating the assignment of local managers versus expatriates

Criteria	Advantages of local managers	Advantages of expats/managers from other countries
1 Personnel costs	• No foreign service pay or reimbursement of costs for trips home, etc. • Low salary level in low-wage countries	• With managers from other countries: potentially lower salary level
2 General skills and qualifications	• Better knowledge of local customer requirements and business practices	• Skills and qualifications conform to standards and mgmt techniques in the home country • Better availability of qualified and experienced employees
3 Product and manufacturing know-how	• Better knowledge of manufacturing processes suited to local circumstances (training, supplier structure, etc.)	• Knowledge of existing processes (important in ramp-up phase and when taking over production of products from other plants) • Knowledge of company-wide best practices
4 Integration and communication	• Retention/development of local specifics, and supply of information about local requirements to company HQ/home factory	• More efficient communication with corporate HQ and home factory • Better contact with central departments
5 Management and control	• Stronger tendency to develop independence	• More efficient involvement in centralized decision-making processes • Typically high loyalty level because intend to return home
6 Personnel development	• Creation of a local management team	• Targeted further development of most suitable staff possible

Source: McKinsey/PTW (ProNet analysis)

ramp-up phase, and is likely to be an ongoing process in the case of high-tech manufacturing techniques. The transfer relates to product features, equipment operation and maintenance, and planning and control systems, as well as company-specific management processes. The quickest method is to send in experienced staff from existing factories.

4. Where the managers come from plays a crucial part in determining the direction, efficiency, and intensity of **communication**. Filling executive positions in production and distribution with local managers promotes interaction with employees and customers. However, in the interests of smooth coordination with the corporate center and the home factories – particularly in the initial stages – it is advisable to staff controlling and works management positions with expatriates.

5. Assigning expatriates to key positions achieves closer central **control and direction setting** (especially in production planning and control). This enables the corporate center and parent company to gain greater transparency on the activities of the foreign branch, avoiding any risks and disadvantages,

most notably in countries where informal business relationships are common and corruption is rife. However, strong central control entails greater organizational complexity. Having an excessively high proportion of expatriates in the management team of a location abroad also usually impedes the location's ability to tap into the local market.

6. The decision on how to fill skilled and executive positions is an important **personnel development tool**. Staff within an international team can enhance their own skills and help define the company's image. Posting employees to foreign production locations contributes towards greater intercultural competence and employee experience. Filling executive positions with local employees right from the start, on the other hand, strengthens local skills, means you can scale down on expatriates faster, and has a positive impact on the company's image in the local labor market.

The following example highlights the careful analysis an international automotive supplier puts into assembling its management team for a foreign location.

Revving Up for Excellence: Global Executive Staffing at an International Automotive Supplier

An international automotive supplier was expanding its activities to a new location. Previously this location had mainly served as a distribution office; it also had a small production department. However, following a decision to relocate production of a series of products for the local market to this location, the company had to decide how to fill the management positions.

Local factory leadership always consisted of a technical and an administrative manager. One of the positions is usually held by an expert, the other by a local manager. Given that a major upcoming in-

vestment would require increased alignment with the home factory to realize synergies, and new products would need intensive coordination with corporate R&D, the supplier decided to make the **technical manager** an expatriate (Figure 6.10).

The position of **administrative manager** was given to a local manager since the site had been used primarily as a distribution office up to then, with very little actual production. The company had already implemented a reliable accounting function for foreign locations and completed the installation of SAP at the facility, so there were no great changes needed in this function.

The products to be relocated had high quality assurance requirements. When a sample run involving comparable products revealed significant quality defects, the company decided to put an expatriate **in charge of quality assurance**, at least for the initial two to three years. This was in keeping with the company's normal approach: Quality assurance is always managed by an expatriate if the quality is unsatisfactory and cannot be improved by deploying skilled personnel for a brief period (from several days to a few weeks).

For the position of **production manager**, the company usually assigns an expatriate only while the products are being transferred from the home factories (one to two years). After that, the expatriate is replaced by a local manager. In this case, the new site already had staff with production experience, and the position of technical works manager was already staffed by an expatriate with knowledge of the products and company-specific manufacturing processes. This meant it was possible to assign the role of production manager to a local manager. The advantage was easier communication and coordination with the production staff.

As the (inbound and outbound) **logistics manager**, the company chose a manager from a branch in another country where logistics processes had recently been reorganized and improved. This manager transferred the knowledge he had acquired during this process to the new location. The position of **controlling manager** was initially staffed by an expatriate to support the local commercial manager. Given that the plant's primary role was to manufacture products for the local market, the position of **sales manager** was assigned to a local manager from the outset.

Bottom line: This company kicked off at their new plant with a balanced, synergistic management team. The search to staff these positions was uncompromisingly objective. Yet the ideal solution, once each profile had been established, was surprisingly close to hand.

Only selected executive positions were not filled by local employees

Fig. 6.10: Organization chart of the location abroad <u>EXAMPLE</u>

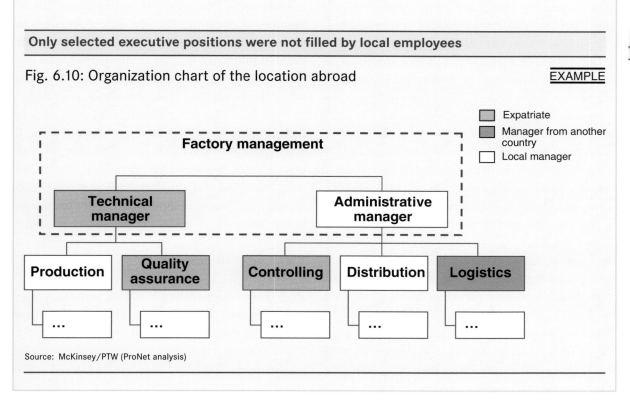

Source: McKinsey/PTW (ProNet analysis)

6.3.1.3 Selection of Suitable Staff for Work Abroad

The requirements of skilled workers and executives to be posted abroad exceed those for assignments in their native country. Alongside outstanding functional expertise, these executives need to have exceptional intercultural and social skills. Overall, an expatriate should fill the following profile:

Functional expertise: Setting up a new production location requires executives with broad technical expertise, especially if sophisticated or innovative technologies are being relocated. They also need basic knowledge of the main corporate functions, such as purchasing and quality management, because executives at the new location will not have the same degree of access to experts from functional departments as those in the larger home factories. Long tenure and allegiance to the corporate culture ensure that these executives will act in line with company strategy and values, whatever the circumstances.

Company network and relationship building: Executives must be able to actively build new relationships that smooth the transition to their new sphere of responsibilities. They also need suitable contacts with suppliers and the company's own functional departments that they can expand on in their new position.

Intercultural skills: The executive needs high-level intercultural competence and knowledge of both the corporate language and the language spoken at the new location. Vital, too, are one-to-one discussions with staff, local suppliers, and customers (only possible with appropriate language skills).

Personal situation: The executive's personality and family situation are further criteria to consider. They must be able to work as part of an international team, be flexible, fiercely independent, and have an appetite for new experiences. High stress tolerance is another important quality, as an assignment abroad is always a great strain. The employee's family situation can also add to this: moving the entire family

can somewhat impair an executive's performance (at least during the initial "teething problems"). This issue should be considered together with the employee (especially since these postings are ideally short-term rather than long-term anyway).

Companies can draw up a **job profile** for a specific position abroad by carefully weighting the requirements outlined above. They can identify specific strengths and weaknesses of potential candidates by comparing existing employee profiles (formal qualifications and proven skills) against the job profile. The results can then be used to initiate any training required to prepare the employees.

At smaller companies, staffing management positions at new locations abroad is often fraught with problems. Many employees appear unsuitable for a foreign posting or cannot be persuaded to go. The few that are appropriately qualified and motivated are often considered indispensable at home.

The difficulties of finding suitable employees for foreign postings must be solved or at least alleviated by implementing a **targeted recruitment and HR development** program. The problem of apparent indispensability needs addressing via **systematic succession planning**. Both these measures have a long lead time, which once again underlines the importance of a forward-thinking globalization strategy.

Experience shows that staff are more willing to spend time abroad if this is part of the overall career plan, and the subsequent reintegration is organized well. The secret is to transform the image of a foreign assignment from a disciplinary transfer to a career accelerator.

6.3.1.4 Expatriate Skill-Building for the Assignment

High-quality **preparation** is crucial to the success of any assignment abroad. Training programs should iron out any weaknesses identified during the selection phase. External institutions can provide language courses and information about the country. A collection of documented best practices for setting up locations abroad may also be very useful for the prospective managers. Meetings with expatriates in the region are also extremely helpful, as are regular exchanges between expatriates (maybe in an "expats blog").

Once someone has been selected for a specific position abroad, this should be followed immediately by a discussion with the candidate. The issues that need clarifying together with the relevant functional departments include: What are the objectives of the foreign posting – is it intended as a personnel development initiative, to transfer specific know-how, or for some other purpose? What is the intended time frame of the posting? What are the basic terms of the assignment (compensation arrangements, working and living conditions, school situation, health care system, etc.)? What are the arrangements for subsequent reintegration? It is also a good idea to encourage employees and their families (if accompanying them) to travel to the intended destination to familiarize themselves with the situation there – the standard of living, working conditions, surroundings in general, as well as to prepare for the move.

> **Preparing expatriates for the heavier functional and cultural demands they will face is a worthwhile investment**

An action plan for **acquiring location-specific knowledge** should be drawn up during the discussion. This will include specific knowledge of products and manufacturing processes, key customers and suppliers, organizational procedures, and national specifics. The employee must also be given access to information about the country and provided with the opportunity to exchange ideas with returning expatriates. Providing support at the location itself can help improve the expatriate's availability and efficiency and may therefore be cost-efficient. This applies both to business services and the supply of specialized information, such as country-specific peculiarities, addresses, and contacts.

Preparing for a posting abroad ideally starts around three months before departure. The strict time frame (covering the employee's family, too) includes func-

tional training requirements, learning the language, gathering information about the country, and any intercultural training that may be required (Figure 6.11).

6.3.1.5 Mixed Teams and Transition to Local Management

Companies are generally interested in assigning their foreign production sites to local managers. This usually means getting local skilled workers and executives up to the company's standards first. They have to acquire experience in the company and take part in training programs. Some companies address the transition plan at the same time as they prepare the location plan, developing specific concepts for immediate implementation.

An innovative concept for filling executive positions at foreign locations is tandem management. This concept, implemented successfully by Volkswagen in its joint venture with the Czech auto producer Skoda at the beginning of the 1990s, involves temporarily allocating positions to two people simultaneously – one an expatriate, the other a local manager.

Back then, Volkswagen found itself in a bind. On the one hand, it wanted to break down the socialist behavioral patterns in Skoda's organization and modernize manufacturing processes and business operations. However, it also wished to retain Skoda's expertise for the joint company. It vowed to retain and strengthen the Skoda brand with its high level of awareness and positive image in the reformed states of Eastern Europe.

The tandem management approach covered around 50 relevant positions, each filled by both one expatriate and one local manager for a period of two to three years. At first, it was mainly the managers sent by Volkswagen who had the authority to make decisions and who familiarized the Czech managers with modern management methods. Responsibility gradually shifted as the local managers progressed.

> **Temporary tandem management means local knowledge can be retained and know-how transferred from other factories, both at the same time**

A similar approach is practiced by the BOC Group, a provider of industrial gas solutions with a head office in Singapore and over 40,000 employees worldwide. The BOC Group sends in a substantial number of expatriates at the start of any foreign engagement. To ensure rapid transition to local management, the company has stipulated that all employees posted abroad should literally "make themselves superfluous" by passing on their knowledge within no more than three to six years following setup of the location. Part of their performance-based pay is linked to this objective.

6.3.2 Recruiting and Training Operational Staff

Employees can be recruited in highly developed countries and regions using standard HR management initiatives. It takes far more effort to attract suitable employees in developing and newly industrialized countries. The general **standard of education** is usually lower. Even more highly qualified candidates tend to be unfamiliar with advanced technologies and **management processes** in international companies. High **attrition** can also be a further issue.

An efficient HR function looks at the selection and training of staff from the viewpoint of the underlying production strategy. Does the company wish to train employees for complex manufacturing processes, de-

---

It takes around three months to prepare expatriates for their posting abroad

Fig. 6.11: Preparing staff for a posting abroad   <u>EXAMPLE</u>

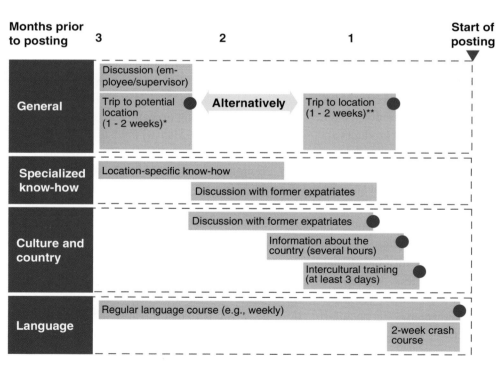

* As basis for the decision
** Following the decision, to make settling in easier

Source: McKinsey/PTW (ProNet analysis)

Fueling for Performance in China: Bosch Wuxi Uses Cross-Site Know-How Transfer to Ramp Up for Sophisticated Injection Valve Production

In the mid-1990s, Bosch Wuxi founded a new plant near Shanghai. The location was set up as a fifty-fifty joint venture with the Chinese company SAIC Holding, and employed around 175 people. The aim was to supply automotive OEMs in Asia and Europe with gasoline injection valves, pressure regulators, and common rails. Injection valves are among the most critical components of a combustion engine. They are subject to tough demands in terms of dimensional accuracy and are made in a highly automated production process. A successful, mature product with a tried-and-tested manufacturing process was chosen for the new production facility in China.

Training for the Chinese engineers began 18 months prior to the start of production. They were divided into several groups. Each was trained for three to five weeks at the lead plant in Bamberg, Germany, and at various equipment suppliers. The training dealt with the production machinery, production line organization, products, and manufacturing process. The new production line was first set up in Germany, in collaboration with the Chinese engineers. Samples were manufactured, and the line was preapproved. It was then transported to China. One of the main reasons why knowledge transfer was so fast is that dismantling and reassembly of the line was performed in mixed teams of German and Chinese production experts.

Around ten expatriates went to China, mainly from the lead plant in Bamberg. Some remained for a few weeks, others for up to five years, ensuring that production was ramped up smoothly. In 1998, two years after production had started, the quality of the Wuxi location achieved the same standard as that of the lead plant in Germany, and has remained at that level ever since.

Bottom line: When setting up complex production lines in newly emerging countries, it is worth stabilizing the manufacturing process beforehand and working in mixed teams to promote knowledge transfer.

veloping initiatives to minimize churn? Or to simplify production so that staff can deliver efficiency and high quality even if their training is only brief? Selection and hiring processes, skill-building activities, and retention programs will all depend on this decision.

6.3.2.1 Methods of Recruiting Local Employees

The approach to hiring personnel at a location abroad in newly industrialized countries is not that different from the procedure used at home. Companies use every possible form of enticement – from word of mouth to a targeted approach to applicants. Approaches vary depending on the target group.

People with academic qualifications such as technicians and engineers like to work for renowned companies – this applies anywhere across the globe, regardless of a country's economic development. It is therefore important that the company invests in reputation and image as much in emerging countries as in the highly industrialized world.

BASF and Motorola, for example, fund university research and award scholarships in China to **raise awareness**. Procter & Gamble hold regular company presentations at universities and have a recruitment coordinator at every college attractive to the company. In Japan, the pharmaceutical company Eli Lilly addresses specific target groups directly and manages to attract loyal and motivated employees by. For example, hiring more women.

Companies should exercise caution when **taking on staff from a partner**, such as in a joint venture, as a partner may not necessarily propose or assign employees who are the best fit for the positions to be filled.

When **advertising for staff**, companies must comply with the design requirements and legalities of the country concerned. In China, for example, companies sometimes need the approval of the labor office responsible before they can publish job advertisements. The content and layout of ads differ greatly from one country to the next.

6.3.2.2 Employment Strategies and Staff Retention

Recruiting and training local employees is expensive, so it is important to develop country-specific strategies to achieve long-term staff retention and lessen the impact of high attrition.

A **retention policy** usually leads to higher labor costs but can definitely be worthwhile in the long term. Anyone willing to pay higher wages will find it easier to hire and retain better educated employees. Hero Honda in India pays its skilled personnel wages around twice the local average for the industry. The company also offers extensive social security benefits and attractive opportunities for further training. A net result of this approach is very low churn. This keeps recruitment costs down and builds up staff experience, which, in turn, boosts productivity and quality (see box below).

Another way to achieve long-term staff loyalty is by implementing bonus schemes coupled to employee tenure. Companies in China, for example, have found factory residence schemes a great success. They make homes available at a discounted price to staff with a specific tenure. Training programs, foreign travel, and language courses also promote long-term

employee retention. Other approaches are long-term contracts and attractive career development opportunities. It is also sometimes possible to generate loyalty with surprisingly cost-efficient measures, such as competitions and awards.

If a company wishes to **retain its top performers**, it has to promote tailored development and market these schemes internally. It should be possible for any employee to climb the career ladder to a senior management position given the right performance. The automotive supplier TRW in Poland is a good example, holding management seminars run by the world-class business school INSTEAD.

> **Companies should make a conscious choice between long-term staff retention and replacement of turnover**

Long-term employee retention may justify additional spending when the goods manufactured are complex but not necessarily if they are unsophisticated. In the latter case, a **strategy of minimal labor costs** that accepts higher churn is often advisable (a replacement policy). This is commonplace with items such as garments and toys. Processes tend to be simple and staff can get up to speed quickly, making it more cost-efficient to manufacture with high attrition and lower wages. Productivity may not be very high, but this is offset by the lower labor costs for staff with limited skills.

This strategy makes it essential to divide up the work into increments to reduce its complexity and content, so new staff can be trained in a matter of minutes. Employees need not have any prior skills if the operating systems are structured accordingly. This type of employment strategy enables companies in China (and elsewhere) to achieve labor costs of around EUR 0.5 an hour. Costs are slightly higher in the economic hubs of Shenzhen and Shanghai, but even these cities have a plentiful supply of low-skilled workers for wages of no more than EUR 0.8 per hour.

Staffing for Success in Developing Countries: Hero Honda Achieves High Productivity via Low Attrition

The Hero Honda factory in Gurgaon, south of Delhi, had around 3,500 employees and a capacity of roughly one million motorcycles in 2004. The factory largely uses Honda's proven manufacturing processes.

Workers are required to have good know-how of the functionality and operations of the machinery but also to work relatively independently – only possible if staff turnover is low. The same applies to high quality and productivity. The cost would be considerable if Hero Honda were to design its manufacturing processes to suit a high churn, and some of Honda's technical experience would get lost.

Consequently, Hero Honda opted for a long-term employment policy. This strategy is not only consistent with the corporate cultures of the two partners, Hero and Honda: it is also in line with the specific conditions of the North Indian location. Here, despite the reform efforts of the national and local government, it continues to be virtually impossible to make adjustments to the workforce at short notice.

Hero Honda worked out its strategy for long-term employee retention in detail and implemented it with great success. In 2004, the undesired attrition was below one percent (excluding retirees). The costs of this long-term employment strategy are outweighed by its economic benefits. The initiatives that have resulted in this high staff loyalty include:

Financial incentives: Hero Honda rewards its workers with variable pay depending on factory productivity. When production was at capacity limits in 2004, this resulted in wages of around 25,000 rupees (roughly EUR 455 a month, which corresponds to nearly twice the normal average wage there for that industry. Although the variable nature of the salary leads to higher costs when the order book is well filled, it also protects Hero Honda against losses when there is a drop in sales. When the level of incoming orders is high, employee loyalty is exceptional because the salary is very attractive, while staff turnover rises slightly when the production volume drops (bringing a fall in wages), making it easier to cut back on staff.

Non-monetary incentives: The company inspires further loyalty via both practical (but non-financial) and symbolic incentives. The first category covers activities such as employee-centric design of the working environment and support of staff members' families. HR representatives are readily available at a stand near the production hall exit, the company organizes courses for workers' wives, etc.

Symbolic activities include planting a tree for every new employee, regular information on key indicators, leisure activities and company contests (such as Sanskrit competitions), with award ceremonies in front of the entire workforce. Management experience has shown that these low-cost symbolic gestures have perceptibly improved staff motivation.

Managers report another positive side effect of this HR policy: the staff have not become unionized. They believe the factory runs more smoothly and effectively as a result. Hero Honda achieves high productivity and quality standards due to its proactive personnel initiatives and variable compensation.

Bottom line: Additional spending on higher wages, social facilities, and events can be cost-effective if the knock-on results are reduced training costs due to lower staff turnover and greater experience, leading to higher productivity.

Overdrive on all Cylinders: A European Hydraulics Manufacturer Fails to Differentiate its Skill-Building

A company based in Germany planned to open a new plant in Europe to manufacture hydraulic cylinders. It opted to use equipment already successfully in use at the German location. Training on the job was provided for all new factory staff at the home location, beginning around five months prior to the launch.

The **production planner** came to the home factory for three months' training. The aim was to acquire company-specific know-how on planning processes, benchmarking methods, and visualization methods, as well as detailed knowledge about products and production processes. He also saw the application of quality management tools. A dedicated sponsor discussed his training progress with him once a fortnight and put him in touch with other staff within the company.

The new plant's **maintenance engineer** was given three months' training at the home site. During that period, he also visited the key machine tool manufacturers. His training objectives included learning about TQM[2] methods. The **equipment operator** came to the home factory for just two months to gain specific product knowledge, expertise in manufacturing processes, and visualization methodology. The home plant's production planning manager was his sponsor. The **machine operators** and assembly personnel – ten staff in all – were sent over to the home factory for four weeks' training on the job, during which they learnt about the products and production.

Skilled workers (from quality management and production planning) from the home factory supervised the production startup process for a month. It was a success but came at a high cost. At around EUR 180,000, the training costs (travel expenses, accommodation, personnel costs, etc.) corresponded to roughly a third of the new site's annual manufacturing costs. One practice that proved particularly useful during startup were the regular discussions held every evening and facilitated by skilled staff from the home factory to assess target fulfillment (unit volume, rejects, rework) and define initiatives.

Bottom line: Recruitment and training should vary by skill level. Exchanging ideas and experience with staff at home may help ensure a smooth start but can be very costly.

_____

[2] _Total quality management._

6.3.2.3 Training of Production Staff

The more complex the processes, the more expensive it is to train local production staff. To familiarize these employees with the processes and tasks involved, many companies send selected skilled workers and foremen to the home factory for a brief period of training. These then train the entire production staff on site with the help of top experts from the home factory (Figure 6.12). This training should form a fixed part of the planning and be coordinated with the start of production.

A mid-size Western European electronics company, for example, managed to transfer all its production-relevant know-how to a Hungarian factory within just six months. It formed mixed **training teams** for every production department. The company made absolutely sure that the trainers from the home plant had no reservations whatsoever about the new plant. The trainers received a bonus of 10 percent of their salary for taking part and a further 10 percent upon successful transfer of their knowledge. The training courses were mainly held at the home factory.

6.4 Production Ramp-Up

A new facility will aim to ramp up to full capacity and achieve the company's usual quality as fast as possible. The swiftest companies manage this within just four months, while the slower ones take over three times as long – around 14 months. Why the big disparity? The delay usually has to do with quality problems. The quality standards are not met immediately and rework is needed time and again. For months on end, technical resources and management capacity are tied up sorting out ever-changing sources of problems.

Global leaders take a different path. They think through production ramp-up beforehand as carefully as they do all the other aspects of site setup. After making a precise evaluation of the situation on site, they choose a ramp-up strategy that minimizes complexity.

What is the best way to organize the start of production? Sequentially with individual products and manufacturing steps or with the entire product range and

multiple processes? Which equipment should be adopted from other locations? Who will be responsible for the transportation of machinery? How can delivery failures be avoided? The winners detail their answers to these questions well in advance.

6.4.1 Ramp-Up Strategies

Companies should expect **production downtimes**, **ramp-up delays**, and **loss of production**, especially if the ramp-up is being implemented with new personnel, new suppliers, and possibly even a new product (Figure 6.13). These deviations from plan can result in a delay in reaching the targeted maximum output (full ramp-up). The results of the ProNet survey show that quality costs averaging 5 percent of total investment in the new location need to be added to the opportunity costs of the delayed ramp-up (cf. section 6.1).[3]

The causes of these deviations can lie both inside and outside the company. Careful preparation cannot rule them out altogether but can reduce them con-

---

Training initiatives should begin a year before ramp-up of complex production activities

Fig. 6.12: Recruitment and training of production staff for locations abroad   <u>EXAMPLE</u>

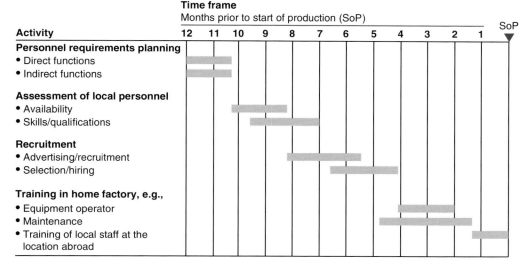

Source: McKinsey/PTW (ProNet analysis)

---

[3] *A tried-and-tested metric for ramp-up speed is the ramp-up factor. It is defined as Ruff = (quantity actually produced)/(quantity that can be produced in theory) x (product lifecycle – ramp-up time)/(product lifecycle). An ideal startup has a maximum ramp-up factor of 1.*

siderably. Causes within the company may include deficits in the skills of machine operators or mechanics, design changes at short notice, and problems in coordinating production lines. External causes include problems with the quality of the materials supplied or delays in the supply chain.

A targeted **sequential introduction** of products and manufacturing steps reduces the risk of deviations from plan. Faults can still occur, of course, but can be pinpointed more easily and resolved with greater focus. **Complexity management** is also an important prerequisite (cf. section 6.2.1). With the sequential introduction of the product portfolio and production technologies as complexity drivers, it is possible to achieve a significant reduction in the total complexity of the ramp-up phase and to adapt this complexity to existing capabilities. **Ramp-up variants** fall into four main categories:

In **ramp-up Mode 1**, sequential introduction of the product range enables staff and, if applicable, sup-

pliers at a new location to prepare themselves successively for new, complex products and their requirements (Figure 6.14). This is particularly suitable for mass products and production lines with long setup times. It means that product-specific lines can be set up right from the outset.

Honda's approach is an example of method 1. Honda uses the production of motorcycles as a beachhead at new locations – as was the case when it began production activities in the US in 1977. The company used the comparatively simple production of motorcycles, which required relatively little investment (approximately USD 35 million), to familiarize itself with the location, identify suitable suppliers, and train a skilled workforce. Two years later, in 1979, Honda announced the ramp-up of automobile production (with an investment of around USD 250 million). It moved its top performers from the motorcycle to the automobile plant and placed contracts for parts with suppliers that it already knew were reliable. In 1985, Honda also relocated

---

Severe deviations from plan are common in the ramp-up phase

Fig. 6.13: Planned and actual ramp-up curve

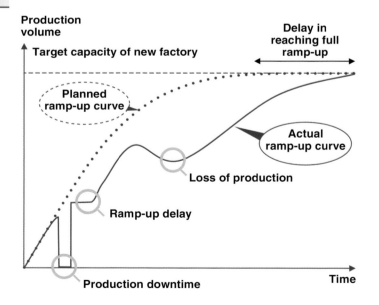

Causes of delays: examples
Loss of production
• Rejects due to incorrect operation of the NC machining center
• Interrupted supply of utilities (electricity, gas, water, etc.)
• Logistical obstacles to the supply of materials (e.g., accidents, delays at customs)
Ramp-up delay
• Delays in coordinating with additional production line
• Training of machine operators scheduled at short notice
• Delay in commissioning transferred machine
• Sample production for customer approval
Production downtime
• Design changes
• Poorly reassembled machine
• Insufficient supply of spare parts
• Materials held at customs
• Late discovery of quality defects in materials supplied

Source: McKinsey/PTW (ProNet analysis)

capacity for engine manufacture to the US. Here again it could rely on experienced staff and suppliers from the motorcycle and automobile plants.

Robert Bosch GmbH adopted a similar approach in 1995 when it expanded a facility in Mexico. It found moderate-quality expertise available and built on that. After Mexico joined NAFTA, the company developed the site into a premium-quality location within a good twelve months, and doubled the workforce. This was only possible because the location already had a core group of well-trained workers and experienced managers.

Mode 2 involves introducing the entire range of products and manufacturing processes simultaneously (Figure 6.15). This method is only suitable if products and manufacturing processes are fairly simple, or staff are extremely skilled and highly trained. Products for which this method applies include sports shoes, household appliances, and simple plastic parts.

This mode was used by the mechanical engineering company in the first case example – and contributed to their downfall. The problem was not just that the local staff were unfamiliar with the manufacturing processes imported from Europe. The company did not do any advance testing or preparation of the territory for its products. The Asian market proved unreceptive, a fact that the company did not realize until it had already invested in production of the entire product portfolio.

In **Mode 3**, the new location enhances its competence in the most incremental steps possible. Products and manufacturing processes are introduced sequentially (Figure 6.16), reducing the complexity of the individual steps to an absolute minimum. Gaps in employee training can be filled successively. This

Sequential product launch is advisable for a more complex product portfolio

Fig. 6.14: Ramp-up strategy – Mode 1

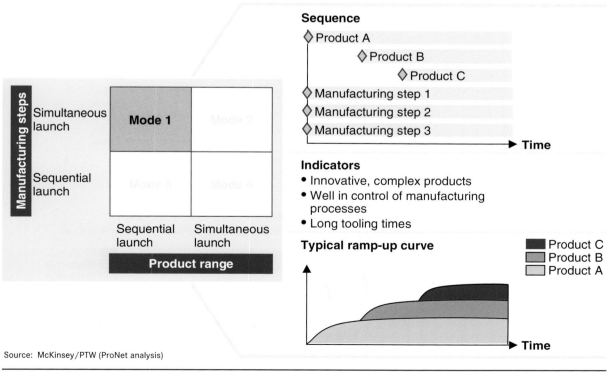

Source: McKinsey/PTW (ProNet analysis)

approach only makes sense if products and processes are very demanding. One disadvantage is the very long ramp-up curve – economies of scale are not realized until a late stage. The approach does, however, provide high process reliability and control over the standard of quality achieved.

This third mode can be divided into two basic variants: (1) introducing processes product by product and (2) introducing products process by process. The optimum choice depends on where the steepest learning curve or greatest economies of scale are expected. As an example, a largely process-oriented approach (variant 1) would likely make the greatest sense for investments in a foundry or a partially automated paint shop.

Ramp-up Mode 4 introduces products simultaneously but production steps sequentially. This method is recommended for very diverse, complex manufac-

turing processes with high quality requirements (Figure 6.17). It also enables the simultaneous market launch of a full spectrum of locally manufactured products. The start of cell phone production in new locations generally uses this method.

Global cell phone manufacturers prefer to start production relocation with the end of the value chain. This has several advantages. The new site is responsible for end-product quality right from the outset, and the logistics chain is not too long during ramp-up (assuming the new location is close to the market). The new facility first takes over final assembly of the products, then preassembly of the modules (e.g., display, casing). The last stage is to add production of the individual parts.

The Mode 4 model can also be used for involving suppliers. It is advisable to introduce technically demanding primary products from new local suppliers

Completely parallel ramp-up is only suitable in very specific cases

Fig. 6.15: Ramp-up strategy – Mode 2

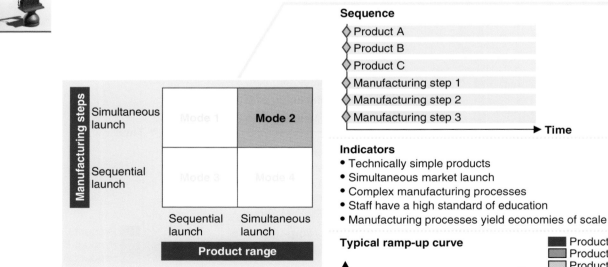

sequentially. This allows you to pace the resolution of any technical problems, rather than risking multiple issues arising all at once.

* * *

Selecting the right ramp-up mode is crucial. The production managers we interviewed also highlighted a number of other valuable levers in the ramp-up process. One is to enlist active support from suppliers. Having local equipment manufacturers that can cover the demand for repairs and spare parts supervise the ramp-up is very helpful. Materials suppliers should also be included in the quality assurance process. Interviewees also stressed the importance of certain organizational activities. Training new machine operators at existing locations using the same or similar products and manufacturing processes is a way to cut corners. Tying wages to reject rates is another. Linking career development opportunities to performance is a third highly effective approach.

6.4.2 Equipment Transfer Versus Purchase

One of the main decisions affecting the costs of setting up a new location is whether equipment should be **transferred or purchased**.

It is impossible to generalize the cost and time involved in relocating machinery. A distinction must be made between standard machinery (no complex peripherals or automation, such as an NC lathe), automated systems (transfer lines in machine-cutting or assembly, for instance), and complex systems in process industries (e.g., chemicals, pharmaceuticals).

With standard machinery, it is usually more cost-efficient to transfer equipment that is no longer needed than to sell it and buy new equipment at the new location. A transfer generally costs roughly 18 to 20 percent of the value of the machinery when new. A sale and purchase operation, on the other hand, would likely cost between 20 to 50 percent of this

---

Mode 3 reduces complexity to a minimum, but levels out the ramp-up curve considerably

Fig. 6.16: Ramp-up strategy – Mode 3

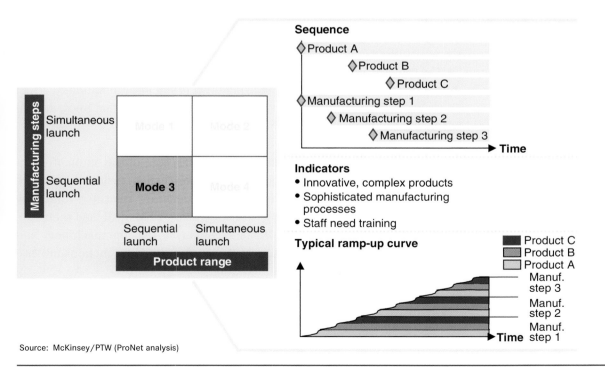

Source: McKinsey/PTW (ProNet analysis)

Ramping up manufacturing processes sequentially makes sense if products are being launched simultaneously

Fig. 6.17: Ramp-up strategy – Mode 4

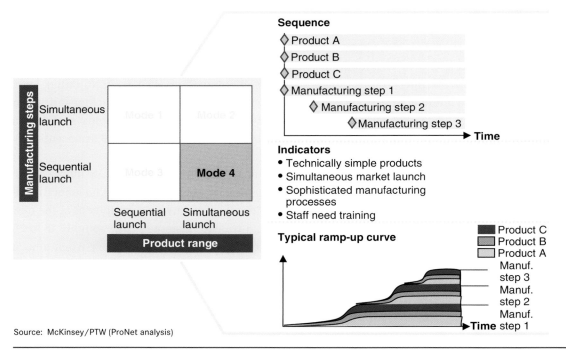

Source: McKinsey/PTW (ProNet analysis)

value. Transfer includes the cost of disassembly/reassembly, customs clearance, and transportation (Figure 6.18). Added to this are the opportunity costs of lost production during transfer. The transfer time depends greatly on the mode of transport and the machine concerned. Transporting simple machining centers over a few hundred kilometers can take several days, while shipping complex production lines by sea can take eight weeks or more.

A number of factors are key for smooth and cost-efficient transfer. These range from planning the optimal export and import sequence, overhauling the machine and procuring suitable packaging/special hoisting equipment, through to monitoring reassembly by experienced staff (Figure 6.19). Industrial assembly experts emphasize the benefits of entrusting several members of the disassembly team with reassembly. The best candidate for remounting a machine is the person who took it apart.

If the company itself does not have any specialists or means of transport for relocating equipment, it may be best to enlist general contractors. As well as transporting the equipment, they will also dismantle and overhaul it, modify it to suit local conditions (correct voltage, etc.), reinstall the production lines, and insure the entire operation.

With certain suppliers, e.g., in the automotive industry and aircraft construction, customer audits are the norm when systems are relocated (and new plants purchased). These OEMs will hold their suppliers liable for ensuring that the machinery and processes in the production lines function faultlessly after relocation.

If machinery currently in use is transferred, this temporarily reduces the production capacity available. The company may plan additional (internal or external) capacity to bridge the loss. Another option often

The transfer of existing machines often costs less than buying new ones

Fig. 6.18: Comparison of machine transfer and purchase (standard machine)* EXAMPLE
Percent of machine value (new)

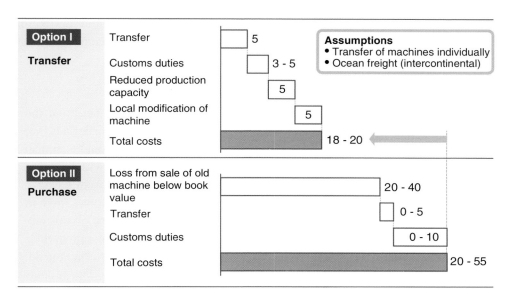

Option I Transfer		
Transfer		5
Customs duties		3 - 5
Reduced production capacity		5
Local modification of machine		5
Total costs		18 - 20

Assumptions
- Transfer of machines individually
- Ocean freight (intercontinental)

Option II Purchase		
Loss from sale of old machine below book value		20 - 40
Transfer		0 - 5
Customs duties		0 - 10
Total costs		20 - 55

* Example of standard machining center, transportation from Germany to China

Source: McKinsey/PTW (ProNet analysis)

Seamless transfer of machinery means production can start quickly

Fig. 6.19: Key factors in successful equipment transfer

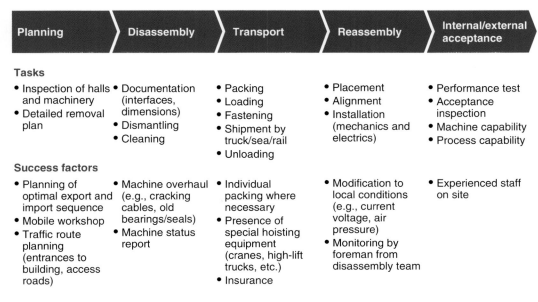

Planning	Disassembly	Transport	Reassembly	Internal/external acceptance

Tasks

• Inspection of halls and machinery • Detailed removal plan	• Documentation (interfaces, dimensions) • Dismantling • Cleaning	• Packing • Loading • Fastening • Shipment by truck/sea/rail • Unloading	• Placement • Alignment • Installation (mechanics and electrics)	• Performance test • Acceptance inspection • Machine capability • Process capability

Success factors

• Planning of optimal export and import sequence • Mobile workshop • Traffic route planning (entrances to building, access roads)	• Machine overhaul (e.g., cracking cables, old bearings/seals) • Machine status report	• Individual packing where necessary • Presence of special hoisting equipment (cranes, high-lift trucks, etc.) • Insurance	• Modification to local conditions (e.g., current voltage, air pressure) • Monitoring by foreman from disassembly team	• Experienced staff on site

Source: McKinsey/PTW (ProNet analysis)

used is to provisionally increase capacity at the existing site before relocation (extra shifts, for example) to build reserves. Outsourcing is usually not an option, whether due to lack of suitable suppliers, proprietary know-how, or complex customer approval processes for new suppliers.

Many manufacturers synchronize short-term capacity downtime with fluctuations in demand. If a new product is to be launched at the same time as setting up a new site, the product lifecycle can be leveraged. This may involve using existing equipment at the old location for a product being phased out and new equipment at the start-up facility for the incoming product.

Navigating a Maze of Requirements: Implications for Bosch When it Goes Overseas

The transfer of machinery for manufacturing automobile components such as injection valves is subject to high process stability and quality requirements. When Bosch relocates production machinery, it has to re-obtain approval from the relevant OEMs before it can start production at the new location.

Even minimal changes to the manufacturing process have to be approved by the OEM according to specific regulations (especially ISO TS 16949, introduced in 2002). This process includes the sequences of production steps at component suppliers and their suppliers, the location-dependent environmental parameters, and even transportation. (The latter covers aspects such as how much plastics swell when transported by sea compared to air). In subtropical and arctic temperature zones, gaining approval for the new process may require a new vehicle endurance run, which takes around three months. Suppliers have to bear the costs and risks themselves if they relocate on their own initiative.

Once the production line has been set up at the new location, one of the acceptance criteria is that all dimensions must match the original technical drawings – around 120 dimensions with injection valves. If the equipment manufacturers have capacity in the new country, Bosch involves them in ensuring 100 percent functionality of the machines and processes. If they only operate in their home locations, Bosch discusses with them in advance how they can guarantee technical support and supply spare parts locally.

Bosch's approval process includes all materials suppliers who supply to the new location. This process is more extensive for new suppliers than for existing ones. However, a low-cost location can only achieve its full cost reduction potential if it uses local suppliers. The best solution is often to convince existing suppliers to relocate their production capacities to the new country as well.

Bottom line: Suppliers should prepare machine transfers carefully to meet their customers' quality requirements, closely involving their own providers of equipment and production materials.

Further reading

Ancona, D. and H. Bresman: *X-teams: How to Build Teams That Lead, Innovate and Succeed.* Boston: Harvard Business School Press, 2007.

Barton, D., T. Hsieh, and J. Sinha: *Becoming a Global Champion: The journey to global leaderships is fraught with peril - and opportunity.* New York: McKinsey & Company, Inc., 2003

Boudreau, J. W. and P. M. Ramstad: *Beyond HR: The New Science of Human Capital.* Boston: Harvard Business School Press, 2007.

Bryan, L. L. and C. Joyce: *Mobilizing Minds: Creating Wealth From Talent in the 21st Century Organization.* New York: McGraw-Hill, 2007.

Eversheim, W. and G. Schuh: Gestaltung von Produktionssystemen. *Produktion und Management*, Vol. 3, Berlin, Springer, 1999.

Huselid, M. A., B. E. Becker, and R. W. Beatty: *The Workforce Scorecard: Managing Human Capital To Execute Strategy.* Boston: Harvard Business School Press, 2005.

RAIMUND DIEDERICHS, TOBIAS MEYER, MARKUS LEOPOLDSEDER, FRANK JACOB

7 Management: Applying Best-Practice Structures and Processes

Summary

The success of a global production network depends on more than just strategic design and systematic setup. Tactical and operational management is just as critical and becomes more challenging as production is increasingly globalized. New locations have to be integrated into the existing network. The reporting structures need to be sensible and incentives in line with corporate objectives, which change over time. Initially, the commitment and direct involvement of senior management are crucial. This should be followed by step-by-step decentralization of decision-making authority until local management is largely independent and directly accountable for financial and operational targets.

As production "goes global," the bar for the supply chain management function rises. Distances between production sites and markets become longer, and service level expectations, the number of product variants, and the cost of holding inventory also all increase. Companies can achieve significant competitive advantage and reduce their costs by optimizing their distribution network and managing transportation intelligently. Specific challenges arise in many developing countries, and applying the right ground rules can make a fundamental difference. Tactical production planning – the assignment of orders to plants across the network – is also increasingly important and a key enabler to leverage a global footprint effectively.

The exchange of expert knowledge, experience, and best practices between different locations poses special challenges, too. The lack of physical proximity makes it even harder to overcome cultural and language barriers. A global production system – a uniform standard for "how things are done" across the network – is an important tool for establishing the sustained use of best practices. It provides a common language for the operations community, and guides frontline managers through standard operating procedures and problem-solving techniques. The right selection of performance indicators and their rigorous tracking is another key element in a successful global production system.

Key questions, Chapter 7

- How should a new location be integrated into a company's organization structure and processes?

- How should decision rights and accountability be divided up between the corporate center and individual locations?

- What controlling tools can be used to promote behavior beneficial to the company as a whole?

- What criteria should be applied for assigning production orders internally?

- What aspects of supply chain management are impacted by the globalization of production?

- How can the distribution network be optimized?

- How can the additional costs and risks entailed by global logistics be reduced?

- How can companies ensure that improvements in production achieved at one site are leveraged worldwide?

7.1 Organization Structure

Once a global production network has been established, corporate management is faced with the challenge of effectively **monitoring** and **managing** it. Needs during the design and transition phase are very different from those of managing a static production network with stable supply chains. Classic organization concepts are generally not the answer to the issues that arise during production network restructuring. Developments are always very dynamic, and organization design requirements change rapidly as a new factory grows.

7.1.1 Form of Organization

In the eyes of most large companies, the best **organization structure** for a stable product business is the divisional model (see box: "The Evolution of Organization Structures: A Retrospective"). However, it is vital to adapt the model used when establishing a new location. Restructuring an organization is a tremendous challenge, and the associated change processes pose difficult tasks for management. It is best to employ a two- to three-stage setup process, particularly with strategically important plants.

The full **support of the corporate center** is usually a vital ingredient when establishing a new location. Local managers do not have the resources to solve all problems, thus running the risk of delays, unmet time-to-market and time-to-volume targets, and budget overruns. However, a new site with low initial volume often does not capture top management attention as much as issues with more immediate impact. This may mean the corporate center is slow to deal with problems, and decision-making requires laborious coordination.

7.1.1.1 Three-Phase Adaptation of the Organization Structure

The solution is to skillfully adapt the organization in advance of each new set of requirements. In parallel with developing the new location, its **organi-**

The Evolution of Organization Structures: A Retrospective

Early organization theory was focused on specific structural principles for dividing up tasks and then coordinating and integrating work with maximum efficiency (Figure 7.1). Examples were grouping similar types of activity, matching decision-making authority with responsibility, and clearly defining accountability. Each person had only one immediate superior, and spans of control were limited. Companies hoped that following these principles would solve organizational problems once and for all, making management a purely routine task.

These principles may have received so much attention because the underlying structures were so rarely discussed. Most industrial companies were structured according to their operational functions, such as sales, engineering, production, and finance, with top management steering their interaction.

After World War II – and even earlier at pioneering companies such as General Motors and DuPont – the increasing number of new products was one of the factors that led to product-group-based profit centers becoming the new basis of the organization structure. Divisional organization now became dominant.

This called for a trade-off between the advantages of functional specialization and the efficiency of making the product group (rather than corporate

---

New organization structures developed over time as industries evolved

Fig. 7.1: History and principles of organization structures

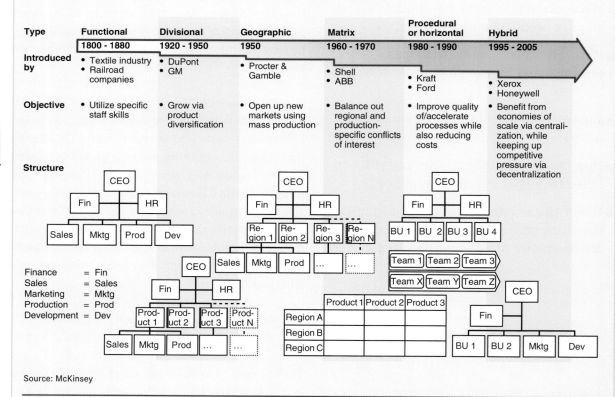

Type	Functional	Divisional	Geographic	Matrix	Procedural or horizontal	Hybrid
	1800 - 1880	1920 - 1950	1950	1960 - 1970	1980 - 1990	1995 - 2005
Introduced by	• Textile industry • Railroad companies	• DuPont • GM	• Procter & Gamble	• Shell • ABB	• Kraft • Ford	• Xerox • Honeywell
Objective	• Utilize specific staff skills	• Grow via product diversification	• Open up new markets using mass production	• Balance out regional and production-specific conflicts of interest	• Improve quality of/accelerate processes while also reducing costs	• Benefit from economies of scale via centralization, while keeping up competitive pressure via decentralization

Finance = Fin
Sales = Sales
Marketing = Mktg
Production = Prod
Development = Dev

Source: McKinsey

unit) the basis of coordination. Divisionalization was a way of decentralizing the integration process in the face of companies' growing complexity. At the same time, however, the intention was to create a structure that pushed decision-making powers and responsibilities further down the organization, thereby promoting the skill building and motivation of line management. The risk perceived by many in this approach was the potential loss of important functional specialist skills, as these were being banished to a less prominent place in the corporate hierarchy. In time, however, it became clear that functional performance was more dependent on organizational skills and value systems than on the organization structure in place.

This trend towards divisionalization was strengthened by another new element of organization theory. At the end of the 1950s, people began to realize that organization structure was an instrument for implementing business strategies, meaning that a company's structure should be driven by its strategy.[1] In the final analysis, this meant that the strategy of a diversified company was better served by a product-based, decentralized structure than by centralization.

The trend towards greater product variety combined with the new organizational theories for handling this complexity led to a general move towards divisional corporate structures. Whereas in 1950 only 20 percent of the US-Fortune 500 had a divisional structure, today the figure is over 90 percent. This trend came somewhat later to Europe and the rest of the industrialized world, but took a similar course, and the percentage of divisionalized firms is now roughly the same as in the US. Other types of organization structure are now rarely found among major companies (though they are neither obsolete nor fundamentally wrong).

Bottom line: A broad portfolio of different organization structures has evolved over the past 100 years, each with its own clear pros and cons. Almost all major companies have adopted a divisional structure.

---

[1] *Chandler (1969).*

zation structure should undergo **three phases** (Figure 7.2).

The organization structure of new, strategically important production sites should transition through three phases

---

1. Production ramp-up: During the ramp-up phase, **top management** is responsible for planning and implementing the setup of the new location. Project management should be assigned to experienced top performers reporting to management at regular steering committee meetings. This direct access to top management has proven its benefits time and time again. It both simplifies and accelerates decision-making processes, as it replaces elaborate co-ordination loops with targeted project management:

"The fact that I can send my reports direct to a member of top management is crucial for driving our activities in China fast and without bureaucracy. It puts me in a much stronger position to take the swift action required."[2] Numerous interviewees in our ProNet survey reported similar experiences.

The optimum interaction of production with the local or global sourcing of components and purchased parts allows much greater efficiency. **Conflicts of interest** between R&D, production, and purchasing are unavoidable. Department-centered attitudes and biases harbor the risk of suboptimal solutions, as each area attempts to achieve its own (cost) goals without considering the goals and interests of other

---

[2] *See Kaufmann (2005), p. 162.*

functions. This problem is exacerbated when the primary goal of setting up the production location is to open up a new market.

Further issues come into play here in addition to the pure cost perspective: how the product needs adapting (R&D), how to acquire customers (Sales), and how to establish the brand (Marketing). Inefficiencies are unavoidable if responsibility is divided. For example, little attention is generally paid to customer requirements outside the sales division during this period due to the more burning issues related to production ramp-up. During this critical phase, it is therefore often vitally important to **bundle responsibility**, putting one person in charge of the entire region.

2. Stabilization phase: Once production has begun, tasks change, as do the requirements on the organization. Any remaining gap versus cost targets has to be eliminated by **lowering scrap rates and improving quality**. Continuous improvement processes (CIP) will be introduced, along with standard corporate functional processes to take advantage of synergies with other parts of the company. Consolidation of the same or similar activities and global cooperation make it easier to implement process innovation and achieve economies of scale. A high degree of specialization and the short orientation periods these experts require have a positive impact on learning and experience curves. These are the classical characteristics of a functional organization.[3] The actual

---

The organization structure should reflect the developing phases of production

Fig. 7.2: Structural transformation: model

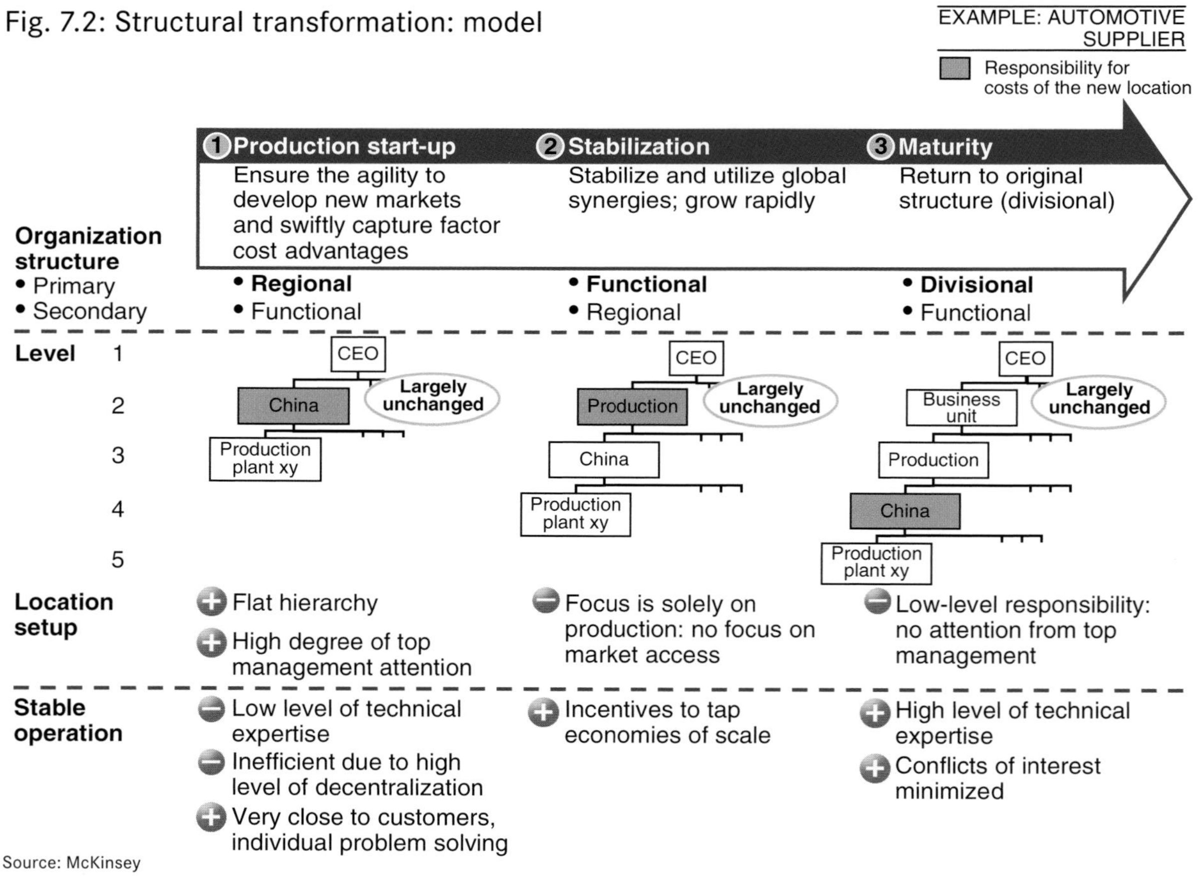

EXAMPLE: AUTOMOTIVE SUPPLIER

▢ Responsibility for costs of the new location

	❶ Production start-up	❷ Stabilization	❸ Maturity
	Ensure the agility to develop new markets and swiftly capture factor cost advantages	Stabilize and utilize global synergies; grow rapidly	Return to original structure (divisional)
Organization structure • Primary • Secondary	• **Regional** • Functional	• **Functional** • Regional	• **Divisional** • Functional

Level	Production start-up	Stabilization	Maturity
1	CEO	CEO	CEO
2	China — Largely unchanged	Production — Largely unchanged	Business unit — Largely unchanged
3	Production plant xy	China	Production
4		Production plant xy	China
5			Production plant xy

| **Location setup** | ⊕ Flat hierarchy
⊕ High degree of top management attention | ⊖ Focus is solely on production: no focus on market access | ⊖ Low-level responsibility: no attention from top management |
| **Stable operation** | ⊖ Low level of technical expertise
⊖ Inefficient due to high level of decentralization
⊕ Very close to customers, individual problem solving | ⊕ Incentives to tap economies of scale | ⊕ High level of technical expertise
⊕ Conflicts of interest minimized |

Source: McKinsey

---

[3] See Vahs (2005), p. 142.

structure the organization should take in this phase depends on the status quo. If – as in most major corporations – the divisional organization is standard for a stable or static operation, then the separate regional structure of the ramp-up phase should be retained. If, however, the usual organizational structure is a functional or a matrix organization, then this should be established during this phase.

The emphasis at this stage is integrating a new production location into existing structures to tap **synergies** and accelerate learning curves. A growing location and its associated problems require the increasing standardization of tasks, competencies, and processes, and the use of supporting management systems. Ideally, the production division will have cost accountability for the new location. The benefits of greater lot sizes and economies of scale can only be captured in a united effort across locations as the site is integrated into the global production network. If this is not achieved successfully and promptly, bottlenecks may result that have to be remedied by corporate management. In the stabilization phase, functional organizations therefore still ensure that top or second-level management remains directly responsible for costs at the new location.

3. Mature phase: In the long term, requirements at the new location will converge with those at established locations. The strategic focus then becomes sustained, profitable growth. To accomplish this, most companies concentrate on developing new products systematically and fast. They also **diversify their product spectrum** to open up new consumer segments. Management needs to shift its focus from stable market supply to a more pronounced entrepreneurial mindset and greater customer centricity. These requirements are often best met by a divisional structure,[4] with the organization transitioning to the structure used elsewhere in the company for mature locations, e.g., the divisional setup. Technologically similar products and production processes are combined to create divisions. The individual divisions are given the necessary autonomy to promote entrepreneurial action. This is the perfect environment for developing cross-functional skills. The division-

al focus on one product segment allows attention to be paid to individual customer requirements.

7.1.1.2 Implementing Organizational Change

The design of this organizational change is a complex task that is critical for the success of the change process. When functions and processes within an organization are restructured, this naturally has an impact on people with their individual attitudes, concerns, and desires. Deliberately conducted **change management** helps to prevent these soft factors from hampering the transition. Instead, it addresses them and makes them a productive part of the process.

There are two **personnel issues** that can cause considerable obstacles if not handled carefully. One relates to how long the original project manager remains responsible for the site. Companies tend to quickly replace individuals responsible for the individual phases of organizational development already described. It is rare for the project manager from the initial planning phase to still be in charge of the plant years after production has begun. This is a natural transition, as cost responsibility is pushed further and further down the hierarchy as a location develops. However, such rapid changes in personnel can have serious consequences should a decision be required that entails **short-term risk** in the interests of long-term success. Standard short-term incentive systems can also have a counterproductive effect. Annual **performance bonuses** encourage on-site managers to duck investments where the payback time exceeds the period they will be in their posts.

> **Performance-based pay should be geared towards the long-term success of a new location, even if the employee only spends a short time there**

The goal must be to create an appetite for risk in the interests of the business. This is often referred to as **intrapreneuring** (a term created by combining the

---

[4] *See Vahs (2005), p. 288.*

terms *intra-corporate* and *entrepreneuring*). Decision makers need to be given as much autonomy as possible to promote an entrepreneurial spirit and creative ingenuity.

The **conflict** is that synergies and economies of scale are best realized via central structures. Corporate management has to grapple with the question of where to draw the line between the two extremes: total local autonomy, on the one hand, and centralized management control on the other. There is no stock solution. The right answer for one set of conditions (in terms of the business environment, competitive position, and organizational skills) may well become suboptimal as time progresses. The pace of change is much faster than it was in the past, and the multiple factors at play make hard-and-fast predictions close to impossible.

This calls for a proactive approach where corporate change is seen as a continuous process. The challenge is to balance a company's current flexibility on the central-decentral scale with greater organizational flexibility over the long term. In today's world of increasing complexity and uncertainty, an organization capable of change has a much greater chance of success than a cumbersome, inflexible entity. Structural solutions alone are not sufficient to create these transformational skills, but can be a vital first step in demonstrating the need for change to the entire organization.

7.1.2 Cost or Profit Center?

An ongoing design task for the management of a production network is thus deciding how to divide up responsibility between the corporate center and local

---

Criteria for deciding on the degree of independence are based on a simple selection matrix

Fig. 7.3: Selection matrix – degree of independence for individual factories

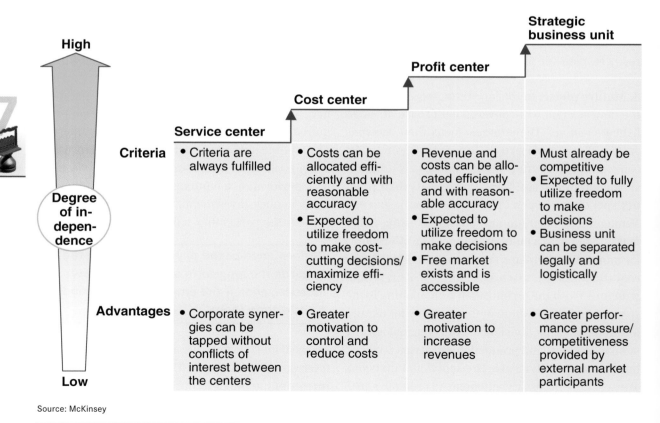

Source: McKinsey

management. The bandwidth ranges from decentralized cost responsibility without any profit accountability to decentralized responsibility for the allocation of profits. Four basic models are used, with varying degrees of independence (Figure 7.3):

Service centers are dependent units that are not accountable for their own income, but provide services for other cost centers (or business units). Although they are subject to budgetary discipline, they have little opportunity to introduce cost-saving initiatives on their own.

Cost centers are measured solely on the basis of costs, i.e., judged by their efficiency, not their own earnings. The objectives are to keep within budgeted costs or to minimize costs for a specific turnover volume. The causes of any deviations from the budget are determined at the end of a budget period. One example of a cost center is the HR department. "Profits" can be made by achieving savings. Performance targets are agreed on with corporate management.

Profit centers are organization units within a company whose own profit or loss is calculated for each accounting period. These profit-based evaluations are then used to improve control of the unit's activities and to review its profitability. The costs are offset against the revenues of the specific corporate unit or department concerned, and a P&L statement is

Boosting Competitiveness and Margins: SBU Pioneers

The concept of the **strategic business unit** (SBU) first emerged in the late 1960s. The basic idea was to define a manageable economic unit and assign it the responsibility and decision-making powers for all the main functional resources it required for success on the market. Business functions were consolidated at the SBU level, generally much further down the hierarchy than previously. The potential advantages were obvious: this allowed more effective integration of tasks closer to the market. The SBU managers now controlled all the variables influencing the performance of their unit, yet could still be held accountable by the corporate center (now at a greater distance). Another key benefit was that top management at the corporate center, freed from operational duties, could now spend more time thinking strategically about the composition of their business portfolio. These substantial changes in roles at every level of the corporation were to have a lasting impact on business in the 1970s and can still be seen today.

The SBU concept is not simply a new face on the old divisional structure. Today, the focus is more on strengthening a company's competitive position in the marketplace and less on merely improving organizational effectiveness. As a result, the size of individual SBUs within a single company can vary from millions to billions of dollars in revenues. The key challenge is to ensure that the constellation strengthens corporate competitiveness.

General Electric (GE) was the first company to introduce the SBU concept. Many of its businesses could be clearly separated by their market characteristics (different products and services, customers, and even technologies). The GE corporate center charged all of its SBU teams with becoming leaders in their market segments. The resources of multiple SBUs were used jointly whenever this could secure a genuine scale advantage, with the managers still retaining responsibility for their share of these joint services. Many other successful companies – among them 3M – created shared resource units to gain cost advantages inaccessible to individual SBUs.

Bottom line: Strategic business units are defined without regard to existing organization units. Their sole purpose is to boost the company's competitiveness and, ultimately, its profit margin.

drawn up. The key idea is that profit center managers should think and act like independent entrepreneurs, and have extensive decision-making authority.

Strategic business units (SBUs) are subareas of a company that operate in a specific market segment independently of other corporate units (see box: "Boosting Competitiveness and Margins: SBU Pioneers"). In contrast to profit center managers, SBU managers are able to determine how profits are allocated. They can also make relatively independent decisions on investment volume and production range. An SBU must have the critical mass to be profitable, which means it rarely consists of just one factory – it is normally an integrated group of factories. This type of center also has the farthest-reaching implications: a decision to employ SBUs affects the entire organizational structure, and usually leads to a divisional organization.

The boundaries between these four types of center are fluid, with various configurations possible within each category (see Figure 7.4). Independence on the scale of central versus local decision-making authority can vary in small increments.

Global production leaders[5] are more centralized to lower their costs via synergies and economies of scale

The ProNet survey also demonstrated the importance of selecting the appropriate degree of independence for a factory (Figure 7.5). Successfully globalizing

---

Borderlines between the various center concepts are fluid

Fig. 7.4: The continuous spectrum between service centers and SBUs

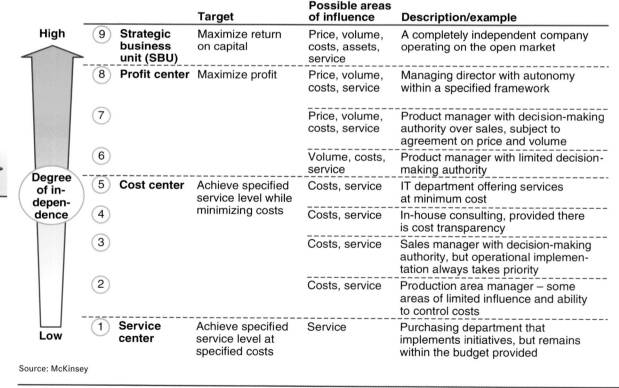

		Target	Possible areas of influence	Description/example
High	⑨ **Strategic business unit (SBU)**	Maximize return on capital	Price, volume, costs, assets, service	A completely independent company operating on the open market
	⑧ **Profit center**	Maximize profit	Price, volume, costs, service	Managing director with autonomy within a specified framework
	⑦		Price, volume, costs, service	Product manager with decision-making authority over sales, subject to agreement on price and volume
	⑥		Volume, costs, service	Product manager with limited decision-making authority
Degree of independence	⑤ **Cost center**	Achieve specified service level while minimizing costs	Costs, service	IT department offering services at minimum cost
	④		Costs, service	In-house consulting, provided there is cost transparency
	③		Costs, service	Sales manager with decision-making authority, but operational implementation always takes priority
	②		Costs, service	Production area manager – some areas of limited influence and ability to control costs
Low	① **Service center**	Achieve specified service level at specified costs	Service	Purchasing department that implements initiatives, but remains within the budget provided

Source: McKinsey

---

[5] See Appendix at the end of the book "A.1.2 Leaders vs. followers: The ProNet methodology" for the definition of pioneers (leaders) and followers.

firms differ consistently from their less successful counterparts in that they are more centralized in steady state (the mature phase), with a tendency towards **central production management**. The disparity is greatest in product design, the procurement of production machinery, and purchasing strategy. The gap between the leaders and followers is statistically significant.

At first glance this result might seem surprising. Surely globalization leaders are optimally placed to make decisions locally thanks to their global diversification? The survey clearly showed that successful companies focus on costs. And it is here – in the synergies and economies of scale – that the advantages of centralization lie. Specifically, leaders prefer to centralize product design to achieve R&D synergies, and machinery procurement/purchasing of supplies to capture economies of scale.

7.1.3 The Difficulties of Internal Transfer Pricing with Cost and Profit Centers

Regardless of which type of center is chosen, a fundamental problem arises: How should internal, cross-center services be charged? The basic idea behind using such centers is to assess performance and provide incentives transparently by measuring the profits achieved or costs incurred. These parameters are set to be largely consistent with corporate objectives, so, in theory, the centers can operate autonomously, aligned with corporate center aspirations without the need for rigid central control. In practice, however, numerous services and outputs in a global production network involve multiple locations, making full autonomy difficult. What transfer price should be used for the input products one site receives from another, within the corporate network? The utilization of services, patents, machinery,

---

Globalization leaders have more centralized decision-making processes

Fig. 7.5: Results of the ProNet survey: degree of dependence on a scale of 1 (dependent) to 5 (independent)

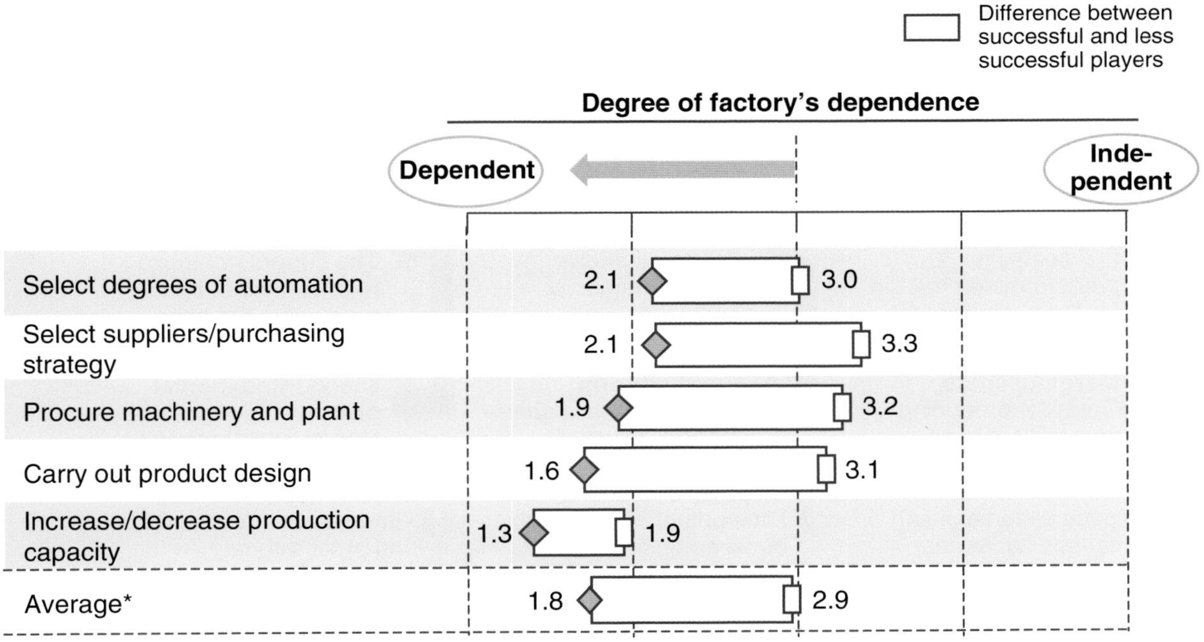

◆ Successful
☐ Less successful
☐ Difference between successful and less successful players

Degree of factory's dependence

Dependent ⟵ Independent

	Dependent		Independent
Select degrees of automation	2.1 ◆	☐ 3.0	
Select suppliers/purchasing strategy	2.1 ◆	☐ 3.3	
Procure machinery and plant	1.9 ◆	☐ 3.2	
Carry out product design	1.6 ◆	☐ 3.1	
Increase/decrease production capacity	1.3 ◆	☐ 1.9	
Average*	1.8 ◆	☐ 2.9	

* Weighted mean of the five sub-items

Source: McKinsey/PTW (ProNet analysis)

and plant can also present internal transfer pricing issues.

7.1.3.1 Objectives of Designing a Transfer Price System

Prices and terms for the internal provision of goods and services are not subject to negotiation between independent market players, so, in principle, they can be determined autonomously. This provides latitude that should be used circumspectly along a number of dimensions.

Value determination function: Transfer prices are used to set the value of a particular item, and customs duties on imports of such items are usually calculated on the basis of this value. This allows companies to minimize the duties payable by specifying low values for the goods in question, or in some circumstances to maximize export bonuses by stating high values.

Profit transfer function: The transfer pricing scheme can be designed to enable differentiated allocation of certain profits among subsidiaries across the network. There are often a number of different options for determining pricing for the billing of intangible services, such as the use of a patent. Booking profits selectively in countries with low tax rates (in line with tax law in the countries concerned) allows companies to maximize returns.

Steering function: The transfer pricing scheme used within a corporation can be tailored to harmonize decisions made by larger subunits. One steering mechanism is to balance supply and demand to achieve optimal resource allocation.

Market-based transfer pricing can increase pressure to perform

Fig. 7.6: Pros and cons of different transfer pricing approaches

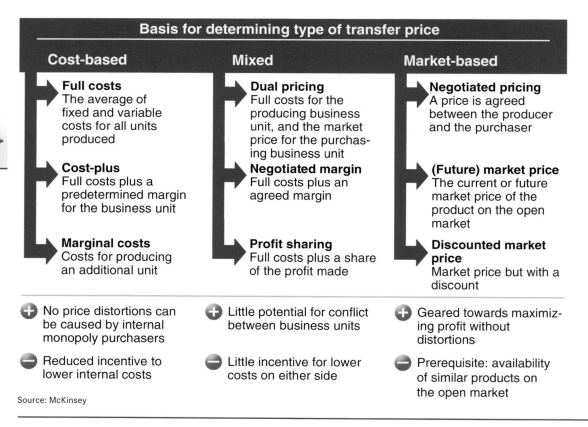

Basis for determining type of transfer price		
Cost-based	**Mixed**	**Market-based**
Full costs The average of fixed and variable costs for all units produced	**Dual pricing** Full costs for the producing business unit, and the market price for the purchasing business unit	**Negotiated pricing** A price is agreed between the producer and the purchaser
Cost-plus Full costs plus a predetermined margin for the business unit	**Negotiated margin** Full costs plus an agreed margin	**(Future) market price** The current or future market price of the product on the open market
Marginal costs Costs for producing an additional unit	**Profit sharing** Full costs plus a share of the profit made	**Discounted market price** Market price but with a discount
➕ No price distortions can be caused by internal monopoly purchasers	➕ Little potential for conflict between business units	➕ Geared towards maximizing profit without distortions
➖ Reduced incentive to lower internal costs	➖ Little incentive for lower costs on either side	➖ Prerequisite: availability of similar products on the open market

Source: McKinsey

7.1.3.2 Methods for Determining Transfer Prices

If the location receiving goods or services is a cost center or profit center, its interests will not be consistent with corporate objectives. It is in the center's interest to get the lowest possible transfer price to improve its own cost situation. This is contrary to the interests of the location supplying the goods or services if it is a profit center. To maximize profits, it must aim for the highest price possible. As a result, the procedure used to set these prices is critical. The center concept can only work if transfer prices are set fairly.

There are a number of approaches for determining transfer prices. These can be grouped into three main categories (see Figure 7.6): cost-based, market-based, and mixed. The primary difference is the degree to which they create **incentives to act in an economically rational fashion**.

Cost-based transfer pricing does not exert any additional pressure on the production location to reduce costs. In fact, it hardly matters how high the product costs rise – the price the production location can receive generally climbs in tandem. As long as the internal supply relationship is not challenged, there will not be any pressure to reduce costs from their side. The location on the receiving end, however, is faced with a real problem: If the transfer prices are high, its materials costs can spiral out of control. This will directly impact its own parameters. Often the only lever is to purchase the parts externally. The corporate center often intervenes, though, as this may be contrary to the interests of the company as a whole. This is a vicious circle that greatly devalues the performance measurement function.

The situation is different with **market-based transfer pricing**. The production location has to compete with other manufacturers and can only make a profit if it manages to keep production costs below the market price. However, this only results in a direct incentive to act if the location is a profit center. As a cost center, it remains unaffected by the prices it receives for its products.

Transfer prices are particularly important in global production networks and can have a great influence on the success of the company as a whole

Setting transfer prices does not just affect how the center's performance is judged. With global production networks, the effects are felt far more widely: Foreign subsidiaries are legally (as well as economically) independent entities. As a result, transfer prices are not only of importance for internal accounting but are also **owed de facto** and therefore trigger cross-border payments. This means that the transfer price has an economic effect on the company as a whole, as well as on the centers themselves. Customs payments are geared to the prices set in just the same way as foreign subsidiary profits are influenced by them. Both of these have a direct effect on the fiscal charges to be paid: High transfer prices can lead to high **customs payments**, but also in some instances to high **export subsidies**. This means that transfer pricing can effectively lead to a **transfer of profits**, affecting the tax burden of the locations concerned.

7.1.3.3 Interaction Between Internal Transfer Pricing and Various Types of Center

The cost center concept and internal transfer pricing are therefore intricately linked. It is not possible to give a blanket recommendation for one perfect combination as the factors influencing the equation differ too widely. Even so, it is possible to sketch a model development (Figure 7.7).

As production locations develop, it is helpful to adapt the center concept and transfer pricing accordingly

The production ramp-up phase is focused on setting up the location's own value-added steps, developing local suppliers, and acquiring new customers in the local market. There are a variety of challenges, and none of the tasks involved constitutes day-to-day operations. In this situation, a setup as a cost center with cost-based transfer pricing offers the appropriate way to measure performance. Requirements change once the site has become established. Complexity is reduced, and workflows become more routine.

At this point, the danger is that the location will become less flexible and lean, and rest on its laurels. This should be countered by building up competitive pressure. It is nearly essential to introduce the profit center concept and market-based internal transfer pricing, particularly if a large share of production is for internal customers.

7.2 Supply Chain Management

In addition to creating suitable structures, the management of production networks also includes designing organizational processes. Supply chain management plays a key role here. The importance of a well-performing supply chain management organization grows as the production network is globalized. Informal coordination becomes much more difficult in global contexts, and it is harder to find pragmatic solutions to malfunctions or errors.

Demands on supply chain management increase as the network's footprint grows

---

The primary task of supply chain management is to **coordinate the flow of materials and information** between the raw materials sources, internal and external production steps, and the customer. As a result, supply chain management essentially concentrates on the organization of supply relationships between factories, i.e., the logistical structure, transport management, and the exchange of data within the network. Cross-factory production planning is also generally considered a supply chain management task, as are demand forecasting and inventory management.

In the short term, new factories should work to optimize their own value added; in the long term, they need exposure to competitive pressure

---

Fig. 7.7: Model development of a production location

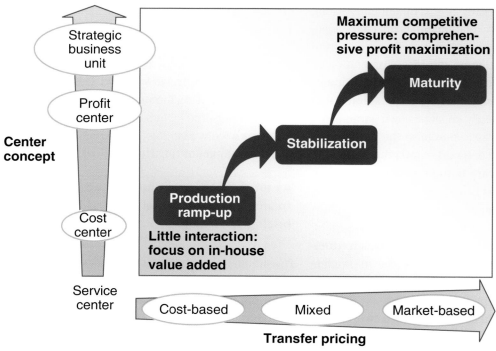

Source: McKinsey

Few topics related to global production have been covered as extensively as supply chain management and logistics. After a discussion on the general challenges for supply chain management on a global scale, this chapter focuses on three approaches as examples for setting up and optimizing a global supply chain. The last section looks at the specifics of logistics in emerging markets.

7.2.1 Challenges for Global Supply Chain Management

There are three main factors driving the ever growing demands on international supply chain management:

The significance of supply chain management is growing in multiple ways due to the distances involved in global production networks. This has a financial aspect as **transport costs rise** and ever more capital is tied up in inventory. And it plays an increasingly important role in the value proposition to the customer. Customers are making ever higher demands on manufacturers' delivery capabilities, while transport times between plants are increasing, making the supply chain less flexible and robust. Logistics is becoming a critical success factor.

Second, the **complexity** of operational control and planning of the logistics chain is increasing due to the varying conditions in different countries. More loading/unloading events when the mode of shipment changes, more transport stages and more cross-border routes all lead to a greater number of business transactions. The logistics chain is becoming more prone to disruption.

Third, **language and cultural barriers** can hinder rapid, unbureaucratic exchange between the parties involved, yet this is vital for finding pragmatic solutions, particularly when difficulties arise.

Globalization tends to lead to longer transport distances and times, which lead directly to **increased inventory** in transit. This is also problematic because the life cycles of many products have become so much shorter over the past few decades. This means a product can lose a great deal of its value during a multi-week transport period (Figure 7.8).

Products are becoming obsolete faster as are inventories. At the same time, demands on delivery times are increasing. Particularly fashion items – a category that now includes items like digital cameras and consumer electronics – come under price pressure within just a few weeks or months of launch. This lowers the margins that can be earned, making reliable supplies critical.

The long periods for which parts, components, and finished products are stuck in transit, i.e., during transport, customs formalities, or unloading and reloading, also make **supply chains less robust**. While it is relatively easy to make up for a missed delivery due to a breakdown by adding an extra shift if the two parties involved are geographically close, this is not possible for intercontinental shipments by sea. As a result, higher **safety inventories** have to be held, particularly in the warehouses where parts are received for the next stage of production, to maintain manufacturing productivity and be able to guarantee customer supplies. Market trends are worsening this problem. Because companies are offering ever more product variants, keeping safety inventories on hand at a variety of locations is expensive. To avoid this (and save warehouse handling costs), more and more manufacturers are demanding just-in-time deliveries, thereby passing on part of the problem to their suppliers. Suppliers must take this into account when choosing their production locations and designing their distribution systems.

A country-by-country comparison demonstrates the increasing importance of logistics as a cost factor in entering new markets and using factor cost advantages by producing abroad. This particularly reveals the varying quality of different countries' infrastructures.

> **Logistics as a share of total landed costs is increasing by more than the transport costs alone**

On average, **logistics costs** comprise between 5 and 10 percent of sales revenues in Europe and the US. For products produced in India, this figure is 13 to 15 percent; in China, 16 to 20 percent.[6] In developing and newly industrialized countries such as China, it is not only the poor infrastructure: government regulation also results in increased logistics costs and lower transport reliability (Figure 7.9). Environmental conditions are improving as developing and newly industrialized countries become more industrialized. Logistics costs as a share of China's GDP dropped from 23 percent in 1997 to 17 percent in 2004 and are projected to fall to approx. 15 percent[7] by 2010, yet they still pose a considerable challenge when attempting to create a reliable supply chain.

The second key parameter for the increasing importance of supply chain management is **economies of scale**. These affect logistics in two ways. Economies of scale in production are a major driver of increasing transport volumes. They also make supplying an

---

Globalization increases cost pressure and the need for flexibility in logistics

Fig. 7.8: Changes to logistics resulting from globalization of the supply chain

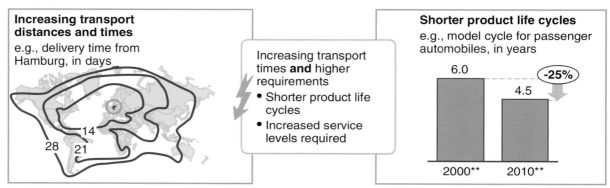

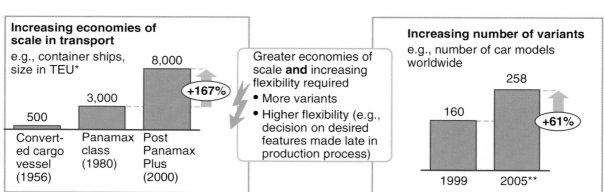

```
  *  Twenty-Foot Equivalent Unit
 ** see Abele (2004d)
```

Source: Abele (2004d), McKinsey

---

[6] See Rodrigue (2005).

[7] See Hammond (2004).

Global logistics has to deal with country-specific challenges

Fig. 7.9: Examples of hurdles to fast and efficient cross-country transport in China

Complicated procedure for granting licenses			Complaints about protectionism
Processes	**Authorities**	**Requirements**	
National truck operating license	• Transportation ministry	• Required by every service provider	"Many cities will not allow a truck to enter without a lengthy registration process. Did you know that a truck from Tianjin cannot enter Beijing until it has waited hours for a temporary delivery certificate?" *Executive of a shipping agency joint venture*
Provincial truck operating license	• Provincial government	• Required at the corporate center	"If a truck comes from a different city, the local authorities often do their utmost to impose a fine and hold it up." *Executive of a shipping agency joint venture*
Municipal truck title	• Municipal transportation office	• Required for the purchase and registration of trucks (e.g., in Shanghai)	"Even though we have a national license, local regulations hamper our business activities in many regions." *Executive of a large port handling company*
Municipal truck operating license	• Local police	• Required to offer transport services in a city	"We prefer to send shipments by truck as bonded goods, as no one will tamper with a sealed truck. Otherwise you need a 'Gong An escort' to travel to places such as North East China." *Executive of a trucking company*

Transport times vary considerably – this variability leads to high safety inventory and poor delivery reliability

Source: Interviews, published documents

Economies of Scale in Logistics: Advantages of Company-Wide Integration

Large companies can achieve competitive advantages via economies of scale – and this particularly applies to logistics. A North American company with annual revenues of USD 30 billion provides a good example of what needs to be done to make this work in practice.

The group comprised almost 100 individual companies, grouped into 18 strategic business units. These companies operated roughly 200 production and value-adding plants in 70 countries and performed research and development in approx. 100 laboratories worldwide, employing altogether around 100,000 people, of which one-third were outside the United States.

The company's global logistics team was faced with the challenge of meeting the individual supply chain needs of all its business units, while taking advantage of the total company's considerable bargaining power in the purchasing of over USD 1 billion in freight volume annually. This power could only be leveraged by cooperation. It was realized that exceptions and individual contracts concluded by the business units were counterproductive, and they were largely abolished.

The successful central negotiating strategy of the logistics team helped the group to meet its cost-cutting targets. The company's distribution costs for finished goods dropped from 5.33 percent to 4.4 percent of sales within only three years. The company managed to achieve this during a period when distribution was undergoing transition, requiring ever more long-distance international shipments that were generally more expensive.

A major reason why it proved possible to use the supply chain profitably was indirectly related to the fact that centralized bargaining power could be used for procuring of transport capacity. Domestic freight distribution, which used to be organized by the individual factories, was centralized and placed under independent management. All business units now booked their shipments through the corporate center, where specialists optimized the movement of freight between approx. 40,000 source-destination pairs, taking into account the transport firms preferred by the indi-

vidual business units. Each BU was informed of the price and service options that the corporate group had negotiated and then made a list of approved transport firms. The group companies were given a range of options from which they could choose the best fit with their needs. Their choices also entailed varying costs, of which the corporate group informed them clearly in advance.

Eighty percent of processes at the corporate center were automated. The objective was to create a system in which entry of an order by any of the BUs would automatically trigger a dispatch note and a freight booking.

Bottom line: Leveraging economies of scale in logistics does not happen automatically. Only an integrated, group-wide process can offer substantial savings, especially when procuring transport capacity.

entire region from a small number of factories – or even just one – an attractive option, especially as **economies of scale in transport and inventory management** can considerably reduce costs. Integrated transport and inventory management allows the capture of corporate synergies regardless of whether transport operations and warehousing are conducted by the company itself or by independent service providers (see box "Economies of Scale in Logistics: Advantages of Company-Wide Integration").

7.2.2 Three Approaches to Setup and Optimization

This section looks at three important aspects of setting up and optimizing a global supply chain. The first, designing distribution networks, is inherently complex. Many companies fail to optimize their costs in this area due to the many transport connections and shipments that need dovetailing. Transport management involves the broad field of tactical planning and operations. We particularly examine the benefits of parallel multimodal transport for increasing effi-

ciency. The third example, tactical capacity planning, looks at how to break down strategic planning to the

Three examples of approaches to setting up and optimizing a global supply chain

Fig. 7.10: Topics related to supply chain management

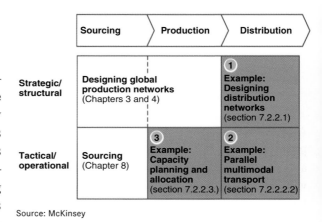

Source: McKinsey

very practical level of capacity planning and allocation (Figure 7.10)

7.2.2.1 Design of Distribution Networks

Almost all value chains contain production steps with significant economies of scale. Sheet-metal forming is a good example in automotive manufacturing. But even with simpler products – from the production of printed materials to consumer goods, such as toothpaste – economies of scale lead to the **concentration of production** in a few locations. The design and management of an efficient, high-performance distribution network to bridge the distance from the factory to the customer are essential to ensure proper market supply. Even if some companies farm out logistical processes either completely or partially – whether intra-company logistics, finished goods warehouses, distribution centers, transportation (line haul), distribution (pickup and delivery), or coordination – it is still necessary to perform meticulous strategic planning and evaluation of the various options. This is particularly true for companies with multiple endpoints in their supply chain (e.g., retail stores), to which a large number of product variants have to be supplied extremely quickly.

Analysis of existing distribution networks typically reveals cost reduction potential of 15 to 30 percent. With rapidly growing networks, transport capacity is often poorly utilized, while in more mature segments, the number of warehouses and distribution centers is either too large or the number of levels (e.g., three-level structure: central warehouse, regional distribution centers, local distribution centers) is not justified by requirements or the costs entailed.

7.2.2.1.1 Service Level – An Indicator of Logistics Performance

A well-performing distribution network is one that provides a **high level of service**. The service level is a measure of how closely customers' deadlines and volume requirements are fulfilled. It is one of the key indicators by which customers implicitly or explicitly rate a company. Unlike costs, where customers are almost always seeking reductions, customers expect their suppliers to meet certain minimum requirements when it comes to delivery capability. Exceeding these requirements is rarely of much importance to them, but failing to meet them is often a knockout criterion.

Key to defining what is required of a distribution network are the **customer's desired delivery lead times** and the optimal **service level**. α and β service level definitions are the most frequently used (see appendix). How much extra would the customer be willing to pay, for example, to improve delivery reliability from 95 percent to 98 percent? What is the most that such optimization can cost and still be worthwhile? In addition to conducting customer surveys, it is also helpful to analyze competitors. What levels do competitors achieve? Where is their performance better? And what will the customer reward or explicitly object to?

7.2.2.1.2 Volume Structure, Economies of Scale, Delivery Requirements, and Product Characteristics

There are a multitude of ways to cover the distance from manufacturer to customer. Direct deliveries from the production site to the customer using the company's own trucks is one. Another is bundling products for dispatch to the target region in full containers, where shipments are then deconsolidated and distributed. In such a setup, it is typically advantageous to use a logistics service provider, either for the local deconsolidation and distribution or with responsibility from end to end.

The key question is which products are best delivered to the customer via which distribution mode. The two main characteristics of this **volume structure** are:

Firstly, the point at which the sales order is assigned to the component (**order decoupling point**). In the case of make-to-order production, transport time from factory to customer is the key determinant of delivery time, whereas with make-to-stock production, delivery may take place from centrally or locally held inventories. With **make-to-order** production, short delivery times mean either a regional production site or airfreight is essential. With **make-to-stock** production, central manufacturing is possible even when delivery times are short. Customers are supplied from warehouses and distribution centers close to the marketplace; inventories are replenished from the factory. When delivery times are longer, there is greater flexibility, eliminating the need for local production or holding inventory.

Volumes per customer: The volume of products to be transported from a factory to a customer determines whether special vehicles can deliver direct, or whether several shipments need to be combined to obtain sufficient volumes to utilize vehicle capacity, ensuring cost-efficient transport. While special trucks may make sense for supplying a large wholesaler ex-works, direct delivery using dedicated vehicles is not really worthwhile for smaller products and multiple delivery points.

An evaluation of order and supply data from previous years can provide a good starting point for determining the volume structure for a specific company (Figure 7.11). This analysis can reveal details such as typical order structure, fluctuations in demand over a year, and regional distribution. This data provides initial indicators as to which options might be best suited to each product, from direct delivery using in-house vehicles to warehousing and distribution by third parties. An analysis of volume per sales order and lead time until delivery deadline offer a starting point for determining the most suitable distribution structure.

Analyzing the order books and identifying requirements are key to optimizing the distribution network

Fig. 7.11: Segmenting transport volume
Percent

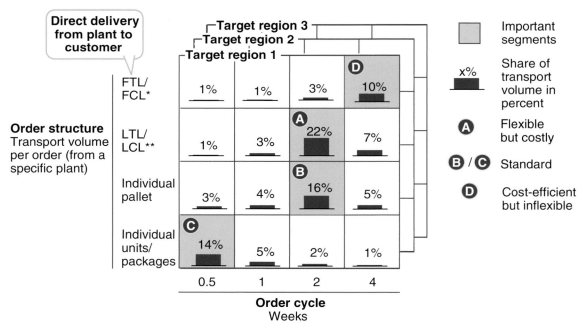

* FCL: full container load, FTL: full truck load
** LCL: less than container load, LTL: less than truck load
Source: McKinsey

The substantial **economies of scale** of logistics networks are an important factor influencing the strategic design of a distribution system. As volumes increase in the network, unit costs drop while performance improves. Capacity utilization of vehicles, ships, and aircraft rises as volumes grow, and more direct transport connections reduce distances to be covered, the number of transfer points, and transport time. It is therefore often advisable for smaller companies to make use of the logistics networks offered by logistics service providers.

It can be worthwhile for large companies to operate their own warehouses or even transport networks. An effectively designed and well-organized distribution network leads to high service levels, allows customized delivery and packing, and lowers costs.

While the economies of scale of logistics networks are a driver for bundling volumes into either one or just a few transport networks, the demands placed on delivery and the product characteristics often have the contrary effect. A transport network must be designed to meet the **highest requirements** of all the products, so including small volumes of products with specific demands can be uneconomical. Goods requiring temperature control, for example, are typically dealt with in their own transport networks. It is equally advisable not to feed small quantities of hazardous materials into the general cargo network, as otherwise it would be necessary to make provisions for handling and transporting hazardous goods across the entire network.

It is thus vital that manufacturers determine which distribution structure is best suited to each product. Cluster analysis should be used to assign as many products as possible to a particular structure to achieve maximum economies of scale.

7.2.2.1.3 Determining Consolidation Levels and Modes of Transport

The segmentation of products into clusters is followed by the definition of the basic distribution network structure, i.e., the number of transfer points and

> **The first step in designing a distribution network is to perform a cluster analysis of the products and their key characteristics**

warehouses. The initial focus should be to determine the **number of consolidation levels** and the fundamental **inventory strategy**.

The next steps are to determine the exact number of company-owned warehouses and transfer points, choose locations accordingly, and then select logistics providers for transport, transshipment, and warehousing. Figure 7.12 outlines the options available for the criterion "product volume per customer."

In the case of **make-to-order production**, the design of the distribution structure has a direct effect on the selection of production locations. If it is not possible to ensure sufficiently quick delivery times using the existing or planned production location structure, then plans must be reviewed and additional assembly locations added.

With the production of mobile phones, for example, one possible solution is to supply customers from a single factory per subcontinent. Demand for mobile phones is very volatile, however, and devices vary greatly – particularly their software and housings, depending on network operator requirements. As a result, mobile phones are produced on a make-to-order basis. Small manufacturers with only one factory can only achieve effective distribution using direct airfreight to the customer's local distribution points. They should use a logistics company due to the high frequency of deliveries and sometimes large number of endpoints. These specialists can further subdivide shipments and deliver direct to individual sales locations.

When designing distribution networks, it is quite legitimate to use several basic structures, though care should be taken not to use too many. For example, if a single factory is supplying all markets, it is conceivable to deliver direct to key accounts' central warehouses using dedicated vehicles. However, complexity increases significantly in line with the number of basic types used, making it harder to optimize the operational efficiency and reliability of logistical processes. Even if each of the individual structures are efficient for the products concerned, combining a variety of structures makes it more difficult to plan for the entire distribution network and achieve transparency, or to achieve the required level of operational excellence.

With **make-to-stock production**, design of the distribution structure is less dependent on the selection of production locations. What is vital is the warehouse structure for safety stock of finished products. Delivery times can be shortened by maintaining a larger number of local warehouses, but decentraliz-

ing safety stock leads to greatly increased inventory overall (Figure 7.13).

A **pharmaceutical example** highlights some of the parameters for designing the warehouse structure in a distribution network. This case study analyzed the effects of shortening the delivery times for critical medicines from five hours to three. Previous process optimization efforts had already tapped the full potential from order transfer, delivery packing, and receipt of delivery to reduce this process to one hour. This meant the maximum permissible travel time radius from the storage location to the customer had to be lowered from four hours to two. As the travel time radius for all warehouses needed to cover the entire area, it became necessary to quadruple the number of

---

The smaller the average delivery volume per customer, the more complex and expensive the distribution network will be

Fig. 7.12: Options for designing the distribution network for make-to-order production

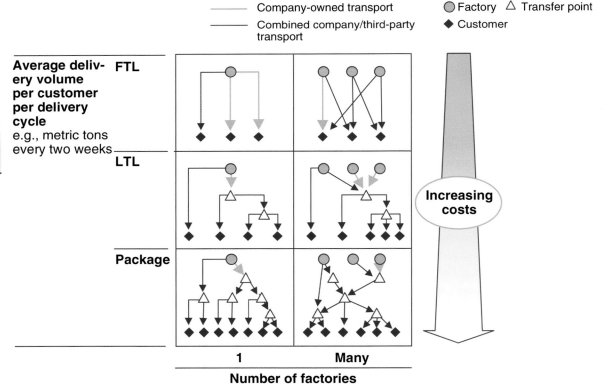

Source: McKinsey

storage locations, which meant a corresponding drop in revenue for each storage location. To maintain service levels, the relative safety inventory (= the safety inventory for each unit of throughput) had to be increased, which resulted in total inventory in the distribution network being roughly doubled.

Optimum planning of **multilevel systems** means minimizing the number of categories/clusters. In pharmaceutical distribution, for example, it may be advisable to limit local storage to medicines critical to life support, ensuring high service levels and very short delivery times on those. Products that are not as time-critical and are likely to require a low service level at the point of sale can then be supplied from a central location with a lower delivery frequency. Critical medicines could be delivered by courier from local warehouses within a few hours, while the wide spectrum of non-time-critical products – from tissues to cough drops – could be supplied at intervals of a number of days (Figure 7.14). These considerations apply equally to mechanical and automotive engineering companies for the supply of spare parts (in contrast to supply of their regular products).

Design of the consolidation structure[8] is of primary importance to companies that operate their

---

The optimum consolidation structure is largely determined by average delivery volume and service requirements

Fig. 7.13: Design options for distribution with make-to-stock production

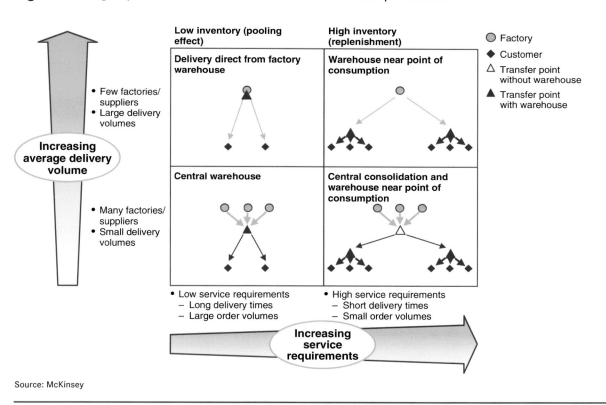

Source: McKinsey

---

[8] Consolidation refers to the pooling of two or more shipments, aimed at capturing economies of scale. This is also referred to as make-bulk consolidation. Splitting consolidated consignments after shipping is known as break-bulk consolidation.

own transport networks. Even so, companies that make use of logistics providers' transport networks for transporting intermediate products and distributing finished products can also gain valuable information from examining the consolidation structure to determine which product volumes can be handled most efficiently by which network. The more inhomogeneous the distribution of shipment volumes is, the more hybrid the consolidation structure has to be.

If the shipment volume remains constant at about a truckload, then direct delivery will be the preferred choice for local distribution. With intercontinental transport, consolidation will only take place at the port of departure, as value-adding companies can rarely (e.g., in the case of automotive manufacturers) fill an entire ship.

In the case of longer continental shipments, it may be worth consolidating consignments into so-called "full trainloads" to make the most efficient use of rail transport.

With general cargo (volume amounting to one pallet), it is usually best to consolidate transport volumes locally. This allows the use of larger vehicles with greater load capacity in line haul (i.e., for transport between distribution centers). The smaller and less regular shipment volumes are for a specific customer, the more consolidation levels will be required to reach high capacity utilization of the vehicles and achieve efficiency. However, consolidation also means that the number of handling events is increased, while the route used will deviate further from the shortest one possible (also referred to as a higher detour factor). This both increases costs and

---

Optimized distribution networks assume complex structures when demands on them are high

Fig. 7.14: Optimized distribution network for pharmaceuticals <u>EXAMPLE</u>

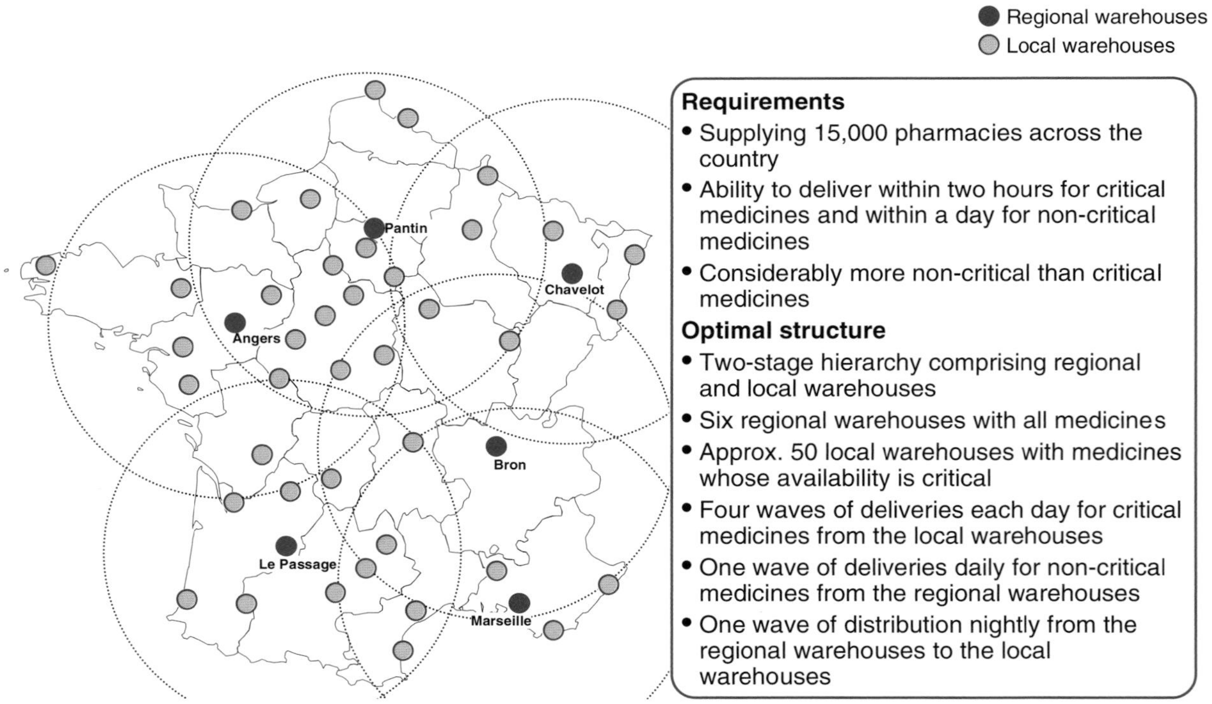

● Regional warehouses
○ Local warehouses

Requirements
- Supplying 15,000 pharmacies across the country
- Ability to deliver within two hours for critical medicines and within a day for non-critical medicines
- Considerably more non-critical than critical medicines

Optimal structure
- Two-stage hierarchy comprising regional and local warehouses
- Six regional warehouses with all medicines
- Approx. 50 local warehouses with medicines whose availability is critical
- Four waves of deliveries each day for critical medicines from the local warehouses
- One wave of deliveries daily for non-critical medicines from the regional warehouses
- One wave of distribution nightly from the regional warehouses to the local warehouses

Source: McKinsey

lowers performance of the network, as the handling events and detours result in higher time requirements.

Vehicle efficiency, capacity utilization, the detour factor, and the number of handling events are the four indicators used to **measure the efficiency of a transport network**. The targeted comparison of these indicators with those of two other transport networks quickly shows which is best suited to a specific product shipment.

Determining the optimal number and location of **distribution centers and warehouses** is a highly complex calculation. Two issues are key:

Transport networks have strongly **non-linear characteristics**. Transport costs are particularly non-linear

when calculated by volume (the transport of 200 kg does not cost twice as much as a 100 kg shipment). Given the importance of the commitment to deliver a shipment to a customer on time, a dedicated vehicle might have to be dispatched to make good on the promise. This means average transportation costs cannot be used to optimize distribution systems. The costs of each specific vehicle in the network are relevant, and the actual cost for the distribution of a shipment depends on the capacity utilization of the vehicle it is delivered with. It is essential to take this into account during optimization.

The total costs of the network compared to the number of consolidation points in a certain range generally demonstrate a relatively continuous function within a **flat minimum** (Figure 7.15). Within this flat solution space of near-optimal costs, there is a

When a network is optimized, several solutions are often equally good

Fig. 7.15: Dependence of achievable cost improvements on the number of regional warehouses

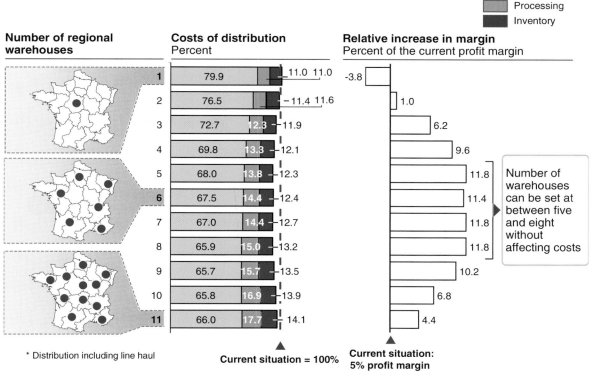

* Distribution including line haul

Current situation = 100%

Current situation: 5% profit margin

Source: McKinsey

great deal of scope for optimization in line with other criteria. Once the solutions with the lowest total costs have been identified, the additional expenditure for migration (investment and restructuring costs) needs to be minimized for these solutions if a warehouse or distribution center structure already exists.

7.2.2.2 Transport Management

Transport management involves tactical planning and operational control of the transport links between the transport nodes.

With global distribution networks, the costs of tied-up capital are often as high as the actual transport costs

---

The costs of a distribution network comprise more than the direct operating costs (transport, handling, and insurance costs, etc.). The **costs of tied-up capital** and the depreciation of the value of goods during transport and storage also represent a significant expense. The example of a relatively heavy, bulky automobile gearbox (see Chapter 4) reveals the importance of these factors. Transport by seafreight and truck from Southeast Asia to Europe costs approx. EUR 30 per gearbox. Assuming total transport and storage time of two months (transportation itself takes approx. six weeks), capital costs of 12 percent per annum, and value depreciation of 8 percent per annum, these costs amount to around EUR 16 – more than 50 percent of the direct transport costs. The relationship between the costs of tied-up capital and logistics costs is greatly influenced by the **value density** of the cargo. If it is possible to

---

Optimizing the distribution network should include the selection of transport routing for cost reasons

---

Fig. 7.16: Regional distribution – an example

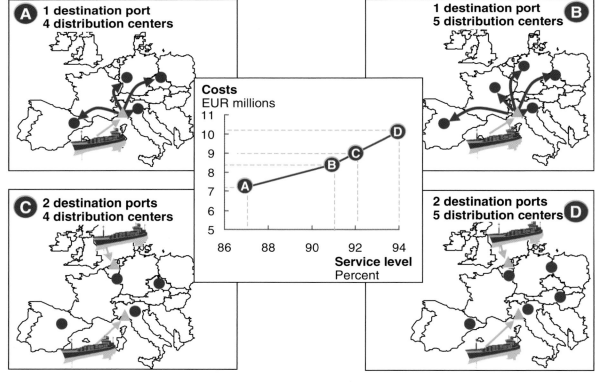

Source: McKinsey

ship multiple units of a product in a sea container, this increases the value per container and therefore the share of indirect transport costs via tied-up capital, value depreciation, and obsolescence costs.

The following subsections examine how manufacturers should organize transport management, and an innovative approach they can take to reduce logistics costs.

7.2.2.2.1 Organization of Transport Management

Transport management tasks can be divided into two categories: tactical planning, which involves **routing** the shipments (i.e., the transport links and transfer points that are to be used as a rule to get from location A to location or customer B), and the operational level, which entails dispatching, monitoring, and billing the shipments for each segment and handling at each transport node. Nowadays, both of these categories can easily be passed on to an external service provider. The organization and **operational control** of individual shipments usually only makes sense for companies with substantial volumes per customer. If general cargo is all that is usually sent, it is not normally practical for one specific company to undertake these operations, as other transport volumes would have to be acquired to ensure high utilization of transport capacity.

It is important to know the available transport links to determine the best **routing** of shipments. Then the most suitable transportation modes and routes can be determined for each product type, possibly with the help of optimization models. Routing can also influence the warehouse structure and transfer points for specific goods.

Figure 7.16 shows an example of the integrated optimization of transport routing, with the corresponding warehouses and distribution centers. The decision is whether to use one or two ports for supplying the region. In this example, using a second port of destination results in lower volume per surface transport route used, leading to higher transport costs per unit due to the decreased economies of scale.

Operational organization and control of transport activities chiefly involves scheduling vehicles to transport all shipments within the time allotted at minimum costs. Transport networks with regular volume flows work with schedules for this that only require minimal adaptation in the event of fluctuations.

If shipments are only needed sporadically, however, vehicles have to be individually assigned to shipments. This poses a dual logistical problem, making planning very complex: the goods have to be picked up and delivered on time, but the vehicle's capacity also has to be utilized to the maximum throughout the round trip.[9] This type of dispatching is only possible with high volume density, i.e., within a transport network with less-than-truck-load volumes per customer being supplied, and a large number of shipments for each day and region. If the volume density is low, it is better to use transfer points for load consolidation. This then means compiling optimized schedules adapted to actual volumes for transportation between the transfer points.

7.2.2.2.2 The Benefits of Parallel Multimodal Transport

For long distances, optimization of the transport modes used can offer substantial savings potential. For distances spanning the globe, the cost of airfreight is approximately 25 times as much as sending the same shipment by sea. Airfreight does, however, offer the advantage of more direct links, particularly for areas far from the coast, and is much quicker overall.

In general, companies can make use of two different modes of transport when optimizing transport networks (also see section 2.5.3). **Sequential multimodal transport** involves using multiple modes of transport one after the other (e.g., truck, rail, and ocean-going vessel) and is widespread. With **parallel multimodal transport**, multiple modes of transport are used at the same time. If, for example, trans-

---

[9] *This can be described as the "capacitated, time-constrained traveling salesman problem."*

portation of a small share of the material by airfreight instead of seafreight allows a disproportionately high reduction in safety inventory, then parallel multimodal transport offers an effective improvement lever. The parallel utilization of air- and seafreight for intercontinental transport links can make the production network both more efficient and more robust.

The parallel use of two modes of transport can both save costs and raise service levels

The increased efficiency is primarily due to the nonlinear relationship between the share of goods sent by air and the required safety inventory. The shipment of even a small percentage via airfreight (less than 5 percent of total transport volume) can result in a reduction of the safety inventory needed in the receiving storage location for the next transport

stage of up to 50 percent. At the same time, this increases both **process and delivery reliability**. Even product quality can be improved by the use of airfreight for a portion of the total transport volume. This reduces the time lag between the production of parts and their assembly at the customer and subsequent quality control. If quality problems occur, remedial action can be taken faster. The difference can be significant. If a quality problem is not discovered until after six weeks at sea, everything produced during that time will be lost or at least require rework. If a portion is sent ahead by air, the fault will be discovered within a few days.

Air transport should be used to cover peaks in demand. The decision regarding the cut-off point, i.e., the level of demand in any given period after which air shipment should be used (demand peak), is determined by the value density of the goods, the desired

---

Multimodal transport is worthwhile for a wide range of products and value densities

Fig. 7.17: Optimum transport policy

SIMULATION MODEL OUTPUT

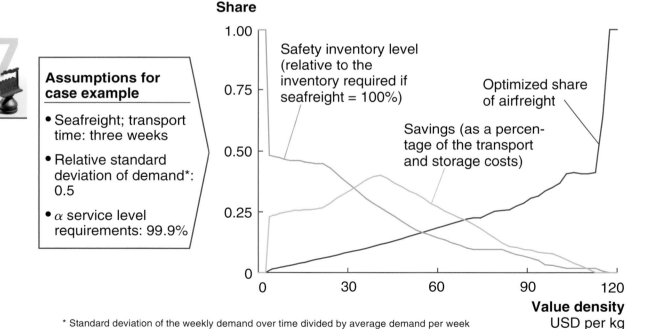

Assumptions for case example

- Seafreight; transport time: three weeks
- Relative standard deviation of demand*: 0.5
- α service level requirements: 99.9%

Share

Safety inventory level (relative to the inventory required if seafreight = 100%)

Optimized share of airfreight

Savings (as a percentage of the transport and storage costs)

Value density
USD per kg

* Standard deviation of the weekly demand over time divided by average demand per week

Source: McKinsey

service level at the destination, and fluctuations in demand. This cut-off point and the resulting target figure for the average percentage of airfreight have to be determined separately for each customer being supplied and for each product, but can be approximated very quickly using tabulated values.

Thanks to the speed of air transport, even partially using it shortens the risk period. The transport volume does not have to be determined until shortly before – or even after – receipt of the order. The benefit of this relative information gain/**shortening of the risk period** is substantial. It is easy enough to optimize the quantity to be sent by airfreight as planners will know how demand and inventory levels have developed since shipping the base volume by sea.

Depending on the value density, desired service level, and standard deviation of demand, this approach can save **up to 40 percent of the costs** vis-à-vis the best unimodal transport option (Figure 7.17).

Companies can use parallel multimodal transport in two ways to increase efficiency:

Proactively determining the share of airfreight and the cut-off value from which airfreight will be used in a particular demand period can be anchored as a permanent feature of transport management.

Ex post analyses of shipments to customers to see where too little or too much use was made of airfreight. This procedure is particularly advisable when first using this approach, as there are often multiple factors influencing the use of airfreight rather than seafreight that should be taken into account in company-specific guidelines for transport management.

7.2.2.3 Tactical Capacity Planning

Chapters 3 and 4 focused on planning the future development of production network locations. This planning is based on a long-term view, which always entails a degree of uncertainty. And the result of this

planning is often relatively highly aggregated: production capacities are usually only shown by year and product group. The strategic planning figures require further detailing and adaptation to become really operational.

Tactical production planning (also called production program planning) is concerned with when and for how long the different factors of production should be combined. The main production factors to be considered are labor, plant and machinery, and materials. These planning steps are essential for the operational workflow. Planning errors can quickly lead to deadline slippage, cost hikes due to poorly chosen lot sizes, unnecessary inventory, or untapped cost advantages. Integrating strategic targets from previous planning steps into production planning is vital to achieve the projected results.

The hierarchical planning system (see box: "Managing Complexity: Global Production Planning at Mercedes-Benz") distinguishes three **planning horizons:**[10]

1. Strategic planning over a number of years or decades (also referred to as location planning or target planning)

2. Tactical planning over a number of months (also known as production program planning)

3. Operational planning for periods of less than one month (also called job shop scheduling or detailed planning).

The detail becomes ever finer as you transition from overarching strategic planning to operational planning. The production window in focus grows ever smaller, the planning horizon gets shorter, and decisions take the form of specific production activities.

Strategic (location) planning has already been detailed in Chapters 3 and 4. It sets the long-term goal for the further development of production capacity at different locations in the production network.

---

[10] See Drexl (1994), p. 1022.

Managing Complexity: Global Production Planning at Mercedes-Benz

As a premium manufacturer, Mercedes-Benz offers its customers highly customized vehicles with a wide range of innovative standard and optional features. This results in a huge number of variants and extreme complexity, further accentuated by the differing legal regulations across the world. Mercedes-Benz installs roughly 110 different rear windscreens for the C-Class, for example, depending on the type of antenna to be attached, the tint, and other features. Optional features for the seats result in over 10,000 variants, covering the variety of upholstery, additional functions, types of controls, option of left- or right-hand drive, and much else. All the parts for these variants

The number of production ramp-ups is growing

Fig. 7.18: Average number of new product lines per year

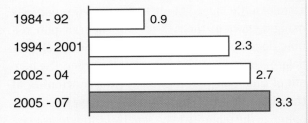

1984 - 92	0.9
1994 - 2001	2.3
2002 - 04	2.7
2005 - 07	3.3

Source: DaimlerChrysler

need to be delivered to the production line just-in-sequence. Production takes place after the customer order is placed, with short lead times.

Hierarchical planning systems allow the effective management of global production networks

Fig. 7.19: Global production program planning at the Mercedes Car Group

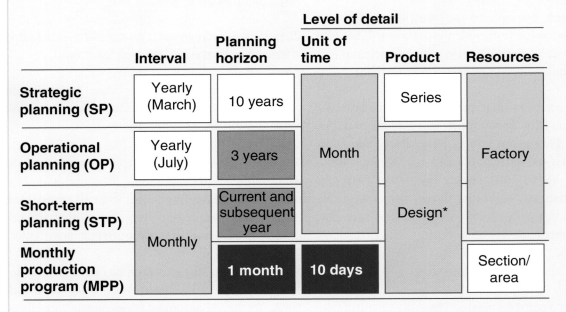

	Interval	Planning horizon	Level of detail		
			Unit of time	Product	Resources
Strategic planning (SP)	Yearly (March)	10 years	Month	Series	Factory
Operational planning (OP)	Yearly (July)	3 years		Design*	
Short-term planning (STP)	Monthly	Current and subsequent year			
Monthly production program (MPP)		1 month	10 days		Section/ area

* e.g., C-Class 320 sedan, left-hand drive

Source: DaimlerChrysler

Mercedes-Benz is spearheading a product drive in which it intends to use new niche models to cover most of the increasingly heterogeneous demands of its customers. However, it faces major challenges due to the large number of additional product lines required (Figure 7.18). Integrated production planning is helping it to overcome them.

The global production program planning process at the Mercedes Car Group (MCG) is organized in line with the principles of hierarchical planning: general strategic planning takes a relatively broadbrush approach and has a long planning horizon. Details are added via operational and shorter-term planning, as well as in the monthly production program (Figure 7.19). This helps to ensure that long-term planning is implemented more or less consistently on a day-by-day basis.

The planning process and execution of the world production program are supported electronically. MCG uses a central planning system that provides all data related to a production program (unit volumes and types, remp-up projects, key data, etc.) from a consistent set of information resources in house. It is also possible to take start-up curves into account and monitor levels of fulfillment using a variety of indicators. Based on the global production program plan, customer orders are allocated to available production capacity and slots. For slots that are not filled by existing customer orders, virtual orders are created based on expected demand. Those orders are then used for finetuning capacity plans and for material requirements planning.

Bottom line: MCG's use of a hierarchical planning system in its global production program allows it to coordinate its worldwide production effectively and produce cars tailored to customer requirements.

The backdrop for **tactical (capacity) planning** is rather different. Production capacity is already in place and only has limited leeway for adjustment within the planning horizon. The primary task of tactical (capacity) planning is to assign customer orders to locations: this is order scheduling at a multi-plant level.

The importance of this varies from industry to industry. In the automotive supply industry, such assignments can remain unaltered for years, as every change in production location requires a new audit by the OEM and entails high costs.

In other industries, such as mobile phone assembly, production can be relocated at short notice to smooth out regional fluctuations in customer demand or take advantage of exchange rate differences to increase returns (see Chapter 2).

In day-to-day activities, operational (monthly production program) planning is responsible for distributing incoming orders to the machine/personnel level and triggering production and purchasing orders through a materials requirements planning (MRP) process. It steers materials flows and inventory to ensure that all order delivery deadlines are met at minimum costs. In this scheduling process, orders are sequenced and resources are assigned, such as production capacity on an assembly line. The prerequisite for cost-optimized scheduling is detailed knowledge regarding the underlying cost structure. It is almost impossible to influence production capacity at this stage, apart from scheduling maintenance, determining lot sizes, and shift planning. This is the domain of MRP and enterprise resource planning (ERP) systems, i.e., computer-aided systems for short-term planning and control of production activities.

A global production network barely has any influence on the approach used for short-term planning, which is almost identical to that for an individual site. It is a different story for tactical planning. Strategic planning specifies the development of production capacity over time at the process step level, and this has to be taken into account in the tactical planning process. If one assumes production capacity is a given, there is still the question of allocation: Which factory should an order be assigned to? This leads to a more fundamental question: What exactly should tactical planning optimize? What parameter should be used to compare two alternative plans? How can you decide which of the two is better? Clearly these questions are vital.

The choice of target parameter for capacity planning is fundamental, but companies select it in very different ways

Nevertheless, companies sometimes have very different answers. Tactical planning can be used, for example, to ensure that **capacity utilization** is as high and as constant as possible, to always supply a customer from the same factory (presenting "one face to the customer" – see box: "Managing Global Production: Hella's Network"), or **minimize costs**. None of these target parameters is fundamentally

wrong, and selecting the most important very much depends on circumstances. Even so, some pitfalls need outlining in general terms.

Particularly at small to medium-size enterprises, **capacity utilization** (and thus the capacity available) is often the criterion of choice. If a factory is being underutilized, it is given the next order received. This might seem an obvious course of action, as underutilization can quickly lead to losses at this location due to the **fixed costs**.

But what are the long-term consequences? What it boils down to is that locations with insufficient performance to prolong and expand on their existing orders are artificially kept alive, while the best locations are given the fewest new orders. Naturally, the underutilization of capacity is not always to be equated with below-average performance, but setting even capacity utilization as a goal risks **concealing weaknesses**.

The fixed assignment of customers to factories – the "one face to the customer" concept – can have a similar effect. Factories are no longer subject to internal

Managing Global Production: Hella's Network

The automotive components supplier Hella generates more than EUR 3 billion in revenues with 23,000 employees – of whom approximately 10 percent are in R&D. This is spread over Europe (85 percent), NAFTA (10 percent), and Asia-Pacific (5 percent). This family firm's production is distributed across 37 locations around the globe, 17 of which are low-cost locations. The company is divided into three business divisions: Light, Electronics, and Aftermarket & Special OE.

The product portfolio of the Light business division comprises all elements of vehicle lighting: headlamps, tail lights, signaling lights, and interior lights. Hella's value added focuses on the following core competencies:

- Injection molding using thermoplastics and duroplastics

- Surface finishes for synthetic materials (metallization, coatings, paints)

- Module assembly and final assembly (welding, adhesive, and screw connections).

Globalizing with the Customer

The overseas activities of key customers were instrumental in triggering the globalization of Hella's production network. Hella went with VW to Mexico, China, and Eastern Europe, where it founded production companies near the factories for which their components were destined. Now Hella has a presence in Eastern and Western Europe, NAFTA, and Asia-Pacific with an emphasis on China to produce complex products in growth regions. This presents Hella with the challenge of effectively controlling and managing a global network.

Setting up a New Location – Example: a Factory in the Czech Republic

Hella's first move into Eastern Europe took it to the Czech Republic in 1992. By setting up production capacity locally, it was able to meet its goal of entering this market, which was dominated by its largest customer, Skoda. An important factor in successfully setting up a location in the Czech Republic was the employment of local managers to run the company. The local management team was given far-reaching entrepreneurial freedom to build up the location under the conditions laid out in the business plan.

This early entry into the Czech market proved advantageous for Hella. Its negotiating position vis-à-vis the authorities was good, as there were few Western firms operating there at that time. There was also a sufficient pool of well-qualified and motivated personnel. Products for Skoda and a number of manufacturers of commercial vehicles could be produced using proven production technology that could be transferred from Germany without difficulty. This meant it was possible to rapidly achieve a high level of productivity and quality.

The location saw further expansion in the late 1990s. Besides additional capacity, new production technologies were implemented, including injection molding and the coating of plastic lenses. No new products were started up to avoid added complexity. Instead, products were transferred that were already being produced in German factories. Before the machinery and plant were transferred, the Czech personnel were trained at the German plants. The transfer was successful, and supplies were never interrupted throughout the process.

The Czech factory very quickly became one of the best-performing locations, and today it primarily supplies the VW Group and Ford. It is on a par with the German lead factories in terms of product quality, scrap rates, productivity, etc. Hella is also successfully using this location as a bridgehead into the Russian market.

Planning, Controlling, and Managing the Production Network

Hella is pursuing the strategy of supplying customers from local locations. It supplies the Skoda plants from its Czech location and the Peugeot plant in Trnava from its site in Slovakia. It maintains maximum possible continuity in the assignment of customers and products to locations, even across model generations.

As a result, each location is equipped with all the necessary technology, and no factories are used purely for manual assembly work (despite their cost advantages). Deviations from this allocation strategy are made only to balance capacity between locations to avoid underutilization.

Technological development of the locations is rigorously managed. Innovations are developed and tested at the German lead factories, integrated into the production process, and then introduced in the other locations. This procedure is used for process and production innovations alike. The improvements cover every field: production facilities, production concepts (such as integrating injection molding and surface treatments), production technologies (substituting vapor coating for painting, for example), and materials (e.g., evaluating and selecting more cost-efficient materials).

Bottom line: Moving into new markets early using established processes can create a sustainable local competitive advantage. Adopting the lead plant concept and encouraging intensive exchange of experience within the network also means know-how advantages from other locations can be used.

Achieving Global Technology Leadership – An Interview with Dr. Rolf Breidenbach, CEO of Hella KGaA Hueck & Co.

Dr. Rolf Breidenbach studied Mechanical Engineering and Economics at The Rhine-Westphalian University of Technology in Aachen, and received his doctorate in Engineering in 1991. Following management positions in industry and at a technical services firm, he joined the management consulting firm McKinsey and rose to become a partner. He became CEO of Hella in February 2004.

Dr. Breidenbach, what is the importance to Hella of its global production network?

Our customers are steadily continuing to globalize, and we are seeing increasing consolidation among automotive manufacturers. It is absolutely essential for Hella to have a global network.

Today, vehicles developed in Europe are also produced in Mexico and China. The simplest solution for an automotive manufacturer is to source its parts in all regions from a single supplier that also produces locally. A supplier without locations abroad can no longer play a leading role. Customers expect local supply, regardless of whether they are in the US, China, or Germany.

What opportunities do you see for Hella to further expand its network over the coming years?

The traditional markets in North America, Europe, and Japan are currently experiencing moderate growth at best, while growth markets such as China, India, and Mercosur are increasing in importance. Any supplier aspiring to a global presence has to expand its network to include these markets.

What factors do you think need considering when setting up a new location to ensure success?

Hella has a great advantage in that it has three high-performing business divisions. This gives us the ability to bundle activities in new markets, if we need to, to create a broader base for a new location. In our experience, it is particularly important for new locations to quickly gain access to the know-how they need to get up to speed on all the different stages of production. This should include close contact and extensive exchange of expertise with the specialists in the global network. The abil-

ity to create a network of local suppliers to avoid relying on costly imports of raw materials and components is another extremely important aspect.

How would you rate the performance of Hella's foreign locations?

The continuous growth of the Hella Group alone is enough to demonstrate the high performance of all our locations worldwide. Naturally, our locations abroad sometimes require support from the specialist departments in Germany.

The situation is somewhat different with process innovations. Our internal benchmarking procedure is always finding new best-practice solutions that have been developed by staff at our locations abroad. But customers, too, recognize the excellence of Hella's foreign locations. Last year, our factory in the Czech Republic received Ford's World Excellence Award, for example.

How does Hella manage this increasingly complex network?

As a rule, all our foreign locations are independent companies that are run as profit centers and judged according to their results. This structure allows Hella to systematically promote entrepreneurship in its subsidiaries. Our sites abroad also report their key operating indicators monthly, measuring performance in areas such as product quality, productivity, purchase cost trends, and delivery capability. We also use internal competition as a tool to link the assignment of projects with criteria for measuring location performance. This provides additional incentives for local management to continually improve performance.

What is the importance of the German locations in the Hella network?

The German locations serve as lead plants and centers of competence where new production technologies and product innovations are prepared for industrial application. We also use them for further refining existing technologies. These optimizations are then passed on to all the relevant locations, where they lead to improved quality and costs.

Our goal is to further develop these German locations as centers of competence. This, in turn, will enable us to continue implementing our technology leadership strategy in the industry at a global level.

competition for orders, as the assignment is fixed by strategic planning. Even if the factories are actually operating as profit centers, this partially negates one of the goals of this organization structure – to create internal competition.

At first glance, setting **cost minimization** as a goal would appear both objective and unassailable. And this is mostly the case – unless the issue is investment decisions or alternatives the effects of which will not be felt until some time after their implementation. Expenditure on investments is usually largely irreversible. This means that even when a capital asset has not been fully depreciated, its sale would still generally only bring in a tiny fraction of its residual value. Companies therefore run the risk of falling prey to the "sunk costs fallacy" when making investment decisions. Alternative investments cannot be compared by contrasting the depreciation (i.e., costs) on past investments under one scenario with the depreciation on future investments under another. Investments that have already been made must be ignored for the purpose of such comparisons.[11] The analysis of cash flows rather than depreciation already implicitly takes the sunk costs logic into account. A more objective and comprehensive target parameter is therefore the net present value (NPV), i.e., all future incoming and outgoing cash flows resulting from an investment, discounted to the present.

To sum up, there are two main risks: growing sluggish due to a **lack of internal competition**, and orientation towards a purely **short-term cost perspective** at the expense of long-term earnings. There are two approaches that can be helpful here:

- Placing contracts based on proposals from internal and external providers.

- Assigning products to the location offering the highest NPV to the company as a whole.

Conceptually, the use of competition as a **market mechanism** is a far simpler choice. When there are new production volumes to be assigned, the factories are requested to submit proposals. Ideally, external providers are also asked to submit proposals in the case of components unrelated to the core business. This ensures that the product is assigned to the factory whose attractive cost situation and available production capacity allow it to submit the best offer. A side effect of this process is that the factories are given a **benchmark** of their own cost position. If

---

[11] *This fact is often overlooked. In 1971, Lockheed attempted to get a bank guarantee to continue the development of the TriStar aircraft. Lockheed argued that it would be foolish to break off a project in which USD 1 billion had already been invested. Yet this expenditure was irreversible, meaning only the future expenditure and expected future income were relevant for the continuation of the project. Also see Brealey (2000), p. 123.*

a **make-or-buy** decision is to be made, this also automatically provides a transparent fact base. The process of economic evaluation is decentralized, and central management effort is minimized accordingly. The advantages are obvious. However, it is only possible to implement this concept if the factories operate as **profit centers**. As cost centers, they do not have enough of an incentive to achieve a profit contribution and increase their revenues.

The second approach is both more universal and holistic – but also more complicated: assignment to the plant that offers the highest NPV according to a **comparison of economic viability**. It may be necessary to make calculations for a number of different scenarios, depending on the number of potential production locations. The situation becomes even more convoluted when there is more than one product to be assigned. Imagine the complex impact of these decisions on revenues and capacity – their interaction with in-house production, third-party sourcing costs, and transportation. The edifice quickly grows too complex for manual planning. This is where an optimization model to support the decision-making process comes into its own. Capturing what is often a wide range of variants of the end products is not usually advisable here, nor do we recommend focusing on individual processing, monitoring, or transport operations in minute detail. Computer-aided approaches allow the inclusion of additional levers for increasing returns. Medium-term **exchange rate** trends can be used, for example, to achieve additional cost or revenue benefits by shifting production capacity within the boundaries of the flexibility available (see Chapter 2).

7.2.3 Logistics in Emerging Markets

The logistics provider landscape is highly fragmented in most developing markets. A 2004 McKinsey survey indicates that there are over five million registered trucking companies in China, and many more that are unregistered. Similarly, 77 percent of all trucks in India are owned and operated by small businesses that own fewer than five trucks in total (Figure 7.20). Such fragmentation is not atypical even

for developed countries, e.g., in Europe. What is largely lacking in developing countries, though, is a layer of large, efficient intermediates that aggregate and manage trucking capacity for shippers. Add this complexity to the difficulties that come with the size of mainland China or the Indian subcontinent, their underdeveloped infrastructure, plus the traffic congestion in major cities, and you get an idea of the challenges of managing physical flows in these markets.

The infrastructure problems mean transportation is unreliable and damage-prone. And poor industry conduct – whether overloading, complex regulation and taxation schemes, or corruption – is a further obstacle. However, supply chain management in emerging markets still needs to accomplish more than just establishing a geographically far-reaching logistics network. Getting goods into cities and towns is only part of the solution. Street businesses and small shopkeepers still account for the vast majority of retail sales. To reach these businesses, companies need a very competitive cost position and additional

---

Most trucks in developing countries are owned and operated by small businesses

Fig. 7.20: Commercial vehicle ownership structure in India
Share of total trucks in India
Percent

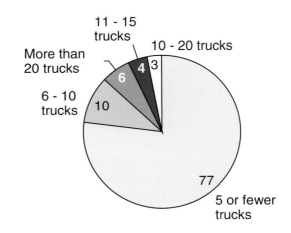

Source: Asian Development Bank estimates

capabilities beyond logistics, with credit risk management heading the list.

Manufacturers need to address these issues by developing the right capabilities and finding suitable partners. Logistics service providers can be crucial to success in a developing country. Finding and developing providers should therefore be a top priority. The following sections will look first at the current logistics situation in emerging markets and then the areas of largest improvement potential for OEMs. Our examples will focus on China and India, as these two countries are of the greatest interest and relevance for most international manufacturers. The issues and trends in China and India are generally comparable to those of other developing countries, such as Indonesia, Vietnam, Brazil, or Nigeria.

7.2.3.1 Coping with a Highly Fragmented Status Quo

The logistics provider landscape is extremely heterogeneous in emerging markets. A Chinese interviewee's comment – "China has millions of trucks and just as many logistics companies" – conveys a sense of this disconcerting market. Companies extending their footprint to these countries need to study the market structures in detail, understanding where it is worth pioneering a new approach, and where it is more advisable to optimize existing structures.

7.2.3.1.1 Multi-Layered, Non-Transparent Distributor Structures

In developed markets, large OEMs typically deliver direct to the retailer's warehouse or outlet. Alternatively, the retailer or wholesaler pulls OEM inventory as required. In contrast to this, distribution from the manufacturer to the end user is a multi-layered system in most emerging markets. Some of the parties involved keep an inventory, and therefore act as distributors. Others – primarily third-party logistics (3PL) providers and carrying and forwarding (C&F) agents – only provide services to the OEM or distributor but do not own the inventory.

Distributors in emerging markets do more than just transport goods and hold stocks. Family-owned retailers or wholesalers that are not incorporated often cannot afford to pay for product upfront. Distributors will therefore advance the goods, only collecting payment once the products have been sold, thus bearing a substantial credit risk. International companies often underestimate the **importance of inventory financing** in the fragmented retail markets of developing countries. Relationships between large and small distributors as well as between small distributors and shopkeepers are typically very close. The well-established business practices at play are completely intransparent to an outsider.

There are several reasons for the fragmented distribution structure in developing countries. The dominant market share of small shopkeepers is one key driver. Tax regulations are another. In India, for instance, intricate structures are often installed to circumvent the payment of interstate sales tax. To sell their goods in another state, companies first transfer inventory across the border into a local warehouse. They are only allowed to sell goods within the state without being subject to the state sales tax after a specific period of storage in the warehouse (a certain number of hours or days), so they hold them there for that time. The downside is an inflated number of stockkeeping points and inefficient transport routes.

7.2.3.1.2 Disparate Carrier Operations

In most major cities in China and India, owner-operated trucks wait in route-specific areas – one for central China, for example, and another for western China. This makes it easy for truckers (or brokers: typically small businesses that consolidate multiple shipments for a trucker) to aggregate volume and fill a truck for a tour. Manufacturers and other shippers are not required to pay for the backhaul – almost a perfect market in action (were it not for the endless negotiations for every single shipment, hardly any quality control, only a rough indication when the shipment will arrive, and often the need to organize the pickup and inner-city transport separately from the main haul).

Case Study: Pharmaceutical Distribution in India

In India, an estimated 800,000 chemists sell pharmaceutical formulations, largely through small independent outlets. To reach such a broad base of retailers, pharma OEMs use a multi-layered network of stockists to distribute their products (see Figure 7.21). While this system allows OEM access to the mass markets, it is highly inefficient and costly, as each intermediary charges a hefty price for their services. This is because stockists have organized themselves into an association that has a de facto monopoly in their sales territory.

Breaking the influence of stockists has long been high on the wish list of pharma OEMs in India. Any attempts they have made to break the cartel-type structure have failed so far. Stockists respond by boycotting the products of pharma OEMs that try to sell direct to institutions and retailers.

The breakthrough is likely to come in the next few years. With organized retail growing in importance and the government taking a more liberal stance, stockists will find it harder to counter a unified pharma OEM strategy. This will require tight alignment among OEMs, however, and they will likely need a logistics provider to establish a distribution network under their joint control. Global management will need to spearhead this process, as local management will shun the high risks.

Drug distribution in India – a multi-layered system

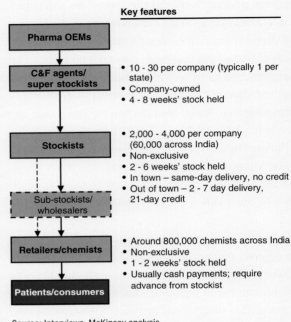

Fig. 7.21: Traditional setup of pharma distribution chain   ILLUSTRATIVE

Key features

C&F agents/ super stockists
- 10 - 30 per company (typically 1 per state)
- Company-owned
- 4 - 8 weeks' stock held

Stockists
- 2,000 - 4,000 per company (60,000 across India)
- Non-exclusive
- 2 - 6 weeks' stock held
- In town – same-day delivery, no credit
- Out of town – 2 - 7 day delivery, 21-day credit

Retailers/chemists
- Around 800,000 chemists across India
- Non-exclusive
- 1 - 2 weeks' stock held
- Usually cash payments; require advance from stockist

Source: Interviews, McKinsey analysis

Operating in this market effectively can, frankly, be a pain. Manufacturers that require substantial full-truck-load (FTL) service capacity have to deal with multiple operators. Even if a single provider is contracted to provide transportation, this company will still use other subcontractors, often with limited control. Enforcing common standards in such an environment is difficult. Large manufacturing companies require an army of people to monitor loading and dispatch, and manage performance of the hauliers. Combined with the poor infrastructure and highly variable transit times, this can lead to chaos around the plants of large manufacturers. Dozens or even hundreds of trucks will often be waiting for their load, turning the streets around the plant into a vast parking lot and camping site (Figure 7.22).

Local trucking companies providing FTL service often develop a presence in road express services. Starting from simple trucking operations, these companies are attempting to grow into higher-value services and to also offer door-to-door service for smaller shipments. While basic transportation can be surprisingly reliable, shippers should recognize that local providers operate with basic or even rudimentary equipment and infrastructure (Figure 7.23). Track and trace is rare, as are bar codes and electronic scanning. Handling is manual, and some companies do not enforce packaging rules, allowing bulky, odd-shaped items. This results in higher shipment damage rates. Manufacturing companies should therefore give some thought to contracting with a premium provider, particularly when in the process of redesigning the supply chain. Damage rates and delivery performance need to be monitored closely at the very least, and also ideally need to be linked to the fees the provider receives. High-tech manufacturers with high-value, fragile shipments particularly need to invest time in selecting providers, designing processes, and developing robust packaging.

A similar situation applies in contract logistics. With consumer goods, for instance, local C&F agents can provide basic logistics services at very competitive rates. The providers typically rely on the OEM's IT systems, to which they require direct access. The capabilities of these companies are typically limited to one region, and the ability to finance larger projects is limited. For nationwide support, companies may have to turn to international providers. These have the resources and can draw on a large pool of experts, but will require adequate compensation for deploying expatriates in emerging markets.

When it comes to air express, road express, and less-than-truck-load (LTL) services, the provider landscape is also very heterogeneous. The air express networks of major international forwarders such as DHL, FedEx, TNT, and UPS extend into most regions of developing countries and guarantee high-quality service. DHL traditionally has the greatest reach, even into the most remote areas. In domestic services, local providers are dominant, but the influence of international providers is also increasing in this field. DHL has a presence in China, for instance, through its joint venture with SinoTrans. In India, the company holds a majority share in Blue Dart, the country's leading domestic provider. TNT has recently acquired local providers in India and Brazil to expand its activities, with a focus on road express. As with local trucking, it is generally still safer to opt for a brand-name provider and stomach the premium rate if speed and reliability are of the essence.

Subcontracting transportation is very common in emerging markets

Fig. 7.22: The trucking and 3PL industry in China observations

- Great majority of trucks owned by individual trucker
- Large 3PL companies subcontract to smaller ones
- Truckers are affiliated to 3PL companies, but are not employees
- Case example:
  - Elee serves 73 Carrefour stores across China
  - Company has only 5 trucks
  - Main contractor uses some 50 subcontractors for everything else
- How subcontracting works:
  - For LTL: Truckers wait in "route-specific" markets (e.g., Shanghai to western China)
  - For FTL: Contractors wait in front of plants (photo)

Source: McKinsey, interviews

In India, pallets and other loading devices are still the exception

Fig. 7.23: Logistics operations of a carrying and forwarding (C&F) agent

Source: T. Meyer

7.2.3.1.3 Much Higher Share of Manual Labor

Handling procedures are rudimentary and largely manual. **Pallets are rarely used**: they are too expensive and take up additional space in line haul trucks. Companies should be wary of introducing them when delivering to their customers. Often customers will neither pay for nor return pallets. Pallet-handling equipment is also rare, though an increasing number of forklifts are being sold in China (Figure 7.24). India's forklift market is expected to reach 4,000 units in 2006 and 8,000 units in 2010, a big leap from the annual domestic demand of only 1,500 units just a few years ago. This growth is expected due to the greater infrastructure and construction activity. Depreciation and maintenance of forklifts and hand trucks, however, is often higher than the cost of additional unskilled workers. These costs can only be justified if the use of forklifts and other handling equipment creates additional benefits, e.g., the storage of goods on multi-level racks. Particularly in locations where an unskilled laborer costs less than USD 5 a day, even larger loads are nowadays packaged in boxes or bags and handled manually. Goods are handballed in and out, with boxes stacked floor to ceiling in the trucks, often with significant overloading.

Level of palletization and automation still very low, but rising

Fig. 7.24: Forklift sales in China
Total shipments, thousands

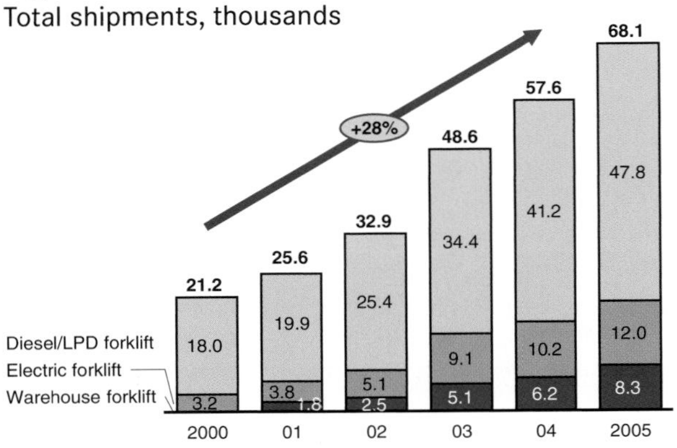

+28%

	21.2	25.6	32.9	48.6	57.6	68.1
Diesel/LPD forklift	18.0	19.9	25.4	34.4	41.2	47.8
Electric forklift			9.1	10.2	12.0	
Warehouse forklift	3.2	3.8 / 1.8	5.1 / 2.5	5.1	6.2	8.3
	2000	01	02	03	04	2005

Source: China Construction Machinery Industry Yearbook

7.2.3.1.4 Severe Issues with Overloading

Overloading is a common practice primarily due to low yields, high toll fees, greater concern over quick truck payback, and tough competition. Many truckers cannot break even without overloading twofold even threefold.

Overloading can become a **serious concern** for international companies. Governments are increasingly aware of overloading as a cause of accidents, road damage, and unreliable transportation. In China, India, Malaysia, and elsewhere, tougher regulations to curb overloading have been put into place in recent years, but enforcement will take a long time. For international companies, the common practice of overloading creates a special issue. Good corporate governance requires that local executives adhere to the new regulations and not do business with vendors that do not comply. However, in industries where transportation costs are a significant part of the retail price, this can be very pricy. As local companies typically do not require their vendors to change their practices quickly and tolerate overloading, the competitiveness of international companies can suffer. It is therefore important to be involved in the process: regulatory management is crucial in this situation. Companies affected by such regulation should try to influence the regulator either directly or through industry associations to ensure an even playing field.

7.2.3.2 Riding the Wave of Transforming Retail Structures

While companies need to adapt to local structures at first, they should be poised to capture opportunities as distribution structures mature. Transformation is taking place across all sectors in developing countries but is most visible in retail, where the consumer is driving change at a rapid pace. We will therefore use the retail sector to describe how the structural transformation of a sector impacts the distribution network.

Consumer preference for **modern retail** structures correlates with increasing wealth (Figure 7.25). Instead of using open markets, street businesses, and small shops, consumers are spending a growing share of their rising disposable income in larger stores and shopping malls. In China, the share of modern retail grew from around 7 percent in 1996 to some 20 percent in 2005. Similarly, fundamental change is expected in India's retail sector. Ever more businesses are adapting to modern retail formats. The success of local companies in multiple formats is exemplified by department stores, such as Pantaloons and Shoppers Stop, hypermarkets, such as Big Baazar and Shoprite, or single-brand outlets, such as Bata. This development is attracting the attention of major international chains, such as Wal-Mart, Metro, Tesco, and Carrefour. These companies will find avenues to enter the market, despite restrictions on foreign direct investment (FDI), accelerating the transformation. Wal-Mart, for instance, recently announced a partnership with Bharti Enterprises, which will run Wal-Mart stores as a franchise, with the backend logistics and purchasing organization controlled by the US-based retailer.

> **Organized retail in India is expected to follow China, which grew from 7 - 20% in 10 years**

Fig. 7.25: Development of organized retail
GDP per capita vs. share of organized retail

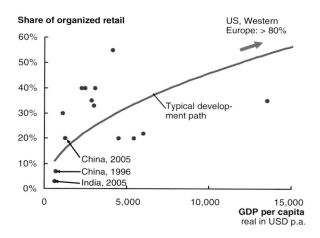

Source: Morgan Stanley, McKinsey

The transformation of the retail sector has important implications for manufacturers, particularly when it comes to the design of their distribution network. Modern retail businesses are much bigger, have easier access to capital, and are more professional in their business practices. They are much more suitable as business partners for international OEMs than thousands of small shopkeepers. This gives international manufacturers the opportunity to become independent of distributors and C&F agents, and redesign their distribution networks. The potential rewards are high in terms of greater efficiency, reliability, and control.

Manufacturers should follow a differentiated approach, remembering that markets mature at different points in time. In China, for instance, multiple route-to-market models must be established and controlled in parallel, based on city and channel characteristics. In the cities, the modern channel structure is already fairly developed. Large manufacturers with dominance in at least one product segment typically have large enough batch sizes to allow direct delivery to modern retail businesses in the cities. This leaves distributors and wholesalers for lower-tier cities, smaller towns, and the more traditional retail formats. Many companies therefore need to run several very different supply chains (even for the same product), often involving many partners, warehouses, and complex logistics flows (Figure 7.26).

7.2.3.3 Controlling the Inbound Supply Chain

For manufacturers, optimization potential lies in inbound logistics as much as in distribution. Realizing this, however, requires more than just improving transport flows and inventories. OEMs have to take control of the logistics spend before they can move to network optimization.

In emerging markets, traditional shipping behavior is still predominant, with the shipper organizing transportation and bearing the costs. This particularly applies to domestic flows. There is little coordination between supplier and OEM other than regular orders specifying volumes and delivery dates well in advance. In contrast to practices in developed countries, OEMs take a pivotal role in coordinating production planning and logistics and pull inventory direct from the supplier, while keeping little or no safety stocks themselves.

International retailers in China have demonstrated how control over their inbound logistics can help reduce costs and improve supply chain performance. Nowadays, nearly all large American retailers use logistics service providers under their control to consolidate volume in China for export. This has multiple advantages for the retailers. They can leverage their buying power and contract direct with shipping lines. They have control over inventory and build container loads to suit their needs, for example, enabling direct shipping to stores in the US. Labor-intensive activities such as labeling and (re-)packaging can be performed in China, and thus at relatively low cost.

Many companies have yet to realize such benefits by taking control over their inbound supply chain.

7.2.3.4 Capturing the Benefits of Growing Infrastructures – Example: India

The logistics infrastructure in many developing countries is still poor compared to industrialized countries. This is set to change, with multiple ambitious infrastructure projects being planned and realized. We will use India as a case example to outline some of the developments, and illustrate how these changes are affecting international manufacturers.

India is renowned for its bad roads, congested airports, and insufficient port capacity. The good news is that large government-led infrastructure projects, privatization, and a more open attitude towards foreign investment and management control are starting to bear fruit.

Many different route-to-market models co-exist in the fast-moving consumer goods industry

Fig. 7.26: Route-to-market in China

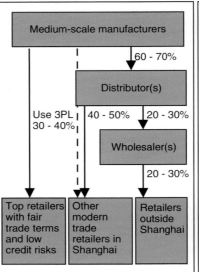

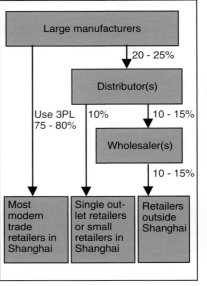

Source: McKinsey

Jawaharlal Nehru port outside Mumbai (Bombay), for instance, has become India's main gateway for seafreight, with multinational port operators in charge of most terminals. Jawaharlal Nehru already accounts for 50 percent of India's container movements. It is India's only port of call for very large container ships – no other port has berths with the depth required to accommodate latest-generation container vessels.

Where airfreight is concerned, Delhi and Mumbai are about equal in size, with Chennai (Madras) ranking third. Kolkata (Calcutta) ranks fourth and has declined in importance in recent years. With the large international airports being privatized, concentration on Mumbai, Delhi, Bangalore, and Chennai is likely to continue. Infrastructure for surface transport has also expanded, but can still hardly keep up with increasing demand. A major highway project is close to completion: the "golden quadrilateral," connecting Delhi (north), Mumbai (west), Bangalore (south), and Kolkata (east). The next stage in the national highway project has started, extending the roads concerned into modern highways with divided lanes. The Ministry of Railways is also planning a significant expansion of rail freight capacity, with dedicated rail freight corridors allowing faster, more efficient transport.

Companies will still have to find workarounds for at least the next decade, with improvements in one area leading to bottlenecks in others, but these infrastructure projects are desperately needed to keep up with increasing demand. With port capacity rising, container handling facilities, for instance, have started to become scarce. Companies that own and operate container rail freight stations (CFS) and inland container depots (ICD) have reported record profits in the last few years. Two of these, Concor and Gateway Distripark, had net margins of 22 and 44 percent respectively in 2006. Despite low labor costs, container handling charges are about two times higher than in most developed countries. In coming years, flat rail cars and rail access are likely to become scarce, particularly around ports and major urban areas. Establishing and operating reliable supply chains efficiently will remain a challenge.

Expected changes in India's regulatory system will have a further and very substantial impact on the availability of the logistics infrastructure. The introduction of a central value added tax (VAT) and abolition of the state sales tax will mean many companies need to redesign their distribution networks. Under the new regulations, having scattered stockholding points across states will no longer be business as usual. Companies will find it attractive to consolidate warehousing near ports and large cities. This will lead to rapidly rising demand for logistics infrastructure in these areas. Anticipating this shift will prove valuable. With land acquisition still a difficult process, major plots in metropolitan centers will become very scarce. The rent for large warehouses will likely increase significantly, as will the cost of services such as container handling, customs clearance, warehousing, and packaging.

Designing and managing effective supply chains is clearly as much of a key success factor in emerging as in developed markets, if not more so. The role of logistics service providers, distributors, and C&F agents is critical. With only a few exceptions, manufacturers have neither the reach nor the capabilities to gain access to the mass market in developing countries. While dealing with the harsh realities of logistics in emerging markets is essential, companies also need to realize that the environment is changing quickly. The transition from unorganized retail to modern trading structures is the most visible indicator that distribution requirements are changing. This gives manufacturers the opportunity to expand their control of the supply chain and reap the rewards of greater efficiency.

Managers should focus on two areas when establishing and **managing supply chains** in emerging markets. Firstly, they need to get access to people and contacts that know how to make things work in a somewhat chaotic environment. This is relatively straightforward, but requires good judgment on how much control can be delegated to local staff over the course of time. Secondly, management needs to be alert to ways to build on changes and prepare local operations for the next phase of development. This

Table 7.1: Logistics in emerging markets – Key issues and best practices

Overloading	Overloading is a hazard and illegal. However, it is difficult for companies to act alone to enforce compliance. Enforcing compliance is generally difficult: It can make distribution costs 50 to 100 percent higher than those of local competition and will still not provide any guarantee that truckers actually do not overload and use the higher freight rates to boost their profits. Also address this issue via industry associations and government lobbying.
Palletization	Start introducing pallets for handling in your own/directly controlled warehouses and movements between your own/directly controlled locations. Be prepared for palletization to increase (e.g., when designing buildings), but **do not try to be the first** to introduce pallets and other loading devices. Customers will not be able to handle them and will often not return empty pallets.
Unorganized retail	Remember that small shopkeepers and street businesses often require credit, and only pay for goods once they have been sold to the end user. This process is very difficult for international companies to handle. Most companies will have to rely on local distributors.
Distributors/stockists	Do not underestimate the role of distributors, which goes beyond logistics in areas such as **credit management**. Develop a plan and gradually eliminate layers of distributors. If distributors/stockists are organized in industry associations, manufacturing companies should align their actions. Trying to break the power of these associations alone can have disastrous consequences. Manufacturers may find their products boycotted.
Exports from LCCs	Aim for "ex-works" or "FOB" trade terms with vendors and consolidate goods in the country of origin. Locate as many value-adding activities as possible in the emerging market to capture the maximum cost advantage, e.g., labeling, packaging. Optimize container loading and routing. **Deliver direct** to the store or customer in the destination country wherever possible.
Imports into LCCs	Aim to locate value added in the country of destination. Make trade terms dependent on which organization can obtain better rates with forwarders/carriers.
Supply chain management	**Categorize goods** and **supply chains** to determine the optimal mode of transport, consolidation structure, and inventory holding points. Use criteria such as value density (the USD value per unit of weight), demand volatility, and criticality to determine the best mode of transport and supply chain structure.
	Define clear interfaces between local supply chain management in the LCC location and the management of international volume flows. Local distribution requires local management on the ground. Better to outsource local distribution as a whole if such resources are not available.
	Ensure that both your own organization as well as logistics service providers communicate well along **trade lanes** at an operational level for international transport flows (e.g., with representatives in the countries of both origin and destination).
Selection of logistics service provider (LSP)	Invest time to understand the capabilities of providers and how they are integrated with their subcontractors. Make sure your provider actually has **control over critical parts** of the transport chain.
	Use a scorecard with multiple decision criteria, e.g., rate, reliability, ease-of-use, speed/inventory reduction, innovativeness of provider/helpfulness in developing new solutions that assist in reducing costs.
	Place greater emphasis on reliability and innovativeness than price when bidding for comprehensive logistics solutions in LCCs: a provider that supports you in optimizing inventory and flow, handles your goods carfully, and prevents them from getting lost, can be worth a lot.
	Companies that require special logistics services such as cold chain logistics or the transport of valuables should plan way ahead. **Testing the solution** before "going live" should be a common practice in these cases.

Contracting	**Monitor spending** by service type, e.g., local warehousing and distribution, airfreight, ocean freight, air express, trucking, etc., and by trade lane.
	Bundle volumes and bid for them by region and service type.
	Enforce compliance with the framework agreement to remain credible in negotiations with logistics service providers and realize lower rates.
	Allow local organizations to opt out of the framework agreement only to **realize innovative logistics concepts**, e.g., the use of international trucking instead of airfreight. Otherwise, strictly enforce compliance.
Expand your reach	Use an LSP to **jointly enter attractive areas** of the product value chain, e.g., spare parts, return and repair services
	Combine OEM brand with the **LSP's reach** and capabilities to achieve a unique competitive advantage.

requires both a good understanding of the local market as well as the ability to draw parallels to the past development of other marketplaces.

7.3 Production Systems

The exchange of expert knowledge, experience, and best practices among different locations is particularly challenging in a global production network. It is not just the physical distances: considerable cultural and language barriers also have to be overcome. Diverging levels of automation and staff training at different locations also mean best practices need to be adapted to local circumstances. Management's role as an intermediary between factories thus becomes increasingly important to ensure that process improvements developed at one location are also implemented at others.

One of the main ways of accomplishing this is by introducing a **global production system**, an integrated set of principles, and methods for achieving lasting improvements in efficiency and quality. The basic principles and methods of modern global production systems are often derived from the concept of "**lean manufacturing**," i.e., they aim to eliminate all non-value-adding activities (waste, or *muda*) and to do so systematically in the framework of targeted kaizen activities.[12]

Production systems largely originated in, and remain particularly important to, the automotive industry.

They have meanwhile firmly established themselves in other manufacturing companies across the industrial spectrum as well. Since the mid-1990s, European manufacturers have increasingly pursued lean production. Even so, implementing this concept is not always as smooth as one would wish.

For one thing, the improvements developed do not always achieve the effects hoped for. Reducing inventory, for instance, improves capital productivity but can also lead to lower service levels and customer satisfaction if the root causes of overstocking are not tackled in parallel. For another, many companies find they are unable to implement the improvements sustainably at a worldwide level, and that improvements previously implemented in some locations gradually fall into disuse.

To succeed at introducing a global production system, a company has to complete three main tasks: develop the production system itself (taking into account the technical requirements of the relevant processes and production technologies), build sufficient skill

---

[12] Kaizen *(Japanese:* kai *= to change;* zen *= good; literally: "improvement")* is a Japanese management concept developed by Taiichi Ohno. In a narrow sense, it refers to continuous improvement involving management and workers alike. According to the kaizen *philosophy, the path towards improvement is not through leaps of innovation but through the step-by-step optimization and perfection of proven products. The focus is not on financial gain, but rather on continuous effort to improve the quality of products and processes. Many companies in the West have implemented the* kaizen *philosophy under another name: the continuous improvement process (CIP).*

and experience in applying the production system methods, and implement the system throughout the global factory network.

A company can increase its chances of success by introducing a production system in the framework of a lean transformation project, which typically has two key phases: the "design and pilot" phase and the "rollout" phase.

7.3.1 Design and Pilot Phase

The objective of the design and pilot phase is to create a detailed blueprint and working model of new, highly efficient manufacturing processes to be rolled out across the production chain. This requires working through a great deal of operating data and verifying the workings of the production system in detail with the relevant production engineers. The goal is a solution that is both tailored and comprehensive in order to maximize the benefit of the system.

While a bewildering array of concepts for production systems already exists – and although they are generally publicly available – it makes sense for a company to develop its own customized system reflecting its company-specific structures, concepts, and terminology. The production system also needs to be designed from the outset so that it works effectively **across different processes and technologies** through which products flow. The improvement opportunities within the specific processes and technologies themselves are then addressed with the relevant tools and techniques of the production system.

Given the many causes of waste and their interdependencies, there are also many different kinds of tools in a production system toolkit. To make an im-

---

Often, sources of waste can only be permanently eliminated by a combination of techniques

Fig. 7.27: Impact matrix – waste and ways to fight it

○ Suitable techniques

Techniques

Cause of waste	Flow principle	Takt time	Pull system	Heijunka (production smoothing)	Poka-Yoke (error avoidance)	Andon*	Systematic error rectification	Standard work	FLS** design	FLS** kaizen	Organization	Performance management system	Competence management	Built-in quality	Maintenance systems	Workplace organization
Overproduction	○	○	○	○	○	○		○	○	○		○			○	○
Waiting	○	○	○	○	○	○		○	○	○	○	○			○	○
Transportation	○		○						○	○						○
Overprocessing							○						○	○	○	
Inventory	○	○	○	○	○	○		○				○			○	○
Motion	○				○	○		○	○							○
Errors and rework	○				○	○	○	○	○			○		○	○	
Process variability	○	○	○	○	○	○	○	○	○	○	○	○	○	○		
Machine variability					○	○	○		○			○		○	○	
Fluctuations in demand		○	○	○				○								
Variability of work practices								○	○	○			○			○

* Tool for signaling when problems occur
** Flexible labor system

Source: McKinsey

pact, these tools need to be applied not only consistently ("fit for purpose"), but also in combination. This is generally the only way to achieve a lasting reduction in "waste," i.e., inefficiencies in the production system (Figure 7.27).

The key point when designing a production system is to avoid waste of any kind

Before setting out to design a new and improved system, a basic prerequisite is an understanding of the current production and supply chain operations and how they support business strategy. What are the weaknesses? Where is there room for improvement? One place the answers must be sought is production itself, through intensive tracking of the workflows and identification of any form of waste (*muda*), strain (*muri*), or imbalance (*mura*). Finding and eliminating

the "three mus" is the mission enlivening the lean management philosophy embodied in the Toyota Production System. Experienced lean experts are often able to identify dozens of possible improvements simply by observing the movement of employees, materials flows, and the storage of intermediate products. This is a bottom-up improvement approach that, when applied systematically over the long term, gradually leads to improvement and increased productivity.

In addition to observing direct production work, however, designing a production system also requires thorough and meticulous analysis of all production flows to identify weaknesses in the supply chain in which the production function is integrated.

Figure 7.28 illustrates the **MIFA approach**[13] (Material and Information Flow Analysis), which can be

Identifying current waste is a vital prerequisite for designing a production system

Fig. 7.28: Material and information flow analysis (MIFA) <u>EXAMPLE</u>

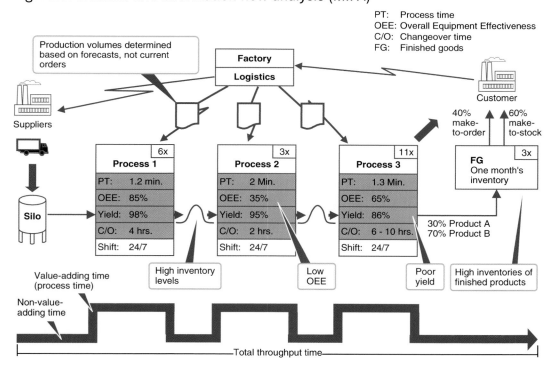

Source: McKinsey

---

[13] *Also referred to as value stream mapping.*

useful for this, and was first used at Toyota over sixty years ago. The objective is to map out the value stream, i.e., the value-added flow for in-house production, then analyze the interaction between the various process flows and individual process parameters to identify any faults or inhibitors. Unlike the "three mus," MIFA is a top-down approach: the system is viewed as a whole, and important potential improvements are identified in an early phase (quick wins).

Once this basic data has been gathered, you can use it to set objectives and design the production system. This step is critical to success. Ambitious **objectives** are important for motivating staff to make radical changes but must be attainable; otherwise, the size of the gap between aspiration and reality will provoke frustration. It is essential to make a solid, clear estimate of what targets are achievable, which re-

quires an in-depth analysis of both the materials flows and the information flows. This preparatory work is also needed to work out the details of the production system: You cannot develop the right guidelines and choose corrective action without knowing the reasons for waste (also see Appendix B, Table 7.B – "Loss/waste, symptoms, possible causes, and key tools and techniques highlighted by the lean concept").

Mercedes-Benz demonstrates how successfully this type of individual production system can be created. Mercedes has been pursuing lean production since the mid-1990s, initiating the Mercedes-Benz Production System (MPS) for this purpose.

Guided by the production principles shown in Figure 7.29, the underlying processes are the foundation for applying lean production methods and striving for

The Mercedes-Benz Production System is a toolkit of best practices with 5 subsystems and 15 production principles

Fig. 7.29: Mercedes-Benz Production System (MPS)

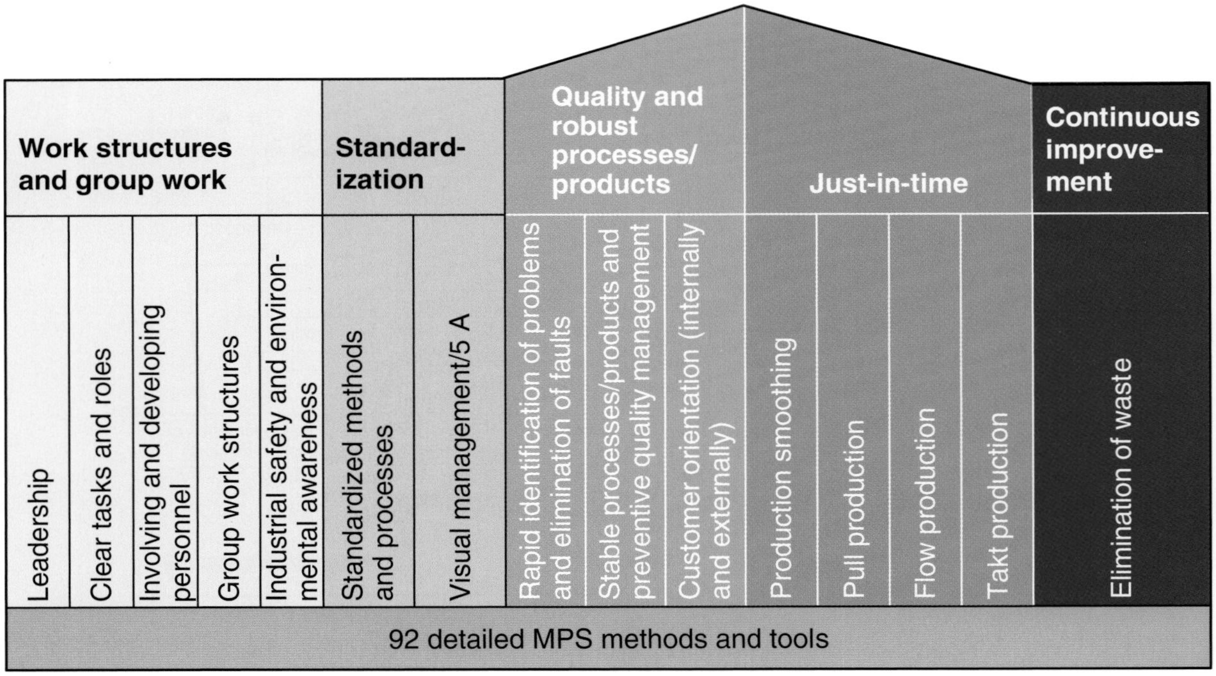

Source: DaimlerChrysler (2005)

continuous improvement – in line with the principle "Operational excellence: Improve and carry on improving forever."[14]

Each of the 15 MPS subsystems has specific methods and tools that existing and new processes have to use and comply with. If, for example, a machine installation is improved in the "Standardized methods and processes" subsystem, the work and the results must adhere to four simple principles:

1. Avoid or minimize storage

2. Keep throughput times short

3. Organize production along the materials flow

4. Keep quality control loops short.

7.3.2 Rollout Phase

If all goes well, the design and pilot phases culminate in rolling out the production system. Depending on the size of the production network, i.e., the number of participating locations, it may be best to perform the rollout in stages.

This means that the changes are not implemented in all factories at the same time, but in a number of "waves," i.e., with a built-in time lag. This offers several advantages: experienced training managers can oversee the start-up at numerous plants, and can themselves benefit from the learning curve. Tracking implementation is also much easier if it is possible to concentrate on a limited number of factories during the critical phase.

> **It is essential to involve the staff during the rollout phase and convince them of the importance of the effort**

The factors critical to the success of the rollout are very different from those critical to the design and pilot phase. Technical analyses and hard facts are not key here but rather the sustained change in **ingrained behaviors**. This involves dealing with the

emotions of employees – reactions to the announcement of changes and reorganization are usually skepticism, anger, and worry. If appropriate countermeasures are not taken, this can lead to an attitude and atmosphere that can endanger the entire rollout. This means a change management process needs to be initiated at an early stage to ensure employees' **identification with the project's objectives** and to create a positive attitude. An important way of dealing with this is systematic internal communication that picks up on experience from the pilot phase and then makes a targeted approach to the different groups of staff. Successes from the pilot phase should be used to demonstrate the advantages of the new approach to key personnel. They can then serve as **multipliers**, passing on positive impressions to their colleagues and subordinates. A central training center at the home factory, for instance, can make a valuable contribution to this, building skills in best practices (see box "Global Lead Factory: Trumpf Transfers its Lean Production Process from its Lead Factory to all its Branch Factories.")

It is also important to clearly understand the benefits of the approaches. Each initiative should be presented at a workshop specially designed for the purpose, where it can be explained to and tried out by employees before they implement the new initiatives at their workplaces. This entails considerable effort, but experience has shown that this is the only way to turn around skeptical attitudes. Creating this understanding includes providing clear, comprehensible documentation (Figure 7.30), an objective description of each action, and tips on its implementation.

Once the methods of a production system have been fully implemented in a factory, the central tasks in ensuring sustainable success are to monitor their continued use and to **quantify the results achieved**. Improvements should be recognized and rewarded to create a performance-oriented culture. After all, the ultimate goal of a production system is to motivate employees to continually **strive to improve** the processes.

---

[14] See Nixdorf (2003).

Global Lead Factory: Trumpf Transfers its Lean Production Process from its Lead Factory to all its Branch Factories

The Trumpf Group, which is run as a family business, achieved revenues of roughly USD 1 billion in 2003/04, with approx. 6,000 personnel in four divisions: Machine Tools, Laser Technology, Power Tools, and Electronics/Medical Technology. In 2003/04, the Machine Tools division had the highest revenues, accounting for 65 percent of the Group's total, with production technologies for laser processing, punching, forming, and bending. Approximately 35 percent of the Group's sales were generated in Germany, another 35 percent elsewhere in Europe, and the remaining 30 percent primarily in the US and Asia.

In 1998, to reduce throughput times and inventories, a lean (*"synchro"*) production system was piloted at the home factory in Germany, where it replaced the static assembly system with flow assembly in tight cycles. This made it possible, for example, to reduce assembly time for the "Trumatic 6000 L" laser machine (weighing over 20 metric tons) from 46 days to only 21, while lowering inventory value from USD 4.6 million to USD 2.2 million.

By 2004, the company had succeeded in implementing the new production system worldwide at all 15 Trumpf production locations. Trumpf now offers training even for staff from other companies in conjunction with the Chamber of Commerce, as well as information days on the *"synchro"* production system.

Bottom line: A lead factory can play a key role in driving innovation in the production system and transferring it to all the branch factories.

Clear instructions should exist for each of the techniques specified in a production system

Fig. 7.30: Mercedes-Benz Production System – Method 2.2.2: Visualization

Description (basis for audit)

- Floor markings, color-coded designations and descriptions are to be used at predetermined workplaces, e.g., for:
  - Storage locations for:
    - Parts to be processed
    - Cancelled parts (scrap and rework)
    - Consumables, auxiliary materials
    - Empty load carriers
  - Tools
- Labels contain more detailed descriptions and are color-coded if errors are possible at this point
- Descriptions and markings show the predetermined maximum inventory level
- The colors of markings and coding are to be uniform throughout the plant
- Markings, codings, and descriptions are to be clear and known to all personnel

Tips for implementation

- Affected area to be informed of measures that have been agreed upon, changed locations, etc.
- Details of the approach should be implemented in a workshop to increase acceptance among personnel

Example from the Sao Bernardo MBBras factory

Source: DaimlerChrysler (2005)

Determining an appropriate portfolio based on key performance indicators in line with corporate objectives and production processes is a key factor in successfully achieving this goal. Well-conceived indicators such as overall equipment effectiveness (OEE), on the other hand, offer powerful solutions at the process level, which are also easy to aggregate across processes (Figure 7.31). Important manufacturing indicators such as scrap rate, capacity utilization, and machine availability are compiled here and serve as orientation parameters for management-by-objectives and performance-related components of compensation.

* * *

Careful, dedicated management is important when implementing a globalization strategy and expanding a company's production footprint. The techniques suggested in this chapter will provide management with crucial support in tackling the challenges of supervising a global production network. Neglecting to apply a systematic approach can mean more than just failure of the new location – it can severely impair the success of the network as a whole.

OEE is a comprehensive indicator that aggregates key performance parameters

Fig. 7.31: Overall Equipment Effectiveness (OEE)
Percent <u>EXAMPLE</u>

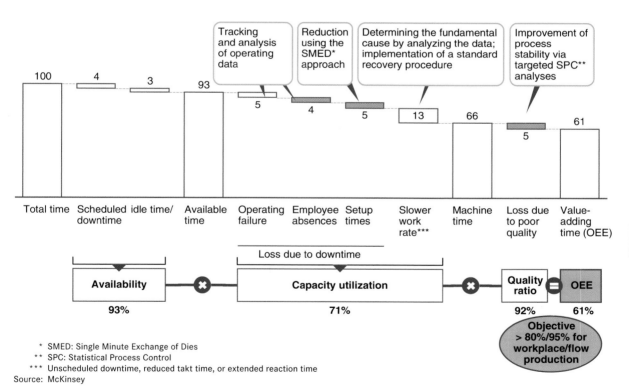

* SMED: Single Minute Exchange of Dies
** SPC: Statistical Process Control
*** Unscheduled downtime, reduced takt time, or extended reaction time
Source: McKinsey

Further reading

Drew, J., B. McCallum, and S. Roggenhofer.: *Journey to Lean.* Basingstoke: Palgrave MacMillan, 2004.

Küpper, H.-U. and S. Helber: *Ablauforganisation in Produktion und Logistik* [Process Organization in Production and Logistics]. 2nd Edition. Stuttgart: Schäfer-Poeschel Verlag, 1995.

Long, D.: *International Logistics: Global Supply Chain Management.* Berlin et al: Springer, 2003.

Mentzer, J. T. et al (eds.): *Handbook of Global Supply Chain Management.* Thousand Oaks, California: Sage Publications, 2006.

Silver, E. A., D. F. Pyke, and R. Peterson: *Inventory Management and Production Planning and Scheduling.* 3rd Edition, New York: John Wiley, 1998.

Simchi-Levi D., P. Kaminsky, and E. Simchi-Levi: *Designing and Managing the Supply Chain.* 2nd Edition, Boston: McGraw-Hill/Irwin, 2002.

Tempelmeier, H.: *Material-Logistik – Modelle und Algorithmen für die Produktionsplanung und -steuerung und das Supply Chain Management.* 5th Edition. Berlin: Springer, 2003.

Thonemann, U.: *Operations Management.* Munich: Pearson, 2005.

Thonemann, U., K. Behrenbeck, R. Diederichs, J. Großpietsch, J. Küpper, and M. Leopoldseder (2003). *Supply Chain Champions – was sie tun und wie Sie einer werden.* Wiesbaden, Gabler [available in English from the McKinsey Operations Practice].

Vahs, D. (2005). *Organisation.* Stuttgart, Schäffer-Poeschel.

Zheng, L. and F. Possen-Dölken: *Strategic Production Networks.* Berlin: Springer, 2002.

Appendix A

The most frequently used and meaningful indicators of logistics performance within a distribution network are the **α and β service levels**:

■ The alpha (**α**) **service level** (also referred to as the *cycle service level* or the *type I service level*) is an event-oriented indicator. It states the probability that an incoming requirement can be met fully from existing inventory, i.e., that there will not be a stockout.

■ The beta (**β**) **service level** (also referred to as the *fill rate* or *type II service level*) is a quantity-oriented indicator that states the percentage of total demand volume that can be delivered within the specified delivery lead time. For the application of the β service level to be meaningful, it must either be possible to partially fill orders or call-off orders or there must be a sufficient number of call-off order per period, i.e., the average required volume per call-off orders is considerably lower than the average total required volume per period.

In addition to these two indicators, there are many others that can be used to specify the service level. For example, the chi (χ) service level includes both the volume component and time component of the service level. Another commonly used indicator is the average time between call-off orders that cannot be immediately filled (*specified average time between stockout occasions*). It is calculated using the number of orders annually that cannot be executed immediately.

Appendix B

Table 7.B – Loss/waste, symptoms, possible causes, and key tools and techniques highlighted by the lean concept[15]

Type of loss/waste	Symptoms	Possible causes	Key tools and techniques
Overproduction – producing sooner, faster, or in greater quantities than is needed by the customer.	Too many parts are produced. Parts are produced too early. Parts accumulate in uncontrolled inventories. Long manufacturing lead times. Poor delivery performance.	Long changeovers driving large batch sizes. Use of an economic algorithm to determine batch sizes. Poor scheduling. Confusion over schedule priorities. Unbalanced materials flow. Prioritization of equipment utilization as a key metric.	Just-in-time (continuous flow processing, takt, pull systems, leveled production). Changeover reduction or SMED (where changeovers drive batch sizes).
Waiting – idle time (for people or machines) in which no value-adding activities take place.	Operators often wait for materials or information. Operators stand and watch machines run. Operators often wait for unavailable machines. Long in-process delays. Low productivity. Long manufacturing lead times.	Large batch sizes upstream causing material shortages. Poor supplier delivery performance or quality. Poor machine condition (low OEE). Poor scheduling. Poor labor utilization. Lack of flexibility in skills.	Flexible labor systems (including standardized work). Just-in-time (continuous flow processing, takt, pull systems, leveled production). Strategic maintenance. Supplier development.
Transportation – unnecessary movement of materials.	Multiple handling or movement of parts. Excessive handling damage. Long distances traveled by parts between processes. Long manufacturing lead times. High indirect costs due to storage space and materials handling equipment required.	Sequential processes physically separated. Poor layout. High inventories; same part often held in multiple locations.	Continuous flow processing and pull systems. Workplace organization.
Overprocessing – effort that is not required by the customer and adds no value.	Performing of processes that are not required by the customer. Redundant approval requirements. Higher direct costs than competitors.	Overengineered processes. Product design. Unclear customer specifications. Excessive testing. Inappropriate policies or procedures.	Production preparation. Standardized work.
Inventory – any parts or materials above the minimum required to deliver what customers want, when they want it.	Obsolete stock. Cash flow problems. Lack of space. Long manufacturing lead times. Poor delivery performance. Extensive rework needed when quality problems are identified.	Overproduction. Poor forecasting or scheduling. High levels of safety stock because of frequent process or quality problems. Purchasing policies. Unreliable suppliers. Large batch sizes.	Just-in-time (continuous flow processing, takt, pull systems, leveled production). Standardized work. Supplier development. Strategic maintenance (where process problems are driven by equipment issues). Statistical process control (where process problems are driven by quality issues).

[15] *Drew (2004).*

Type of loss/waste	Symptoms	Possible causes	Key tools and techniques
Motion – unnecessary movement of people or materials within a process.	Searching for tools or parts. Excessive walking by operators. Double handling of parts. Low productivity.	Poor layout of workplace, tools, and materials. Lack of visual controls. Poor process design.	Workplace organization. Continuous flow processing. Motion kaizen. Standardized work. Visual management.
Rework – repetition or correction of a process.	Dedicated rework processes. High defect rates. High materials costs because of spoilage levels. Low productivity. Large quality or inspection departments.	Poor quality materials. Poor machine conditions. Unstable or unsuitable processes. Low skill levels. Unclear customer specifications.	Statistical process control. Autonomation. Strategic maintenance. Supplier development. Standardized work.
Variability – any deviation from the standard or nominal condition.	High levels of scrap or rework. Large quality or inspection departments. Recurring problems that are patched up with quick fixes. Output measures that show an unacceptable level of variation (e.g., quality).	Unstable or unpredictable processes. Unsuitable processes. Poor quality materials or supplied parts. Low skill levels.	Statistical process control. Autonomation. Supplier development. Standardized work.
Inflexibility – response to demand variability issues that arise as a result of variation in customer demand.	Unable to react quickly to changes in customer demand. High levels of overtime. Periods of underutilization.	High inventories. Long changeover times. Poorly balanced work. Low skill levels. Over-scoped equipment.	Just-in-time (continuous flow processing, takt, pull systems, leveled production). Flexible labor systems. Changeover reduction or SMED.
Working practices – normal working practices that obstruct flexibility in the operating system.	Unable to change ways of working significantly. Work frequently delayed when the right people are not available.	Terms and conditions not configured to facilitate change. Operators are highly specialized and often only one person can do a specific job.	Standardized work. Flexible labor systems.

MICHAEL STOLLE, ULRICH NÄHER, FRANK JACOB, NICOLAS REINECKE,
JAMES HEXTER, MARINA DERVISOPOULOS

8 Sourcing: Extending the Footprint Reconfiguration to Suppliers

Summary

For most industrial products, the biggest cost factor by far is materials. This means companies can only tap the full potential of a global network if, in addition to having their own manufacturing sites, they also source worldwide as cost-efficiently as possible. It often takes these savings in materials costs to actually make foreign locations attractive in the long term. However, nurturing competent local suppliers is a demanding task that requires patience and intuition. Companies should factor in substantial time and resources for selecting and building suppliers in developing and newly industrialized countries. Planning this process is crucial to realizing these savings.

The development of a tailored sourcing strategy begins with segmentation of the materials groups required. The company has to decide (for each category individually) whether to source items locally for the new plant in the medium term or to continue buying from the current supplier. Quality risks and process complexity should be factored in as decision criteria alongside materials costs.

Ideally, local sourcing is built up in two phases parallel to relocating production. First, the company realizes quick wins, sourcing parts from local providers who can supply them reliably without intensive preparation and training. As soon as local production is running smoothly, the second phase begins. The aim is to tap the full potential of local sourcing and ensure that more complex parts can also be obtained locally. To achieve this, the manufacturer needs to systematically refine the capabilities of its own sourcing organization in the target country as well as supplier skills. A local supplier structure can only be built up efficiently with a high-performance local sourcing organization.

Key questions, Chapter 8

- What role do materials costs play compared to the other costs of production?

- How does the share of materials costs differ between industries?

- What savings can be achieved by building up local sourcing?

- What are the main factors influencing the level of potential savings?

- How do the challenges of sourcing in low- and high-cost locations differ?

- Which parts should be sourced locally first?

- What approach can be used to increase the share of parts sourced locally?

- What should the tasks of the local sourcing organization be, and how should it be structured?

- How can continuous improvement of local sourcing be ensured?

8.1 Why Sourcing is so Significant in a Production Network

Materials make up a large share of a product's manufacturing costs. The figure is typically 70 to 80 percent for electronic products such as PCs, DVD players, or games consoles, around 60 percent in the automotive industry, and as much as 20 to 40 percent even in the pharmaceutical sector (Figure 8.1). This means the integrated optimization of production must also include the supplier network.

Only a few companies currently implement concepts that do this systematically. Often, foreign plants obtain most of their supplies (even the simplest of parts) from their home country, showing how greatly companies underestimate the importance of this topic. It makes sense to build up local sourcing for almost all foreign plants, particularly for components with low value density, where any advantages of production at the home base are quickly wiped out by transportation costs. If these components are manufactured using simple but labor-intensive methods that are difficult to automate, sourcing from a low-cost location is generally advantageous.

However, the specific weaknesses of developing and newly industrialized countries may be an even greater obstacle to sourcing than when setting up the production site itself. On-time delivery often suffers from poor infrastructure, and cooperation with suppliers is impaired by the lower qualifications and technology skills of their product and process engineers. As a result, companies frequently require more technical staff than originally planned. Business partners are all too often unfamiliar with international practices, and have divergent expectations and aspirations stemming from the different culture. These difficulties may be exacerbated when it comes to complex parts and components.

Building up local sourcing therefore deserves focused management attention. Opportunities and risks have to be evaluated with the greatest care and folded into network planning.

Global manufacturing locations only become attractive once a local supply base has been established

8.1.1 Cost Potential

The main motivation for local sourcing is additional savings potential. Done right, it can boost the economic viability of the entire optimization project. The potential for reducing the labor costs of in-house production is already limited for many companies due to their low level of vertical integration. In the consumer electronics industry, for example, in-house production is below 30 percent on average, measured by the costs of production. Of this, a large share is accounted for by depreciation, the costs of tied-up capital, as well as other operating costs and overhead. As a result, direct labor costs usually range between a modest 5 and 15 percent of production costs.

Materials costs account for over half of the cost of goods manufactured in many industries

Fig. 8.1: Share of materials costs* for production in high-cost countries
Percent of cost of goods manufactured**, example: Germany

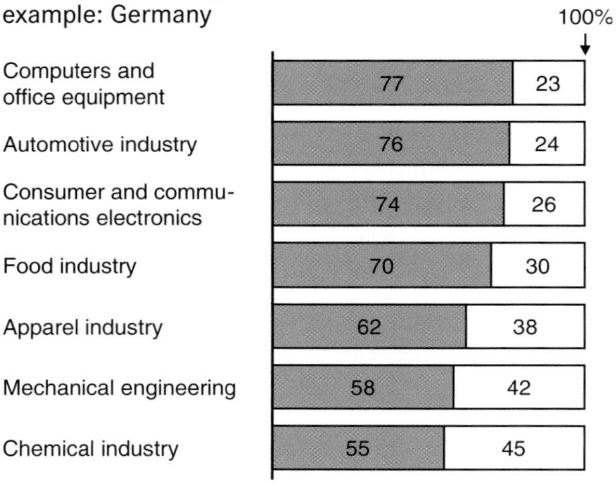

100%
Computers and office equipment
77
23
Automotive industry
76
24
Consumer and communications electronics
74
26
Food industry
70
30
Apparel industry
62
38
Mechanical engineering
58
42
Chemical industry
55
45

* Use of materials and sourced merchandise
** Use of materials, sourced merchandise, labor costs, costs for industrial and trade services, costs for contract work, depreciation
Source: German Federal Office of Statistics (Yearbook 2004)

If manufacturing is relocated to replace the current site, potential savings are often accompanied by sizeable one-off costs. This is because staff reductions can entail high redundancy payments, especially in high-cost countries.

However, incorporating suppliers into the optimization concept (as well as the company's own manufacturing steps) revolutionizes the savings opportunities. The share of labor costs – seen across the entire supply chain – is substantially higher, while the lower factor costs of firms in the new location feed into the calculations favorably. These two aspects explain why, as already demonstrated in the strategic location concept (Chapter 4), an integrated approach can often lead to many times the savings attainable from the company's own production steps alone (Figure 8.2). As a rule of thumb, the higher the share of personnel costs for supplied parts, the higher the potential (Figure 8.3).

An estimate reveals that General Motors and Ford could together save around USD 10 billion each year if they purchased half of their standard supplied parts (e.g., cast parts, simple electronic components) in low-cost countries such as China, India, or Romania. Given the magnitude of these savings, it is astonishing that this lever is so often insufficiently exploited.

8.1.2 Challenges

Why don't manufacturers purchase more of their parts from low-cost countries? One reason is frequently lack of insight into local practices. Companies can fall down badly if they lack transparency on (or sensitivity to) the local business culture, or the resources to overcome these hurdles. Even where mutual understanding develops well, the process can prove more time-consuming than purchasing from familiar suppliers in the home market. Often the choice of technically experienced suppliers available at a low-cost location is limited (and perhaps non-existent). Additional issues are linguistic and cultural barriers, differences in the maturity of technical equipment, or simply a lack of trust on both sides.

Relocation can realize savings potential of over 30%, but only by moving both production and sourcing to low-cost countries

Fig. 8.2: Cost reduction potential for an automotive supplier from relocating to and sourcing in low-cost countries
Cost of goods manufactured per unit, EUR

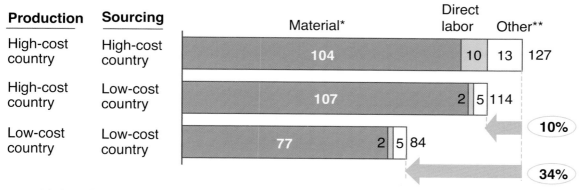

Production	Sourcing	Material*	Direct labor	Other**	
High-cost country	High-cost country	104	10	13	127
High-cost country	Low-cost country	107	2	5	114
Low-cost country	Low-cost country	77	2	5	84

10%

34%

* Incl. materials handling overhead; materials prices franco domicile
** Depreciation, indirect labor in production, other operating costs
Source: Company data, McKinsey analysis

Take the example of a German company that subcontracted the production of a complex cast part to a Chinese supplier. It took this supplier two and a half years and four attempts before its samples finally passed the quality test. In the meantime, the manufacturer had already changed the design, so the part was no longer needed anyway. The OEM had failed to develop sufficiently close contact with the supplier and did not provide appropriate support. The takeaway: much more time needs scheduling to develop new local suppliers than companies would expect from past experience (Figure 8.4).

Average savings rise by 0.6% for every percentage point increase in the share of labor costs

Fig. 8.3: Savings potential for supplied parts from sourcing in low-cost countries

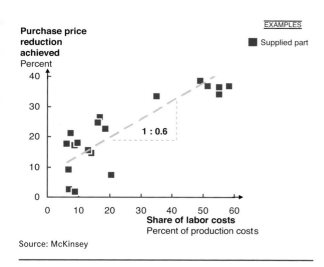

Purchase price reduction achieved
Percent

1 : 0.6

Share of labor costs
Percent of production costs

Source: McKinsey

Building up suppliers in developing and newly industrialized countries takes much longer and costs much more than at home

However, many of the special challenges of low-cost locations can usually be overcome more efficiently from a local production site than via sourcing from the home base. If a company lacks knowledge of local market conditions, it can bring in agents familiar with local suppliers, their cost structures, and decision-making practices. In parallel, the company can broaden its experience base via contact with suppliers. Alternatively (or later), the company can also participate in the local market without intermediaries, encouraging its regional staff to prepare and handle negotiations.

Building up new suppliers in low-cost countries can take much longer than at the original high-wage location

Fig. 8.4: Time frame for building up suppliers

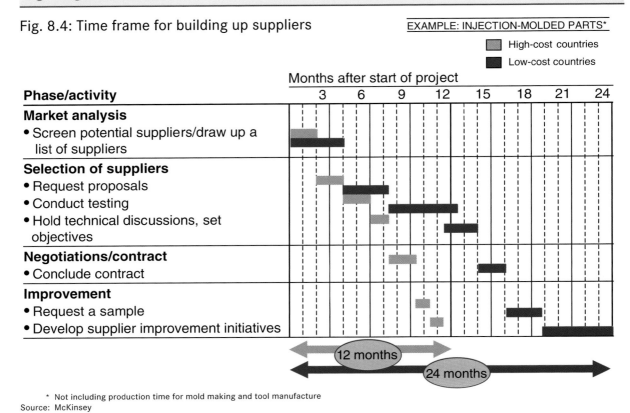

EXAMPLE: INJECTION-MOLDED PARTS*

■ High-cost countries
■ Low-cost countries

* Not including production time for mold making and tool manufacture
Source: McKinsey

If local suppliers have insufficient skills, close cooperation can be arranged with the company's own engineers. This means shorter feedback loops, and any problems can be tackled direct without long journeys and interpreters. Local presence is also an advantage when recruiting highly qualified buyers and managers. The better information on (and contact with) other companies and local universities makes access to suitable candidates much easier.

Definition of Local Versus Global Sourcing

We speak of **local sourcing** if it is in close proximity to the production location. The aim is often to build up local suppliers for product-specific components or components that are difficult to transport. If, for example, a company sources parts for a production location in India from an Indian supplier, this is an example of local sourcing. If, however, the parts for the plant in India are sourced from China or the US, for example, this is **global sourcing**. The aim is to find the best or cheapest supplier in the world for a particular supplied part. Global sourcing can open up very attractive potential, particularly for standard products with high value density. In other cases, quality or know-how may be the deciding factor. The term "global sourcing" is therefore not necessarily synonymous with purchasing from low-cost countries.

8.2 Segmentation of Sourced Parts

A production network's sourcing strategy should be based on segmentation of the materials groups it requires. Which should it source locally in the short or medium term, and which should it continue to buy from the same suppliers? The cost potential in the new region is the key criterion, but factors relating to risk and complexity are also important.

> **Segment materials groups by cost, complexity, and risk before developing a local sourcing strategy**

Using this logic, sourcing strategies can largely be assigned to four segments:

1. Strategic partnerships: Generally it is best to buy technically complex components that relate to the company's intellectual property from established suppliers.

2. Global sourcing: High-volume standard components with low logistics costs should be purchased from whichever supplier in the world can offer them most favorably.

3. Local quick wins: Simple parts can typically be sourced at relatively short notice wherever the new plant is situated.

4. Local supplier development: With parts that are more complex but could still be sourced locally, the best strategy is targeted development of local suppliers.

Assigning parts to these segments is not always easy: the guidelines below can help.

1. Strategic partnerships: If sourcing parts locally poses acute risks – whether due to technical and quality risks or possible loss of know-how – the materials group should be assigned to the first segment. This does not mean local sourcing is out of the question, but the company should venture into it only if local suppliers have proven technically capable and

trustworthy over the course of a long-term partnership, and the transition can be prepared carefully. The central purchasing department and local sourcing team should carefully examine strategic alternatives for this segment. Might current suppliers be prepared to establish their own site in the region? Would insourcing be worth considering?

2. Global sourcing: Materials groups and parts for which local sourcing is not expected to yield any further cost advantages should be allocated to the second segment. These may, for example, be parts that the manufacturer already sources from other low-cost countries. The current supplier is used for items in this segment even if production moves to a new location. The local sourcing team works closely with the central purchasing department to coordinate the purchase of these items as efficiently as possible.

Beware, though, as experience shows it is usually a myth to expect the same prices the world over. Numerous examples – particularly from Asia – show that only the presence of a local sourcing department and insight into supplier structures and negotiation strategies will uncover a region's true optimum. Allocation to this category should therefore be reviewed several times – especially once local production is fully established.

3. Local quick wins: This third category covers technically straightforward components with relatively high logistics costs compared to the value of the parts. They should be sourced locally from the outset as the savings potential can be captured without much difficulty – eliminating the logistics costs alone will make a huge difference. This segment therefore has absolute priority for the local sourcing team until production kicks into action. Appropriate local suppliers should already be available, making it fairly unproblematic to switch.

4. Local supplier development: The fourth segment should include materials groups for which local sourcing is attractive even though suppliers do not yet have the necessary skills. In these cases, the sourcing team should link supplier development closely to setting up the new plant. The simplest scenario is to convince existing suppliers to relocate, too. Alternatively, the OEM needs to acclimatize new suppliers step by step to the technology and quality level required (via increasingly complex components). The manufacturer will continue to source the items in this segment from its previous suppliers until local sourcing is secure and stable.

The example below suggests how this strategic segmentation might play out in practice (see accompanying box: "Producing shower heads in Eastern Europe").

We recommend a sequential approach that focuses initially on Segment 3, "Local quick wins," awarding contracts to competitive local suppliers as quickly as possible. These suppliers must be capable of developing the parts through to production without major support while the OEM sets up the plant. In a second phase, the main focus shifts to the strategic development of suppliers to source complex parts locally. We describe both phases in the following sections, supplemented by examples of **best-practice approaches**.

8.3 First Wave: Sourcing Simple Parts Locally

The relatively simple components in Segment 3 can usually be sourced direct from local suppliers, even in low-cost countries. Difficulties can still emerge, mostly because Western buyers tend to overlook or disregard key prerequisites for successful cooperation with local partners. A systematic approach can avoid these stumbling blocks, realizing cost potential even in the short term. OEMs also need to structure their own organization appropriately and hire local staff to ensure that internal inefficiencies do not diminish these savings.

8.3.1 Setting up a Systematic Process

When relocating their sourcing, companies should set up a clear process defining the end products and roles of everyone involved for each step (Figure 8.5). Any culture-specific factors should be given appropriate attention at each stage of the process.

A systematic sourcing process that respects cultural differences is vital when selecting suppliers

Fig. 8.5: Ten-stage sourcing process for low-cost countries

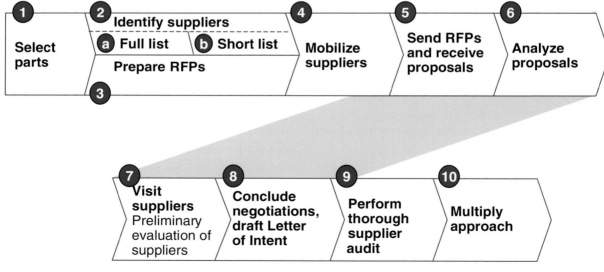

Source: McKinsey

Producing Shower Heads in Eastern Europe: How to Segment Sourced Parts Strategically

A brand manufacturer of shower heads decided to establish a plant in Eastern Europe. The following table gives an overview of the four key components:

Component	Explanation	Supplier to date
Spray former	Dual-component part made from hard plastic and silicon	Western Europe (two suppliers)
O-ring seals	Special rubber compound authorized for use with drinking water	China
Internal plastic parts	Standard injection molding	China (various suppliers)
Housing	Galvanized plastic part, complex geometry	China (two suppliers)

In high-quality products, the spray former with nubs that the water comes out of is a dual-component part made of hard plastic and silicon. This is a genuine differentiating feature between branded shower heads and cheap products from low-cost countries, as customers can clearly tell the difference between the two-component spray former and other products. As a result, the manufacturer considers this intellectual property. The spray former therefore belongs in the "Strategic partnerships" segment, and the manufacturer continues to source this part from its established Western European suppliers.

O-ring seals are largely made manually in China. They require a special rubber compound that the supplier has to develop, but once an official license has been obtained for the compound (and this was the case), only the production costs remain. Provided these conditions are met, manual work in China is superior to European automation. The seals are also small and light, so transportation costs are very low. It makes economic sense to assign the O-ring seals to the second segment, "Global sourcing," and continue purchasing them from China.

The internal plastic parts are injection-molded products of little value. The share of labor costs is relatively low: materials make up most of the costs. At the same time, tool payback is fast due to the large unit volume. Eastern European injection molding suppliers combine the advantages of a low-cost country with highly developed technology and the resulting low reject rate. Together with lower transportation costs, this clearly makes Eastern European suppliers a better option than the current Chinese suppliers. The internal plastic parts are therefore allocated to the third cluster, "Local quick wins," and local bids should be requested as soon as possible.

The chromium-plated housing of the shower head is relatively large, suggesting local sourcing could lower transportation costs significantly. The share of labor costs is comparatively low, so Asia has barely any advantage over Eastern Europe. The total costs would be lower if this part were sourced locally in Eastern Europe rather than in China. But market analysis reveals that very few companies galvanize plastic in Eastern Europe – and none could chrome-plate such a large part perfectly. So the housing is allocated to the fourth cluster, "Local supplier development." The manufacturer will continue to buy it from China until an Eastern European galvanizer can deliver the right quality.

Bottom line: Use precise differentiation to segment sourced parts optimally, analyzing strategic significance, know-how requirements, and cost structure.

1. Select pilot parts: The sourcing team should first select a few parts for pilot projects from the third segment. They should be representative, cover all relevant production technologies and levels of difficulty, and be produced in large quantities to enable high levels of savings from the very start. The team should run through the entire sourcing process for these parts, testing possible suppliers.

Pilot projects like this reduce the risk of unexpected problems shortly before production kickoff. They also shorten the preparation time needed for second-wave parts as suppliers already understand the manufacturer's quality requirements. Naturally, testing suppliers takes time. This process should therefore be tackled immediately after deciding on the new location to ensure that production can begin punctually.

Suppliers of Turned Parts in Eastern Europe: Preliminary Screening Based on Simple Criteria

In a global sourcing project, a West European industrial company was looking for suppliers of turned parts in Eastern Europe. Initially it identified around 400 suppliers using Internet searches, lists from embassies and trade associations, and public databases.

Project members verified the contact details of all these companies, and requested the following information:

- ISO certificate

- Revenues and number of employees

- Product range and reference customers

- Materials processed in metric tons per year

- Number/types of machines and maximum diameter for workpieces on single-/multi-spindle lathes

- Existence and, if applicable, size of own design engineering department and own tool shop

- Interest in collaboration

The number of relevant suppliers was reduced from the initial 405 to below 50 based on certain

criteria: good fit of product range, processing of the materials required, and minimum revenues (Figure 8.6). This preliminary screening allowed a very targeted approach of the shortlisted suppliers.

Bottom line: You can greatly reduce the number of potential suppliers by applying formal screening criteria, thus simplifying the subsequent selection steps.

> **Preliminary screening based on simple criteria quickly narrows down the number of possible suppliers**

Fig. 8.6: Investigating suppliers of turned parts in Eastern Europe
Number of suppliers

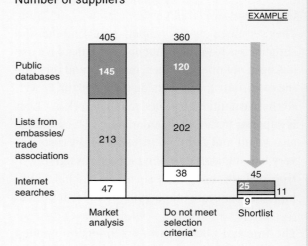

* Technology/raw materials, size of parts, unit volumes

Source: McKinsey

2. Identify potential suppliers: The next step is to identify relevant local suppliers. This process is normally difficult without a good grasp of the local language. Databases, consulates and trade associations can provide initial advice. Suppliers may be located in technology clusters. In China, for example, the majority of manufacturers producing zinc die-cast products are located around Guangdong and Xiamen. It can also be useful to work together with external scouts. Many now operate in countries such as China and India (also see section 8.3.2).

Qualitative criteria are then used to prepare a short list of the companies identified. The sourcing team will need to gather information from potential suppliers both in writing and by telephone. The box below illustrates the selection process using the example of suppliers for turned parts in Eastern Europe.

> **Targeted preselection of suppliers using technical and economic criteria increases the response rate and quality of bids**

3. Prepare requests for proposal: The next critical step is to prepare clear, comprehensive documents for the request for proposal (RFP). Ideally, all documents should be available in the local language. The request should contain key data and the requirements of the component to be produced. It is vital to include:

- Volumes required

- Technical specs and engineering drawings

- Quality requirements and guarantees

- Schedule and delivery specifications

- Additional service requirements.

4. Mobilize suppliers: Before sending out RFPs in low-cost countries, it is essential to be aware of circumstances there. Many suppliers in China, India, and South America are growing at an annual rate of 30 percent or more. This is a seller's market: the position of new customers is not very strong. Cold-callers' inquiries often end up unread in the wastepaper basket at even the best suppliers in these regions.

> **Western companies need to pull out all the stops to attract appropriate suppliers in low-cost countries, gaining their trust via personal contact**

Personal contact is the only way to coax enthusiasm from such suppliers and convince them that your RFP is intended seriously. Asian business people view personal relationships between business partners as particularly important. Paying the supplier a visit or extending an invitation to your local office are a must.

Mechanical Engineering Component Supplier in India: *New Supplier Day*

A Western European mechanical engineering company was looking for low-priced suppliers in India for a large share of its sourcing volume. The central purchasing department carried out a detailed Internet search and sent out requests for proposal to all the suppliers it had identified worldwide – without first gathering further information about the firms or making contact with them. The result:

in the first round, not one proposal was received. In the second round, the few proposals submitted were useless. They were either much too expensive, or so cheap that the supplier had obviously not understood the complexity of the part.

At the third attempt, the company changed its strategy: 15 of the Indian suppliers that had previously been approached unsuccessfully were invited to a conference. All 15 accepted (in most cases the sales manager). The OEM brought RFP packages to the conference customized to each supplier.

These packages contained a list and photographs of the products concerned, engineering drawings labeled in English, quality specifications, and detailed RFP forms (Figure 8.7). The latter turned out to be particularly helpful for understanding precisely how the suppliers had made their calculations.

The products were also displayed during the supplier conference and discussed in detail with the participants.

During the conference, the mechanical engineering company's sourcing team arranged to visit the five most suitable suppliers during the next few days. These visits were motivating for both parties: they boosted supplier enthusiasm ("they're interested in us!") and reversed the frustration of the Western European company's purchasing staff that had been lingering from the earlier (failed) attempts.

Bottom line: Written requests for proposal are not always appropriate. Personal discussions are often more efficient and successful.

Using an RFP form helps make the supplier's costs transparent

Fig. 8.7: RFP form for assessing suppliers   EXAMPLE

A. Cost of raw materials	Raw materials used	Utilization			Unit costs		Net costs
		Gross	Net	Unit (kg)	Costs	Unit (per kg)	
						Total	

B. Processing costs	Production step	Machines and equipment used	Processing costs		Net processing costs
			Costs	Unit (per kg/per hour)	
	• Casting				
	• Machining				
	• Deburring				
	• Washing				
	• ?				
				Total	

C. Costs of tools	Custom tools	Investment	Depreciation	Net costs per part
	• Casting molds			
	• Machine tools			
	• ?			
			Total	

D. Expected margin	No. of parts per year	Expected margin
	• < 100	
	• 100 - 500	
	• 500 - 1,000	
	• > 1,000	

E. Price offered (add A, B, C, D, net)	....

F. Other costs	Cost type	Unit costs
	• Overheads	
	• Rejects	
	• Packaging	
	• Transport	
	• Taxes	
	Total	

G. Landed costs (add E and F, net):	....

Source: McKinsey

Some firms also find supplier conferences highly successful, as the following example demonstrates.

5. Send out RFPs/receive bids: Manufacturers need to actively support potential suppliers as they draw up their bids to ensure the maximum response rate, even after the RFP package has been issued. Fixed contacts should be appointed to deal with queries, and suppliers approached and offered help with any difficulties. The OEMs should clarify any misunderstandings and ask suppliers to elaborate on their proposals if necessary.

6. Analyze the bids: The next step is careful analysis of the bids submitted. Particular attention should be paid to the following:

■ Bids that do not cover the entire scope requested

■ The supplier's price level as a whole

■ Unrealistically high or low prices for individual parts.

The first price offered is a starting point for further technical and commercial discussion, not the criterion on which to award the contract

---

The first-round bids often provide interesting information on the suppliers' general technological competence. If a supplier only offers some of the parts requested, this may indicate that they do not have certain production technologies or cannot satisfy particular quality requirements. Unrealistically high or low prices often suggest that suppliers have not understood the specifications correctly.

7. Visit suppliers: Sourcing teams should hold detailed discussions on the bids with suppliers – misunderstandings will often come to light. This also gives suppliers the opportunity to improve their offers. Significant price cuts can be achieved in the process, as Figure 8.8 shows.

8. Conclude negotiations and draft a Letter of Intent: Once the final bids are on the table, the teams

can work out target prices for negotiation with the suppliers. They should use what is known as **best-of benchmarking,** i.e., select the lowest-price subcomponents and production steps from several bids. The lowest of all for each cost item is entered on the bid analysis form (Figure 8.9). The entire part is then recalculated on this basis. The resulting target price should be within the range calculated in the *landed cost model* (see Chapter 4) for the location.

Occasionally, suppliers' bids are not below the price level of high-cost countries, or only marginally so. One reason may be substantial productivity disadvantages. This should not be accepted for the long term. Instead, the OEM should give the supplier technical support during production ramp-up to increase productivity. Another reason may be that the supplier has only worked with small batches in the past and therefore has not included possible economies of scale in the costing.

Detailed discussion of the bids helps clarify these issues. It also needs to be remembered that many suppliers from low-cost countries now have excellent insight into the cost structure of competitors from

---

Substantial price reductions can be achieved even in early negotiation rounds

---

Fig. 8.8: Additional price reductions resulting from negotiations
Index (previous costs = 100)

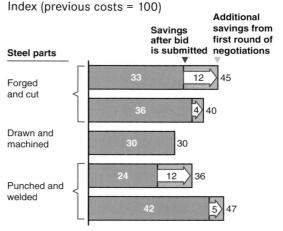

Source: McKinsey

high-cost countries, so often submit proposals just far enough below the price level of the high-cost country for relocation to be profitable. In these cases, it is up to the purchasing department to uncover additional latitude for negotiation by identifying the actual costs.

Once the team has reached an agreement with the supplier, a contract is drafted – preferably a Letter of Intent (LOI) to start with. The final contract should only become valid once the samples have passed all the quality tests. However, it is still important to draw up the LOI carefully, including the following points:

■ Prices and future price development

■ Quality requirements

■ Delivery instructions

■ Milestone plans for producing samples and approving parts

■ Warranties and penalties for noncompliance

■ Requirements on supplier development going forward.

9. Perform a supplier audit: In the auditing and sample testing phase, companies should make the same requirements on new suppliers as on their previous ones. However, it is useful to divide the conditions into two categories: those that are vital from the outset, and those that are desirable but not essential at the start of production. The samples supplied

---

Target price is determined using best-of benchmarking

Fig. 8.9: Form for calculating target prices   <u>EXAMPLE: CAST PARTS</u>

	Supplier A			...	Supplier E	Own assessment	Target price
A. Cost of raw materials							
Consumption – Gross (kg) – Net							
Costs (per kg)							
Net costs							
B. Labor costs	Costs	Unit	Net				
Processing steps – Casting – Machining – Deburring – Washing Total							
C. Cost of tools	Invest-ment	Depre-ciation	Net				
Tools – Casting molds – Machine tools Total							
D. Margins							
E. Price offered							

Source: McKinsey

often point to the need for an immediate program to improve quality (see the following case study).

10. Extend the approach to different parts: Approval of the sample parts concludes the overall process of local sourcing for the pilot parts. The sourcing team can then multiply this process from its experience with the suppliers and successively award contracts for the remaining parts from Segment 3. Figure 8.10 shows a possible process control system for this multiplication phase following strict milestones.

8.3.2 Establishing a Local Sourcing Organization

A top-caliber local sourcing organization is the bedrock of any local supplier structure. In the early stages of developing the location, its tasks include:

- Analyzing the market and selecting local suppliers

- Negotiating and concluding contracts with local suppliers

- Managing the site's sourcing

- Coordinating local logistics, including the export of parts if necessary

- Implementing test plans for central quality management.

> **You can only build up a local supplier structure if you have an effective and technically competent local sourcing organization in place**

Technically competent local quality assurance is especially important when building new suppliers in low-cost countries. One example of an organization structure that has proven valuable in managing these tasks is shown in Figure 8.11. This structure is

---

A process control system is essential when placing contracts for large volumes of parts

Fig. 8.10: Process for monitoring parts contracts
Number of parts

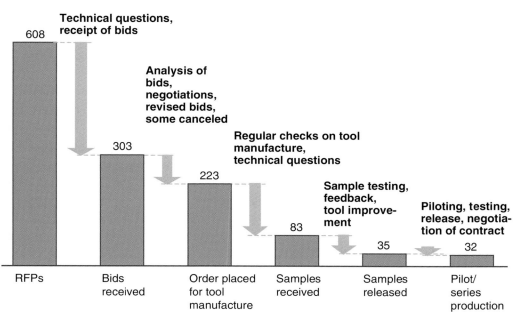

Source: McKinsey

Sourcing an Electric Motor Housing from Slovakia: Selecting and Building Local Suppliers

A Western European company from the electrical and electronics industry set up a new plant for electric motors in Slovakia. They needed to find a new local supplier to manufacture the pole housing for the motor.

A project team of buyers, developers, and quality managers began selecting and training suppliers. In the first step, five potential suppliers were identified within a 200-km radius. Following visits to the firms, two were shortlisted based on their apparent competence and the cost level of their first proposals. The future supplier was selected from these two after a detailed process audit. In the end, the deciding factor was greater tooling know-how.

Together, they immediately made a plan to prepare for production start-up. This was followed by initial training courses for the supplier's employees on quality management just a few weeks later. They jointly defined new guidelines, and the supplier changed its previous quality assurance processes.

In parallel to this, work began on tool development with the support of an expert from the customer's company. Since the first sample parts did not yet satisfy the requirements, they discussed the deviations in workshops and defined detailed improvement initiatives before further sample parts were made. Thanks to this intensive support, the supplier managed to achieve the quality standards in the second round of samples and, 17 months after the selection process first began, the parts were approved for series production.

Bottom line: Companies can source parts successfully even in low-cost countries by rigorously managing the selection process and training their suppliers. It is essential to budget additional expenditure for providing intensive supplier support.

---

also useful for mapping further supplier development activities required later in the process.

Ideally, employees in the local sourcing organization should have experience in their own company and in the relevant branch of industry, know the local market in detail, and speak the local language fluently. As a rule, of course, candidates of this quality will rarely be available. The best solution is therefore a good mixture of expatriates and locals, perhaps combined with locals who have returned from living, working, or studying abroad (and thus have international experience). The box below describes a real-life example.

It can be worthwhile using external service providers for sourcing small volumes or supporting the procurement process

In view of the effort it takes to build a local sourcing organization, many companies wonder whether it would make more sense to call in external agents, at least initially. The answer depends on the specific task in the sourcing process, as well as the type and number of parts that need purchasing.

As a rule, companies should handle the core tasks of local sourcing themselves, right from the start. This particularly applies to the selection of suppliers and contract negotiations, as both tasks presuppose familiarity with the company's requirements. There are also tasks, however, that many companies are already successfully outsourcing to service agencies in low-cost countries today, e.g., market analysis and identification of suppliers, quality audits and tests according to defined measurement protocols, and the execution of logistics.

Other criteria for the use of external agents include the size and type of parts to be sourced. If, for example, a company wants to purchase just a few special

Rigorous local quality assurance is crucial when building up new suppliers

Fig. 8.11: Typical purchasing organization in a low-cost country

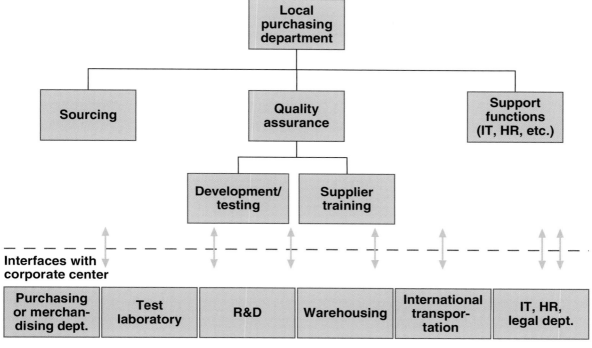

Source: McKinsey

Example: Establishing a Purchasing Office in China

A large European company with a global presence had run a purchasing office in Shanghai with a European manager and two local purchasers for several years. Despite this, the company bought hardly any of its parts in China – unlike its competitors, who were much more successful at purchasing locally. The company decided to launch a project to boost its sourcing from China.

It was key to find the right purchasing manager locally to steer the organization through the difficult setup phase. The company looked for a Chinese national with five to ten years' international management experience in purchasing for the industry in question or a related field. They also needed to be able to attract and retain other good employees. Finally, the candidate had to be prepared to commit to a long-term working relationship. Since the competition for highly qualified purchasing managers in China is extremely fierce, a very attractive salary was essential. Following advice from headhunters, the company put together a package that secured the purchasing manager an annual salary of around USD 250,000, provided the agreed objectives were met.

In addition to the purchasing manager, four local buyers also signed up within the first four weeks. Since usually only four weeks' notice is required in China, they moved into their new positions fast. These buyers were around 30 years old and had already gathered two to five years' purchasing experience in similar industries following their graduation from Chinese universities. All spoke good English, so they could communicate well with the corporate center. Their salaries ranged between USD 6,000 and 20,000 per annum – with considerable differences between Shanghai and Shenzhen. This salary meant the buyers were earning many times the average income for qualified workers.

In spite of these attractive packages, the company was confronted with a problem typical for China – constant staff turnover. Some were hired away after just a few weeks. This meant recruitment had to continue over the next few months.

To support the local staff, the company transferred an experienced engineer from Germany to China for a year. He brought technical experience from the central development department to the work of the local sourcing organization and gave the local buyers systematic training in industry-specific technology issues. He also served as a contact for employees from head office.

Bottom line: Establishing an effective local purchasing office requires critical minimum expenditure to retain sufficiently qualified staff. Compensation packages for high performers have to be attractive enough to fend off local competition.

technology parts, having its own organization would not be economically viable. Experienced agents can be worth engaging in these cases, particularly for small-volume specialist materials groups and parts. Although they can often cost three to four times as much as the company's own employees, the expense is worth it because they achieve the objective much faster due to their contacts. Some specialist agents also have privileged access to certain local companies or special market know-how. It can also be valuable to use agents on a continuing basis for large volumes. Companies have to use their own discretion on this, based on their assessment of the value added.

8.4 Second Wave: Sourcing more Complex Parts

Once local sourcing is on a solid footing, companies should also tackle the issue of purchasing more complex components (Segment 4) locally. Both the local suppliers and the company's own local sourcing organization will need training to achieve the quality, reliability, and efficiency required for the collaboration. It will also be vital for the supplier's production department and the company's own R&D or production department to cooperate closely due to the complexity of the parts. Manufacturers should also at-

tempt to tap into additional savings potential by providing local purchasers with methodical training in this second wave. The immaturity and dynamism of many markets in low-cost countries often make the use of purchasing tools even more successful and worthwhile than in high-cost countries.

8.4.1 Targeted Development of Local Suppliers

Local suppliers generally need their customers' support to build up the additional competences required for producing more complex input products. The company's efforts in the second wave should therefore focus on the targeted development of selected local suppliers. As relevant analyses show, this can result in further significant reduction in materials costs, while also improving quality and on-time delivery.

There is no one-size-fits-all solution for developing suppliers. Their needs will be very different depending on the country, the product, and sometimes even on the individual supplier. This means close cooperation and the systematic identification of possible improvement opportunities are indispensable for the success of the project. The case study below shows how this might be applied in practice. The same mid-size company

Sourcing of Injection-Molded Parts in China: Developing Suppliers Using Quality Management Support

A small to mid-sized industrial enterprise wishing to purchase injection-molded parts from China experienced years of frustration with samples produced by suppliers there (also see section 8.1.2). In the first test, the dimensions were often wrong. The second sample failed to meet the physical requirements. In the third pass, the surface proved substandard. The company therefore set up a project named "2nd Time Right" aimed at fundamentally redesigning the collaboration.

In the first step, they altered the process of sample testing and auditing in four ways:

- The first samples were now checked immediately based on all the criteria. Samples with shortcomings and defects (e.g., incorrect dimensions) were detected much earlier. This meant the suppliers received detailed feedback about all the features they still needed to work on right from the outset.

- Company staff now discussed this feedback with the suppliers locally. Previously, suppliers had received the test results on German forms by fax, generally without any comments.

- The test periods were shortened. Rather than testing all the samples in Germany, external service providers in China performed certain standard tests. The bottlenecks that had previously been the norm in the German test laboratories were avoided by announcing the new samples early and having strict rules for processing. At the same time, the company quickly set up its own test laboratory with local staff in Shanghai.

- The most promising suppliers had already been audited before the order was placed, preventing wrong decisions when the contracts were placed. Each supplier also had sufficient time to remedy any faults parallel to the samples being tested. Chinese staff members were trained as auditors to introduce even greater flexibility.

In addition to these process improvements, the company also gave the local suppliers direct support. It set up a Supplier Quality Team that – following intensive training from the engineers at corporate center – advises suppliers from the sample production stage through to release for series production. The team now spends three-quarters of its time with the key suppliers, helping them to stabilize processes and meet the quality requirements in start-up processes.

These far-reaching changes paid off. Thanks to their steep learning curve, the suppliers managed to dramatically reduce defect rates in the initial sample test, thus shrinking total development times for new parts.

Bottom line: Working closely with local suppliers is the only way to identify needs and tap into improvement potential.

sourcing injection-molded parts in China serves as the example – with a revolutionized approach.

Leading automotive OEMs have also recognized the advantages of targeted supplier development. They cooperate closely with their suppliers across the globe to lower their costs and improve quality and delivery standards. Dedicated teams perform audits, identifying and then implementing improvements together with the suppliers. The results are often impressive, as the example of one Indian supplier shows. With the help of its main customer – a European OEM – this supplier introduced lean management and modern purchasing tools, had its executives trained in modern management methods, and established an incentive scheme for continuous improvement. The result: delivery times fell from eight weeks to five (with increased flexibility), reject rates dropped, and inventories were reduced (Figure 8.12).

> **Companies can often realize additional cost potential by training and developing local suppliers in issues relating to production, quality, and management**

What is the lesson to be learned if manufacturers wish to be successful in low-cost countries in the long term? They have to adapt their supplier development to the specific needs of the situation to realize the full potential of these relationships.

8.4.2 Methodical Skill-Building for Purchasers

As in high-cost countries, skillful negotiation is important for purchasing in low-cost countries – sometimes even more so. Purchasers therefore need sound training in conducting fact-based nego-

> **Developing suppliers in low-cost countries can achieve fast, impressive results**

Fig. 8.12: Tapping into potential from supplier development

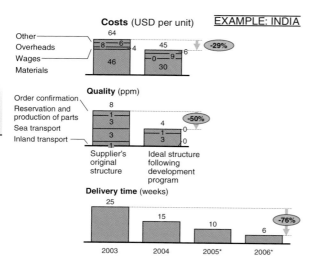

* Forecasts

Source: McKinsey

tiations. Methods range from auctions through to best-of benchmarking and target pricing. Many companies report savings of 10 to 20 percent from these techniques.

> **Modern methods of preparing negotiations create transparency on further cost reduction potential**

Making a correct assessment of the actual savings potential available is a big challenge in low-cost countries. This is because suppliers from India, China, or South America always offer very low prices from the perspective of Western purchasers. Western companies do not always have a firm idea of local suppliers' actual costs. This means negotiations are often based purely on comparison of the proposals they have to hand. Many suppliers in China and other low-cost countries have been achieving impressive profit margins for years as a result.

This was a painful lesson for an American IT company. After merging with a Chinese competitor, it found that the Chinese were paying up to fifty percent less for comparable parts from the same local suppliers. Insight into the nuances of business in the part of the world you are sourcing from is the key. Sophisticated purchasing tools are available for creating the necessary fact base.

Two examples of these tools are explained in greater detail in the appendix: clean-sheet costing and linear performance pricing (LPP). Every company should train its local purchasers in the use of such tools for preparing negotiations, and ensure that they are applied consistently. Having this kind of head start will definitely pay off – both directly in supplier negotiations and indirectly by giving companies an edge on their competitors.

Globalization in the Automotive Industry: BMW

In 2006, the BMW Group, manufacturer of the BMW, MINI, and Rolls Royce brands and premium motorcycles, sold over 1.35 million vehicles and generated revenues of nearly EUR 49 billion. The company employs around 106,000 people worldwide. Its most important markets, each accounting for a quarter of revenues, are North America and Germany. A large share of BMW's production facilities are concentrated in Germany, and the latest BMW plant was opened in Leipzig in 2005.

Around 235,000 MINIs are manufactured in the UK, and more than 150,000 BMW X5 and Z4 models are produced in Spartanburg (South Carolina/US). Additional production plants are located in Rosslyn (South Africa) and, since 2003, in Shenyang (China). The Group also has eight assembly plants.

"Take Established Suppliers With You if Possible" – Interview With Dr. Klaus Richter, Senior Vice President Production Materials Purchasing, BMW Group

Dr. Richter, the BMW Group has expanded globally, especially in the last decade. Established suppliers are also going global, and new ones are emerging in developing countries. What effect is globalization having on BMW's sourcing and its suppliers?

Source: BMW AG

Dr. Klaus Richter was Senior Vice President for Production Materials Purchasing at the BMW Group from 2004 to 2007, after having served as Vice President for Purchasing Strategy and Controlling since 2003. In his role als Senior Vice President he was responsible for a purchasing volume of over EUR 25 billion and contacts with around 1,000 suppliers. In November 2007 he became Executive Vice President for Procurement at Airbus. We spoke to Dr. Richter about the globalization of the automotive supply industry and BMW's sourcing practices.

Globalization has two very different drivers where suppliers are concerned: first, cost potential, and second, the opportunity to conquer regional markets.

The BMW Group has set up market-oriented facilities twice following a standardized approach: in South Africa and in Spartanburg in the US. We typically establish a market-oriented supplier base in three phases.

The first step is to set up a certain supplier base parallel to our own production. Local sourcing is often essential, due to customs requirements, if assembly does not cover the minimum value added necessary to achieve the status of a local vehicle manufacturer. In a country like South Africa, this is genuine pioneering work. Initially, companies find hardly any suppliers there, and those that are in place are generally unable to deliver high quality at low cost – by global comparison – at first. It is, therefore, important for BMW's existing global suppliers to build up a local branch. We generally try to take our key suppliers with us to a new location. We then make a joint effort to attain the minimum local value added required by relabeling and local assembly, but increasingly also by expanding local sourcing. This costs money and is not productive, but tariff barriers make it the most appropriate way to get established.

In the second step, a genuine supply market develops that actually does deliver more cost-efficiently to local manufacturers because of low transport costs and customs advantages. In this phase, more and more parts are sourced locally, the total value of which may go beyond the minimum local content required for customs reasons.

In the third step, a level of productivity and quality is achieved that allows us to export locally sourced parts for use in other plants. In South Africa, we have gone through all three phases, and today we supply many leather products from there for our worldwide production.

How can the other objective be achieved: more cost-efficient sourcing?

This is the major task of our tier-one suppliers – the manufacturers of parts and components who supply the BMW Group direct. We frequently purchase entire systems – brake systems, for example – on a just-in-time and just-in-sequence basis, sourcing them from a plant relatively close to the BMW production site. The flexibility we offer our customers – they can change the car's fittings up to five days before production begins – often means packing and assembly of the systems have to be kept within a very small radius.

This year we've opened a number of international purchasing offices in order to accelerate our global sourcing. We already had offices in Spain/North Africa, Singapore, China, and Japan and have now added BMW purchasing offices in India, Turkey, and Eastern Europe. We naturally want to tap tier-one suppliers for established technologies, but, at the same time, support them in optimizing their value chains and optimizing their component footprints.

Subcomponents and smaller parts, however, come from a much larger catchment area. The obvious one here is Eastern Europe, particularly for products such as wiring harnesses that are labor-intensive and simple to produce. We know the relevant regions well and use them systematically. At present, for example, we buy parts worth around EUR 3 billion from Eastern Europe.

Suppliers that do not use low-wage countries for the production of parts and components often cannot offer competitive price structures in their proposals. This becomes a problem for these suppliers if we are forced to look for alternatives and develop other suppliers for economic reasons.

What experience – positive as well as negative – has BMW had with suppliers and parts from emerging countries, particularly in Eastern Europe?

The BMW Group has not had any negative experience with parts from low-wage countries. These suppliers do need to be carefully supervised and audited, though. The BMW Group has not worked directly with Chinese suppliers a great deal so far, for instance. There is no doubt that China harbors certain risks, partly because it is so far from our home location, but this is a subject that we must and will address.

How do you react if suppliers – e.g., Bosch, Delphi, or Faurecia – reorganize their plant networks? How do you interact?

As a rule, suppliers have to inform us if they wish to change their production locations. The audits we demand for every start-up create a certain transparency, and naturally we also reap a share of the positive cost impact, as we generally do our own costings for our suppliers' services. The sourcing potential of low-wage countries often also comes from cooperation with local second-tier suppliers. The potential of such markets therefore goes beyond the labor cost advantages of our direct suppliers. All in all, I consider this a practicable approach because we, as the manufacturer, cannot handle every step ourselves. With regard to cycle time, reliability, and on-time delivery, we need partners who can manage their own supply base appropriately and who know what it means to be integrated into our sourcing process.

BMW's strategy varies depending on whether the objective is "opening up the market" or "tapping sourcing markets," doesn't it?

Yes, that's right. When opening up new markets, we have to break new ground – due to trade barriers as much as anything else – and start producing and purchasing locally very quickly. On the efficiency side, however, we benefit from established sourcing markets. We won't be the first to enter Mongolia, for example, and neither do we want to be. All in all, globalization is an evolutionary and very successful process for the BMW Group, with a multiyear lead time.

But having a presence in established foreign markets is important. Japan is a trendsetter in active suspension and infotainment systems, for instance. Spain is interesting in the context of purchasing for small cars – after all, 3.5 million vehicles are built there in this class. Italy is important as a supply market for motorcycle construction.

How long does it take to build up suppliers in new locations?

The process of getting up to speed, where suppliers expand their skills in a new location, takes around two generations of vehicles, i.e., ten to 15 years. The Chinese want to cut down the length of this process, but it is not easy. In Eastern Europe, ramp-up is less critical. The culture is European, and starting up production is not very different to being in Germany if you have a small group of expatriates.

How do you go about setting up the local supplier structure for new plants?

We try to take our existing suppliers with us, if possible. Of course, this is barely viable for them at the start because the unit volumes are still relatively low. This means we have to support them either directly or indirectly to make it attractive and feasible for them to build up production there. In the long term, however, the effect is extremely positive. The value of the materials sourced via Spartanburg, for example, is now around EUR 2.5 billion.

Overall, this process is very fruitful for the entire German industry – despite or even because of production overseas. The BMW plant is the nucleus that persuades even SME suppliers to set up production in the new country. As a result, the suppliers create a base from which they have a better opportunity to open up the corresponding market. A whole range of suppliers who went with us to Spartanburg now also supply the big three in the US, i.e., Ford, GM, and Chrysler.

What are the consequences for Germany, BMW's home location?

Positive. In the past few years, the great success of German automotive manufacturers and suppliers has led to a 15 percent increase in employment in the industry, equating to over 100,000 additional jobs in Germany. To be fair, however, it has almost exclusively been the more highly skilled workers who have benefited from this growth.

It is clear that parts and components can be produced in low-cost countries if processes are stable. But you have a much harder time of it as a systems integrator. High qualifications and direct contact between employees are very important in this field. Germany still greatly underestimates the priority of training and education. Better training is absolutely essential for us to differentiate ourselves from other countries that have low labor costs. In the future, production will gravitate towards the locations around the world that offer the best business environment, and will also depend on where R&D is located.

Appendix

Clean-Sheet Costing

Clean-sheet costing is a tool for calculating target costs that the purchaser then tries to realize in negotiations (Figure 8.13). The parts to be sourced are first broken down into their individual components and cost drivers and then recalculated based on local factor costs. If precise cost data is not available, e.g., machine processing times, a realistic estimate is usually sufficient. Following the costing, the purchaser identifies what percentage of the price offered remains unexplained.

The aim of the negotiations is not usually to clarify how the supplier worked out their proposal down to the last detail. More important is to understand where there is room for maneuver and roughly how much. Targeted focus on these points may enable the manufacturer to reduce the unexplained percentage.

Linear Performance Pricing

While clean-sheet costing sheds light on the supplier's internal cost structures, linear performance pricing (LPP) allows comparison of the prices of similar products. To do this, the product's crucial performance driver (e.g., the tension force of a spring) is identified and the price compared with this performance parameter. In the example of a spring, this results in a price per unit of tension force, e.g., 10 eurocents per kilo Newton. The aim here is to push all prices down to the lowest relative price.

For LPP to be successful, it is essential to choose the right performance parameter. This is relatively easy in the case of an electric motor, as illustrated in Figure 8.14. That there is a linear correlation between the power of an electric motor and its price is an obvious assumption. It may be harder to define the performance parameter for more complex products or modules, although in these cases the transparency gained and resulting potential are correspondingly higher.

Clean-sheet costing means breaking down components to be sourced into their cost drivers, identifying the share of costs that cannot be explained

Fig. 8.13: Clean-sheet costing example – electronic accessories
USD per unit

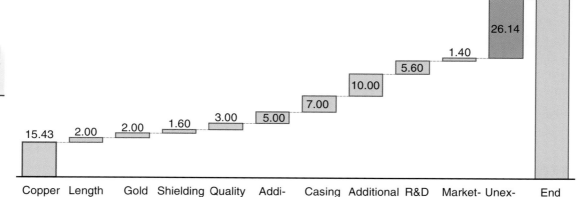

Source: McKinsey

Linear performance pricing analyzes bids based on their central performance parameter

Fig. 8.14: LPP – example: air-conditioning motor

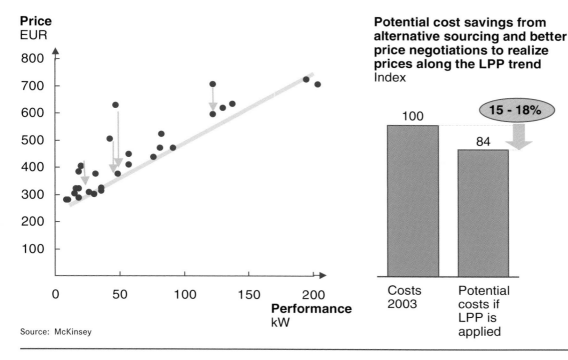

Source: McKinsey

Example: Determining the Performance Parameter for an Automotive Wiring Harness

Every modern automobile has a central wiring harness that connects all the key electronic components. Its complexity varies greatly depending on the type of vehicle, number of electronic systems, and add-ons. For a long time, it was considered impossible to compare them: the lengths of cables, plugs used, and other features were all too different.

However, one of the big OEMs used the LPP method to make its wiring harnesses comparable for all of its group's brands, models, and variants. The performance parameter was specified simply as the number of cables used in each harness. On this basis, the automotive manufacturer was able to negotiate a standardized target price for wiring harnesses with its global suppliers, which they all had to reach within four years. Each purchaser was then able to work out the unit price for each wiring harness simply on the basis of the number of cables it contained.

Bottom line: LPP can even make the costs of complex components comparable as the basis for negotiation.

Further reading

Assaf, M., C. Bonincontro, and S. Johnsen: *Global Sourcing & Purchasing Post 9/11: New Logistics Compliance Requirements and Best Practices.* Fort Lauderdale: J. Ross Publishing, 2005.

Campbell, R. M., J. Hexter, and K. Yin: "Getting sourcing right in China" in *The McKinsey Quarterly*, Special Edition: China Today. New York: McKinsey & Company, Inc., 2004.

Cook, T. A.: *Global Sourcing Logistics: How to Manage Risk and Gain Competitive Advantage in a Worldwide Marketplace*, New York: AMACOM/American Management Association, 2006.

Dimitri, N., G. Piga, and G. Spagnolo: *Handbook of Procurement.* Cambridge: Cambridge University Press, 2006.

Hexter, J. and A. S. Narayanan: "The challenges in Chinese procurement" in *The McKinsey Quarterly,* Special Edition: Serving the new Chinese consumer. New York: McKinsey & Company, Inc., 2006.

Hexter, J. and J. Woetzel: *Operation China: From Strategy to Execution.* Boston: Harvard Business School Press, 2007.

Knapp, B. W.: *A Project Manager's Guide to Contracting and Procurement.* The Project Management Excellence Center, Inc., 2006

Laseter, T. M.: *Balanced Sourcing: Cooperation and Competition in Supplier Relationships.* San Francisco: Jossey-Bass, 1998.

Nassimbeni, G. and M. Sartor: *Sourcing in China: Strategies, Methods and Experiences.* New York: Palgrave Macmillan, 2006.

Nelson, D. R., P. E. Moody, and J. Stegner: *The Purchasing Machine: How the Top Ten Companies Use Best Practices to Manage Their Supply Chains.* New York: Free Press, 2001.

Paquette, L.: *The Sourcing Solution: A Step-by-Step Guide to Creating a Successful Purchasing Program.* New York: AMACOM/American Management Association, 2003.

Vietor, R. H. K.: *How Countries Compete: Strategy, Structure, and Government in the Global Economy.* Boston: Harvard Business School Press, 2007.

SEBASTIAN SIMON, ULRICH NÄHER, MADS D. LAURITZEN

9 R&D: Aligning the Interface with Production

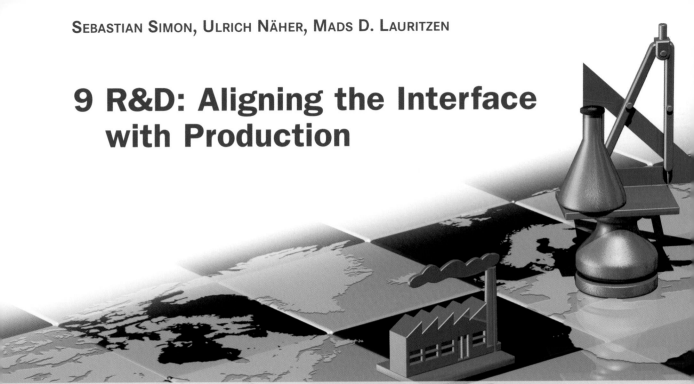

Summary

The optimal site for a company's R&D is not guided by the same parameters that determine its production location. Though factor costs and market proximity also play a role, access to top-caliber engineers and the nurturing vitality of a deep-rooted knowledge cluster are often the key drivers. As a result, the R&D of companies based in developed countries is currently not as globalized as production. The widening geographic gap between production and R&D means management of their interface is becoming ever more important.

To what extent should companies collocate these two functions, with R&D following production abroad? The decision is always case specific. When products involve little R&D and simple manufacturing processes, collocation of production and R&D is not necessary. But with more complex products, detailed analysis is required to determine which aspects of the R&D process should remain centralized and which would benefit from collocation.

Depending on local product design requirements and location-specific production technologies, various constellations are possible – from mere support of production ramp-up to wholesale transfer of product R&D. In the latter scenario, it is advisable to expand gradually from a pure production site to a center of competence with full product accountability.

Future trends suggest that high-cost countries will continue to have high appeal as the locus of R&D. This is mostly due to the established knowledge clusters for specific industries, with their high density of private and public research activities, together with suppliers, customers, competitors, and venture capitalists. This mix creates a rich network that stimulates constant know-how exchange and innovation. Inevitably, however, the landscape will change as emerging markets grow and their education standards rise. In some industries, high-cost countries will continue to dominate product development, but others will see the shift of entire clusters.

Key questions, Chapter 9

■ What are the most important indicators for collocating production and R&D?

■ What different solutions are available for collocation?

■ What options have companies chosen in practice?

■ Under what circumstances will emerging countries develop as R&D locations?

9.1 The Challenge: Finding the Right Constellation

Any company setting up a global production structure has to consider how to integrate R&D. The interface between the two functions needs harmonizing with the product strategy, the needs of the target market, and the production network configuration. When should R&D move with production, and how much should be invested in managing the interface?

R&D locations gravitate towards knowledge clusters and markets, while production locations are more oriented to factor costs

Surveys show that the R&D departments of Western companies are far less globalized than their production plants (Figure 9.1). At German companies in the sample analyzed, for example, around 30 percent of manufacturing is carried out abroad, but less than 20 percent of development work. Even if R&D is set up globally, a high-cost location is usually chosen, while production tends to be based in low-cost countries.

This is because factor costs play a much greater role in selecting production sites, while R&D locations are geared more toward knowledge clusters and markets. This is illustrated by many case examples, such as the European automotive supplier ZF (Figure 9.2). This company has plants scattered throughout the world and has expanded its capacity, particularly in Asia, Eastern Europe, and South America. Yet its R&D locations are largely situated in high-cost countries. Its two platform development locations are still based in Germany and the US.

The intensity of interaction between the two functions is a crucial factor in the decision. There are multiple interactions along the entire product life cycle. A typical example is the development of a new vehicle (Figure 9.3). In the concept development phase, R&D uses inputs from production experts (e.g., technical feasibility studies or calculation of the required investment and production costs). In later development phases, production follows specifications defined by R&D, such as when preparing logistics and manufacturing concepts. In the ramp-up phase, responsibility for the vehicle passes from R&D to the production division. Ramp-up problems and quality defects have to be rectified in short communication loops between R&D and production. The two functions also have to jointly develop and implement any cost-cutting initiatives that are required.

The more intensive the interaction between R&D and production, the more seriously collocation should be considered

Does this interaction mean R&D and production need to collocate? Not necessarily. Practice shows that automotive OEMs often decide not to. A detailed, case-specific analysis is required to find the optimal solution, factoring in location-specific constraints and structures.

Across industries, two indicators influence the intensity of interaction required on a product-specific basis. A high level of **product innovativeness** frequently necessitates complex and detailed knowledge transfer between R&D and production, and close feedback loops between these functions. Product life cycles are often short, resulting in repeated production ramp-ups.

The **complexity of the production process** determines the extent to which product and process development engineers require interaction with production, and vice versa.

The ProNet survey revealed a clear correlation between these indicators and R&D integration (Figure 9.4). Wherever both indicators were found to be highly important, the companies surveyed opted to integrate R&D more closely.

R&D is less globalized than production, but expanding rapidly

Fig. 9.1: Cumulative global foreign R&D investments

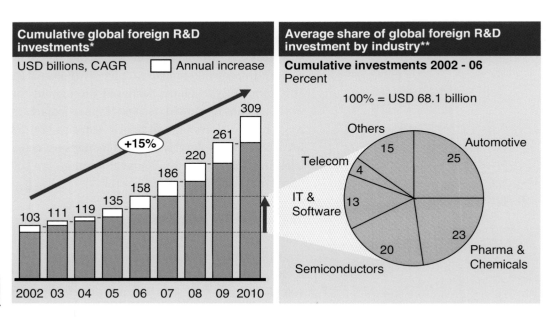

* Since 1982; projection of global annual foreign R&D investment inflows based on R&D investments in 80 countries; 2006: projection based on 9 months of data; 2007 -10: forecast
** Industry sector split based on foreign R&D investments in Brazil, China, Eastern Europe (12 largest countries), India, Japan, Mexico, Russia, US, Western Europe (EU-15, Norway, Switzerland)

Source: Locomonitor 2006, UNCTAD 2006, McKinsey analysis

In general, R&D is more bundled than production

Fig. 9.2: Production and R&D locations of ZF

Source: ZF

Development and production have particularly close ties in the automotive industry

Fig. 9.3: Interface between development and production in the automotive industry – example

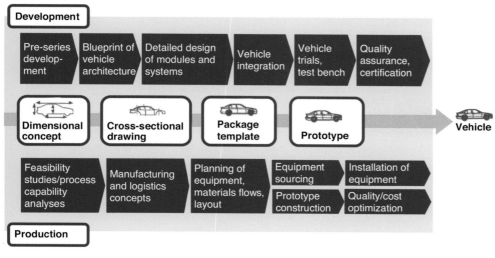

Source: McKinsey

9.2 Design of the Interface with Production

The above indicators provide initial pointers on how far the collocation of R&D and production might be advisable for a particular product. The development of a new product, however, cannot and should not be considered monolithically *en bloc*. The **five typical phases** of R&D (Figure 9.5) make very different demands on proximity to the market and production. A company's basic research activities, for example, are often conducted in central research facilities, in close cooperation with external knowledge centers, such as universities and public research institutions. Its application development processes, on the other hand, require market-specific insight, and may even be closely integrated with the R&D and production networks of key customers.

The need for collocation of R&D depends heavily on production complexity and degree of innovation

Fig. 9.4: Drivers of R&D collocation

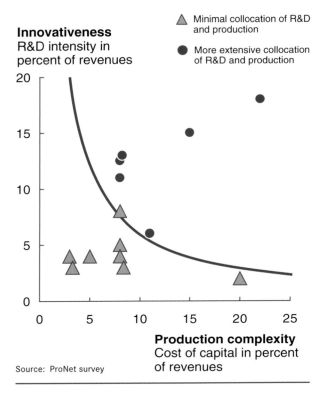

Innovativeness
R&D intensity in percent of revenues

△ Minimal collocation of R&D and production

● More extensive collocation of R&D and production

Production complexity
Cost of capital in percent of revenues

Source: ProNet survey

The various phases of R&D require differing levels of proximity to the market and production

Accordingly there is a whole spectrum of possible solutions between minimal and complete collocation of R&D with production. Choosing the right solution and designing it effectively plays a major part in

---

The R&D process has five phases

Fig. 9.5: Typical phases in the R&D process

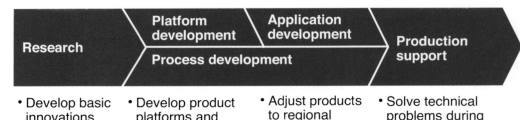

	Research	Platform development / Process development	Application development	Production support

Tasks

- Develop basic innovations (products, technologies, manufacturing processes)

- Develop product platforms and standardized modules
- Develop manufacturing processes in parallel

- Adjust products to regional markets and customers

- Solve technical problems during ramp-up
- Optimize ongoing production

Source: McKinsey

helping to optimize efficiency and effectiveness at the interface between the two functions. Five basic options can be distinguished (Figure 9.6):

1. Independent networks with minimal collocation: In this option, only a few R&D employees are based permanently at the production location. This is particularly suitable for mature products that are already being manufactured at other locations.

2. Collocation of process development: The second option involves developing production processes at the site itself. This is recommended if processes need adapting to local circumstances (see Chapter 5), or are so complex that substantial development support is required even in ongoing operations.

3. Relocation of application development: The next step up is to transfer application development. This approach is often used for highly market- or customer-specific products, and can be combined with option 2.

4. Relocation of platform development: Option 4 is to relocate product platform development (in addition to application development) to the same location, or at least the same region. This is relevant when the target market requires a completely different product architecture, such as a low-cost product platform.

5. Full collocation: Companies usually only opt for full integration at long-established locations that

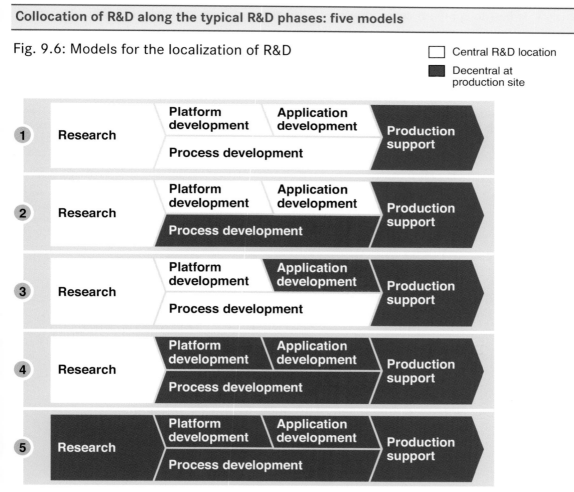

Collocation of R&D along the typical R&D phases: five models

Fig. 9.6: Models for the localization of R&D

□ Central R&D location
■ Decentral at production site

Source: McKinsey

serve as a knowledge cluster for an entire sector. Moving R&D generally means an industry's locus has shifted fundamentally.

The general principle: the easier it is to uncouple R&D from production, the more the R&D and production networks can be optimized independently of one another. However, if there are good reasons for collocating, it is important go the extra mile. It may take up to two years to move process and application development, depending on the complexity and innovativeness of the products and production processes. Redesign of the entire development network, systematically setting up and expanding product development competence through to platform development capability, may realistically take up to five years or more. Forward planning is crucial in view of these time scales. It is best to approach collocation on a step-by-step basis, initially adapting just the production steps, for example, and then moving other development competencies in a second phase. Full integration may take five years or more.

The collocation of development functions with production can take a long time and needs considerable forward planning

The section that follows discusses the five interface design options in greater detail, with case examples for illustration.

9.2.1 Option 1: Independent Networks with Minimal Collocation

If production processes, products, and relationships with markets allow it, companies can plan the limited collocation of R&D and production for new production facilities (Figure 9.7). The plant adopts tried-and-true production processes from other locations, with just a small team of production planners, operations schedulers, and engineers on hand to provide support. During the ramp-up of new products, technical experts may also be brought in from other locations to provide temporary assistance. The product modifications implemented by local product and

Minimal collocation for products where the degree of innovation is low and manufacturing processes are simple

Fig. 9.7: Option 1 for R&D collocation

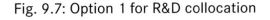

Legend:
- Interface
- □ Central R&D location
- ■ Decentral at production site

Indicators
- Products and manufacturing processes not very innovative
- Low complexity/limited number of manufacturing processes
- Not much adaptation of products to specific requirements required
- Manufacturing processes highly standardized

Source: McKinsey

process engineers are only minimal. Given that the interdependence of the two functions will remain limited, this solution allows general optimization of the R&D network according to R&D-specific location factors without constraints arising from too great a focus on the production footprint.

The ProNet survey revealed that many companies prefer this approach. The leaders, in particular, opt for centralization of fundamental design decisions on products and production processes when establishing global production networks (see Chapter 7). This facilitates integration with existing know-how clusters, the use of economies of scale in R&D, and reduces the number of variants due to greater product standardization.

The creation of R&D-independent plants in peripheral locations is often supported by a lead factory concept. Lead factories carry out the central development of production technologies as well as the transfer to all branch factories. This means they also have to produce a minimum share of the production volume themselves (product platforms and possibly local variants) – either temporarily during the production launch, or even permanently (Figure 9.8).

> **The most successful global manufacturers tend to centralize their fundamental design decisions**

The branch factories adopt production processes that are already stable and reliable. Lead factory employees also help branch factories optimize specialized production processes for local product variants and small series. In effect, they act as centers of competence with a global presence. They assess and optimize manufacturing methods, train staff at new sites, gather and validate optimization ideas, and generally drive continuous improvement.

Lead factories produce some of the volume required

Fig. 9.8: Lead factory concepts: parallel and staggered production

	Indicators	Advantages for R&D integration
Ⓐ Parallel production 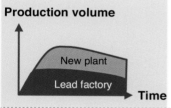	• High migration costs • Short product life cycles • Need for flexibility	• Lead factory closely supervises production ramp-up • Lead factory supervises ongoing product changes • Lead factory constantly optimizes production
Ⓑ Staggered production 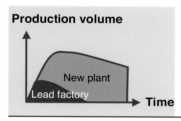	• Wage-intensive production • Long product life cycles • Little need for coordination after stabilization	• Lead factory closely supervises production ramp-up • Lead factory's capacity is only needed for a short period

Source: McKinsey

Transplants of Japanese Automotive OEMs in Europe, with Japanese Lead Factories

The Japanese OEMs Daihatsu, Honda, Isuzu, Mitsubishi, Nissan, Suzuki, and Toyota (Figure 9.9) provide an interesting example of how to divide up R&D and production. They employ a combined total of around 36,000 production workers in their European production transplants – with a bias towards locations with relatively low wages (Spain, Portugal, Eastern Europe). Small R&D functions operate at these sites, but these are mainly for stabilizing and optimizing manufacturing processes. The bulk of product and process R&D is based in Japan.

In addition, over 2,500 staff members in application or design centers are responsible for regional product adjustments. These centers are also located in Europe, but distributed according to a completely different pattern. To be close to what makes their business tick, these companies have tended to choose locations close to their competitors, right in the most sophisticated automobile markets with the greatest purchasing power. The majority are in Germany, Benelux, and the UK.

Bottom line: Japanese OEMs are a prime example of the strict segregation of production and R&D locations. Production tends to be in the low-cost periphery of Europe, while R&D is close to the market. The lead factory concept also holds firm: transplants replicate production processes tried and tested first in Japanese plants.

Japanese OEMs in Europe use different sites for application development and production

Fig. 9.9: Locations of Japanese OEMs in Europe

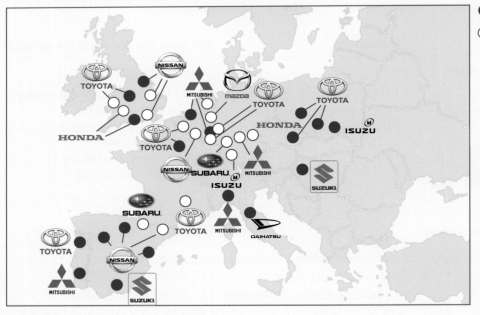

Source: Japan Automobile Manufacturers Association (2004)

9.2.2 Option 2: Collocation of Process Development

The collocation of process development (Figure 9.10) may be advisable if the manufacturing processes are very R&D intensive (e.g., in basic materials and process industries such as steel and chemicals) or have to be adapted heavily to conditions at the location concerned. A typical example is where companies revert to their previous semi-automated or manual production processes in low-cost countries (see Chapter 5).

Although these processes do not need to be redeveloped, they do have to be harmonized with current products. To collocate just process development successfully (without application or platform development), these parallel R&D phases must largely be independent of one another.

Option 2 is often found when production capacity is moved for cost reasons. It frequently makes sense to expand a low-cost location with high production volume into the lead location for a specific production technology. One European automotive supplier relocated production of its air-conditioning systems to Eastern Europe. The company hoped to benefit from the wage structures there while still supplying to customers throughout Europe at acceptable logistics costs. The supplier then decided to enhance its local production technology competence to optimize its production processes in Eastern Europe and supervise the launch of new products. Within a few years, the Eastern European location had assumed the global leadership function for all manufacturing processes related to its air-conditioning systems. The production experts based there are involved in all product development processes with automotive OEMs worldwide from the outset to ensure design-to-manufacture and design-to-assembly solutions.

9.2.3 Option 3: Relocation of Application Development

The expanding markets of Asia and South America require specific products that make close-to-the-

---

Collocation of process development (only) may be an option for low-cost locations

Fig. 9.10: Option 2 for R&D collocation

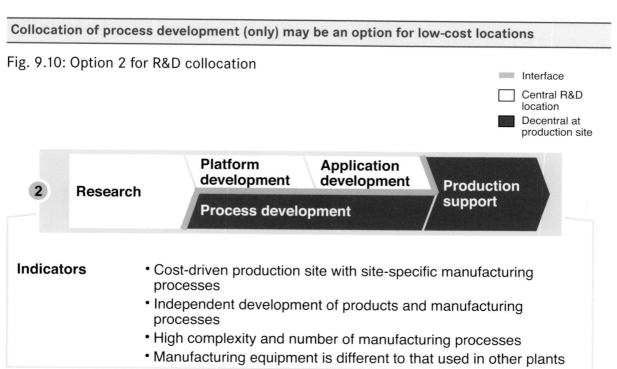

Indicators
- Cost-driven production site with site-specific manufacturing processes
- Independent development of products and manufacturing processes
- High complexity and number of manufacturing processes
- Manufacturing equipment is different to that used in other plants

Source: McKinsey

ground R&D essential. This is not just to lock in certain functionalities: the product also needs modifying to local regulations, language, and consumer preferences. These adjustment needs cannot be fully understood (or forecast) from a distance. Usually the more promising and cost-efficient solution is to adapt the products on the spot with local employees, in close contact with key customers. A local R&D department is highly recommended, often integrated with the local production facility (Figure 9.11). Relocation does not automatically mean having production and application development at the same location, especially since they are driven by different criteria (low labor costs and access to local suppliers versus availability of well-educated engineers). This is a trade-off between reducing network complexity and optimizing these functions individually. Often, though, companies prefer collocation to minimize complexity, especially in emerging countries in the entry phase to a new market.

Geographically distributed R&D tasks require highly standardized processes

In order not to lose sight of platform strategy and the desired level of identical parts, it makes sense to differentiate between a) standardized modules developed centrally and b) customer-specific product applications developed by local subsidiaries. Many of the large automotive OEMs follow this approach, globally standardizing basic auto platforms including chassis, body, and powertrain. Add-ons that vary according to brand and country are particularly the interior and exterior fittings. Such a broad spread inevitably makes high demands on R&D. A must for efficient cooperation between the R&D and production sites are standardized product development processes. Assigning overall project accountability with extensive decision-making authority also helps to stabilize the product development process. An example is discussed in the accompanying box: "Distributed Application Development Hand-in-Hand with the Customer."

Collocation of application development is a good solution for products that require adaptation to local market requirements

Fig. 9.11: Option 3 for R&D collocation

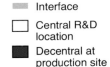

Interface

Central R&D location

Decental at production site

3 **Research** **Platform development** **Application development** **Production support**

Process development

Indicators
- Products need adaptation to specific requirements
- Low complexity/limited number of manufacturing processes
- Manufacturing processes highly standardized

Source: McKinsey

Distributed Application Development Hand-in-Hand with the Customer: Flextronics Builds on a Standardized Approach

Flextronics is the world's largest electronics manufacturing service provider, based in Singapore. As well as manufacturing products they pursue joint application development with customers. A current example is a mobile phone specifically for cardiac patients. It not only functions as a normal phone, but can also take ECG and blood pressure readings and put through emergency calls using GPS tracking. Flextronics locations in several European countries were involved in developing the device. Austria was responsible for the overall design, as well as production and development of the mechanics. Hungary devised the ECG measurement technology, while Sweden developed the software and antennas. The Special Business Solutions Center[1] in Austria handled communications with the customer's corporate center in Germany. The geographic proximity meant that meetings with the customer, whether to clarify product requirements or evaluate prototypes, could swiftly be arranged at any time.

For this cross-site, customer-centered development to work, Flextronics needed highly standardized processes and robust project organization.

Development process: The development process follows a standard split into concept phase, product architecture, detailed development, product and process validation, and production launch.

Every phase follows an established procedure ending at a quality gate. All milestone decisions are made jointly with the customer. The concept phase quality gate, for example, covers much more than just technical specifications. It requires a clear calculation of the target price and number of units. Suppliers submit component development proposals, and together OEM and customer examine alternative production site scenarios. This allows an accurate estimate of the product's profitability even at this early stage.

Project organization: The key contact was a Program Manager who supervised the product through every phase, ensuring overall compliance with schedule and budget. A Design Project Manager coordinated the product development staff (in Austria, Hungary, and Sweden), while an Industrial Project Manager was in charge of staff at the production site (Austria). Both of the latter were involved throughout the project, but responsibilities were gradually transferred from the Design Project Manager to the Industrial Project Manager.

Bottom line: Multinational standardization of development processes across projects allows decentral development, but requires great process discipline and a strong project organization.

---

[1] Special Business Solutions Centers (also called Low-Volume/High-Mix sites) at Flextronics are special sites for the manufacture of products with low unit volumes and high diversity. They have highly flexible production equipment with short setup times. Typical products are measuring instruments, auto navigation systems, and medical equipment.

9.2.4 Option 4: Relocation of Platform Development

Modifying applications is enough to satisfy local market needs for some products, but highly specific market conditions may require an entirely new prod-

uct platform (Figure 9.12). In these cases companies should consider whether it would be better to create an entirely new product line, including platform development, at the offshore site. Clearly this entails considerable effort and limits potential economies of scale in product development. Manufacturers should therefore only choose this option if application development is definitely not enough – either because existing platforms cannot satisfy the needs of local customers, or the effort required to modify or add applications is out of all proportion to the viable sales price. This often applies to emerging markets such as China and India. The primary driver for this option is proximity to the market, not so much the integration of production and development, so whether the two really need to be at the same location is a separate issue (see Option 2).

Whatever the situation, central R&D should be involved in the development of market-specific platforms to ensure standardized interfaces between modules, quality standards, and a high share of identical parts. The two case examples that follow – on flex-fuel vehicles and laser cutters – show how this kind of cooperation can work.

9.2.5 Option 5: Full Collocation

It is usually possible to tap into far more economies of scale and synergies by concentrating all research activities for a particular field in one geographic area than by collocating with production. However, sometimes it may still be advisable to relocate the relevant R&D to a new plant – or not to move production away from the home base at all (Figure 9.13).

Three different scenarios are conceivable for Option 5:

1. Leadership role of the regional market: A new market can develop dominance over the standards of an entire industry (category definition). This is happening with flat screens, where Asian competitors (and thus Asian locations) have gained hugely in importance in recent years.

2. Specific customer issues: Customer constraints of a technical or regulatory nature may call for completely new solutions and product architectures. One example of this is ABB (see the following case example). The dynamics of the energy market in China and the specifications laid down by state and state-dominated customers were reason enough for the company to set up research centers close to its Chinese production locations.

Highly specialized customer needs may require the collocation of platform development

Fig. 9.12: Option 4 for R&D collocation

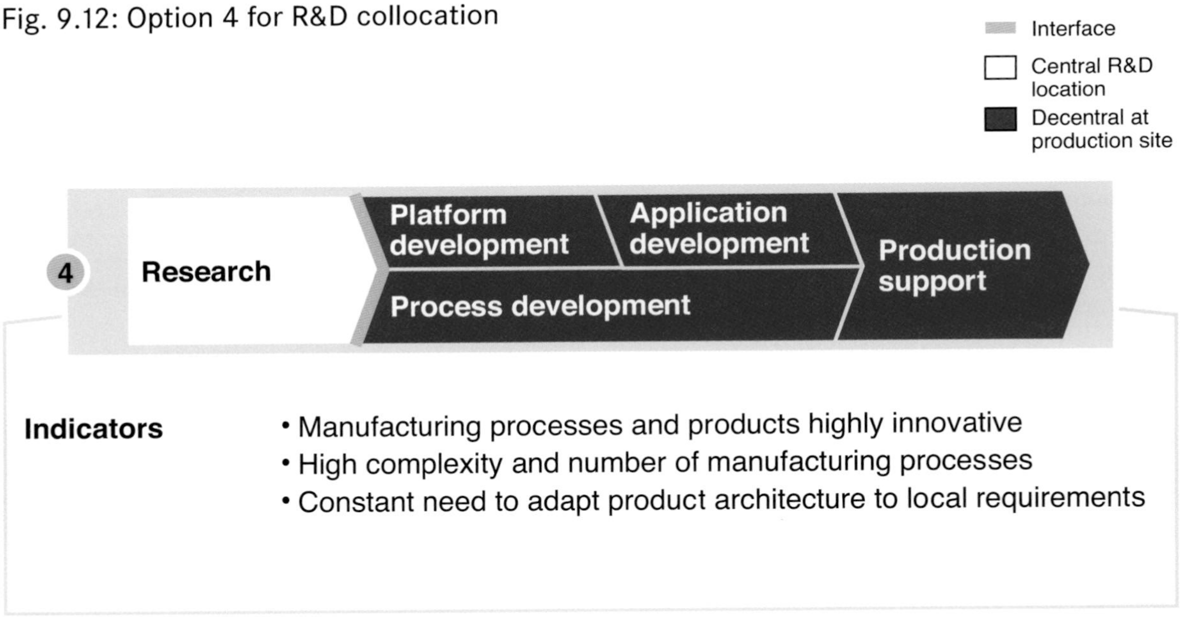

Legend:
- Interface
- Central R&D location
- Decentral at production site

4 — Research — Platform development | Application development | Production support | Process development

Indicators
- Manufacturing processes and products highly innovative
- High complexity and number of manufacturing processes
- Constant need to adapt product architecture to local requirements

3. Extremely research-intensive products: In some (rare) instances a company's products demand continuous and close communication between research and production. An example is the photonics division of JENOPTIK AG, which mainly manufactures its highly innovative products as one-off, make-to-order items. This is the reason it has tended to combine its research labs and production facilities (see the Jenoptik interview at the end of this chapter). If production were moved on its own to a low-cost country, any labor cost benefits would be outweighed by the effort of storing and transferring product data, as well as training production staff who had no research background. Researchers would also miss out on valuable information gleaned from assembling and testing existing products.

Collocation of research is only necessary in exceptional cases

---

Fig. 9.13: Option 5 for R&D collocation

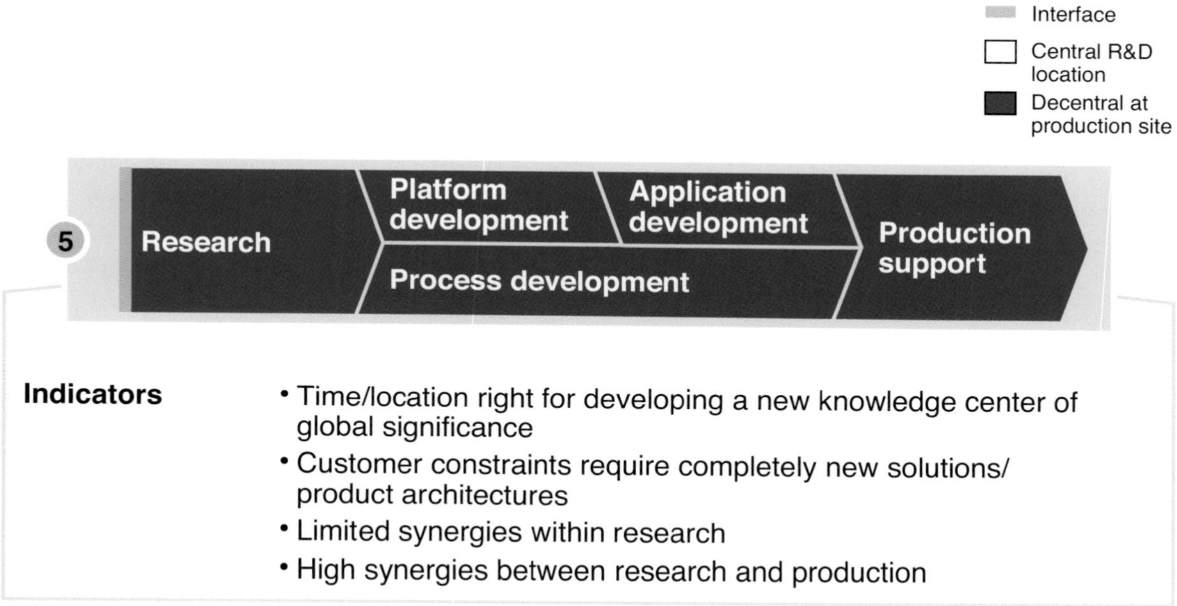

Legend:
- Interface
- Central R&D location
- Decentral at production site

5 | Research | Platform development | Application development | Production support
Process development

Indicators
- Time/location right for developing a new knowledge center of global significance
- Customer constraints require completely new solutions/ product architectures
- Limited synergies within research
- High synergies between research and production

Source: McKinsey

Flex-Fuel Vehicles for Brazil: Volkswagen and Bosch Achieve a Breakthrough Innovation Due to Local Presence[2]

Brazil's fuel price situation is unique: ethanol produced from sugar cane costs consumers roughly 30 percent less per kilometer traveled than gasoline, averaged out long term. This is mainly because the Brazilian government has been encouraging the use of ethanol for decades now by imposing high taxes on gasoline and promoting the cultivation of sugar cane. Mechanization also raised sugar cane yield significantly in the 1970s.

As a result ethanol vehicles had a market share of over 80 percent until the late 1980s. The price ratio was up-ended temporarily when crude oil prices reached a low and the sugar cane crop failed in 1989. The market reacted quickly. The share of new registrations for ethanol vehicles fell to less

---

[2] *Cf. ANFAVEA (Association of Brazilian Automotive Manufacturers) (2005) and Almeida (2004).*

than one percent, and remained low in the 1990s – a sorry situation both ecologically and macroeconomically (Figure 9.14).

The situation has changed again recently due to the initiative of Volkswagen and Bosch. Both companies have had a presence in Brazil for decades now. The VW Gol, for instance, which was developed in Brazil, has been the highest-selling automobile every year since 1987. Bosch began operations in Brazil back in 1994 with the development of a flex-fuel injection technology based on the ethanol vehicles in use at that time. This technology allows vehicles to be fueled with any mixture you wish of petrol and ethanol. The technical challenge lay in analyzing the ratio of ingredients and adapting the injection and combustion process precisely to the mix.

VW Brazil launched the VW Gol Total Flex, the first vehicle with flex-fuel technology, in March 2003. Its success was so resounding that VW introduced further flex-fuel models within a matter of months. Brazilian customers were delighted to benefit from the generally lower price of ethanol, while having the option to revert to petrol if conditions changed. They were happy to pay a premium (around USD 300 - 350) for this innovation.

By 2004, vehicles with flex-fuel technology had a market share of some 24 percent in Brazil, with annual sales totaling 1.56 million units (9 percent of them VWs, and the other 15 percent also largely using the Bosch technology). VW Brazil also exported around 157,000 flex-fuel vehicles in the same year, mainly to China, India, and Australia. Market experts predict that, within a few years, flex-fuel vehicles will dominate the Brazilian market.

Bottom line: Having R&D in the heart of a consumer market can inspire product innovation that would be virtually impossible from a distance.

Flex-fuel vehicles are replacing ethanol-only vehicles, which became unpopular in the early 1990s

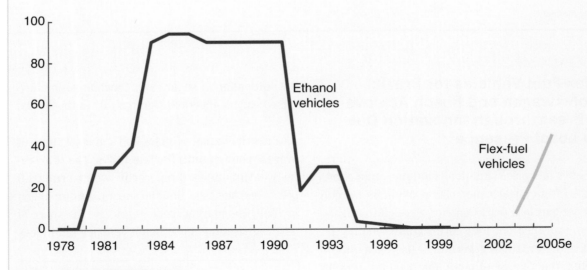

Fig. 9.14: Market share of ethanol vehicles in Brazil
Percent

Source: Almeida (2004), ANFAVEA (2005)

Trumpf Develops an Innovative Laser Cutter for the US Market Using a Unique Collaboration of Local Platforms

In 2002, Trumpf, a global mechanical engineering company based in Germany, began developing a completely new laser cutter for working sheet metal, aimed mainly at the US market. The Trumatic L 2510 has an integrated automated sheet loading and unloading unit and a high-performance laser that combines compact design with low operating costs due to a new coaxial electrode arrangement. Another new feature is a common mechanism to drive the automation and cutting head motion units.

The project, which involved nearly 50 design engineers in total, was a joint venture between the three company locations in Farmington (US), Ditzingen, and Neukirch (both Germany).

Farmington was in charge of the overall concept. It developed and produced the laser, motion unit, and machine frame, and designed the automation concept. The site also involved key US customers in the concept phase.

Ditzingen contributed the cutting technology, supporting table, and machine casing. Its staff were also responsible for compliance with the standardized development process and product documentation (management of product data, manuals, etc.). The automation unit was developed and manufactured in Neukirch.

The development work was completed in just 18 months. The secret of success was collaboration with key customers throughout, allowing the product to be geared precisely to market needs. Cost-optimized construction was chosen in view of the relatively high price sensitivity of American customers, eliminating one drive motor. The technical specifications stipulated a maximum (structural steel) workpiece thickness of twelve millimeters, as it was found that 80 percent of US customers do not work with thicker sheets anyway. Also, the early involvement of key suppliers and production specialists resulted in perfectly conceived design-to-manufacture and design-to-assembly components.

Bottom line: The newly developed device is a good example of the advantages of customer-centered platform development even for capital goods.

9.3 Outlook: Globalization of R&D

Even R&D is facing advancing globalization, despite how different the criteria for its location are from those driving production.

> **Low-cost and high-cost countries are already competing with one another on the labor market for top researchers**

Parameters for new R&D locations should be market requirements and knowledge clusters. The presence of a local application development department is often essential to satisfy the needs of local customers.

Many companies are moving parts of their application development to the vast growth market of Asia for precisely this reason. The best example is the consumer electronics industry. Western mobile phone companies, such as Motorola or Nokia, have set up development centers in China with thousands of staff to design mobile phones specifically for the Chinese market (designed to recognize Chinese handwriting, suit local taste, etc.).

Factor cost differences are important for R&D locations, but not as much so as for production. Wage cost differences become less significant the more educated employees are. In some circumstances, the

ABB Research Locations in China: Sophisticated Solutions in Cooperation with Local Experts

The Swiss-Swedish ABB Group generates global annual revenues of around USD 28 billion (2006) from business in power and automation technologies. Just about half of this – and nearly half of 111,000 ABB staff – are spread throughout locations outside Europe. The company's eleven research centers are based in nine countries: 500 of the 710 or so research employees work in Europe, 150 in Asia, and 60 in the United States. Every center specializes in particular disciplines, such as production automation, energy transmission, or insulation technology.

ABB has stepped up its presence in Asia massively in order to participate in its expected high market growth. In China, for example, experts expect an annual increase in power station capacity of 25 gigawatts in the near future. ABB generated revenues of USD 2.4 billion in 2005 and lifted this to USD 2.8 billion in 2006, making China its largest market in terms of revenues. It plans organic growth of around 20 percent p.a. over the next few years.

In 2006, ABB moved the global headquarters of its Robotics division and three business units – Marine, Power Electronics, and Metal – to China, added a low-voltage generator joint venture in Nanchang, expanded and updated seven existing factories, established eight new sales and service branches, and hired 1,900 new employees bringing the total number of people in China to 11,000.

China makes very specific demands on the power supply industry. Large quantities of electricity have to be transported over long distances – such as the 2,000 km between the hydropower plants inland and the Chinese coast. To date, landlines are designed for no more than three gigawatts. The ABB research department is working flat out to devise innovative technical solutions to increase line capacity. In the spring of 2005, ABB established an R&D center in Beijing that will employ 50 to 100 scientists and engineers and work closely with Chinese institutes. Its primary fields of research will be energy transmission and distribution, new production technologies, and robot control systems.

Bottom line: Companies that operate in demanding growth markets with highly innovative technology may have to set up research facilities there as well. The key is to collaborate with leading knowledge centers and translate local customer needs into successful products.

Automotive Supplier with Production Plant in Spain: Transforming an Extended Workbench into a Center of Competence

At the end of the 1960s, a European automobile supplier established a plant in Spain to benefit from wage cost advantages and compensate for currency fluctuations. For decades, the location served as an "extended workbench" for the main factory. Its activities were confined to the manufacture of simple components, and it was run as a cost center. The original factory would only send out teams of experienced production engineers to provide temporary support when new components first went into production.

Since the equipment used locally was different from that used at the main site, it became increasingly necessary to make location-specific adjustments to production processes (e.g., gluing instead of welding). The Spanish plant successively extended its manufacturing competencies. The high demand from Spanish customers gradually led to investment in new plant capacity. The plant was increasingly able to demonstrate how powerful and competitive it was in the company network.

From 1994 onwards, the product range was expanded to include more complex units, and the site was upgraded: a development department for production technology was installed, and the portfolio of manufacturing technologies was enhanced via major investments in automation and mechanization. The year 1995 saw new products for the global market go into production for the first time.

The next move was to extend its product development capabilities. Initially, the location was only responsible for developing product variants, but in 1997 it received complete R&D responsibility for several product areas. It began to localize supplied materials, and ran the plant as a profit center. It already had the top talent it required. The relatively low labor costs for Spanish engineers helped the location step up from a production center to a center of competence.

Bottom line: Production sites abroad can gradually grow into highly effective lead factories and development centers.

high mobility of top-level personnel even results in low-cost countries competing with high-cost countries for the same resources. In fast-growing low-cost countries, such as China or South Korea, the labor market is very tight, despite large numbers of college graduates, as the demand for well-trained engineers and technicians exceeds supply (Figure 9.15). In some cases, the candidates available also cannot

Very large number of engineers available in some low-wage countries

Fig. 9.15: Global labor market for engineers

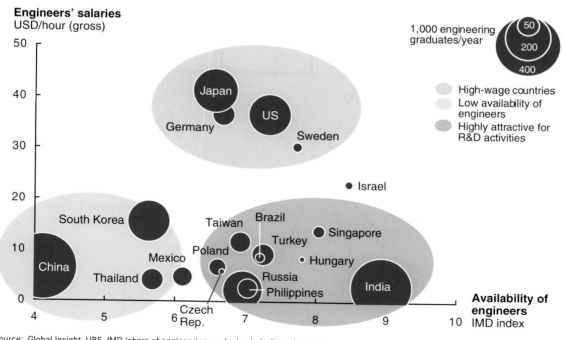

Source: Global Insight, UBS, IMD (share of engineering graduates in India estimated)

(yet) meet the education standards required (see Chapter 2). From this perspective, Eastern Europe and India are comparatively well placed.

As already mentioned, peak-quality R&D tends to thrive most in knowledge clusters, where leading universities, state and private research institutions, specialized industrial zones, and venture capitalists interact to create exponential dynamics. The self-reinforcing feedback mechanisms that form in these clusters are difficult to replicate anywhere else.

High-cost countries have so far offered a better environment for developing knowledge clusters

As a result, world-class knowledge centers are frequently found in high-cost countries. Examples are California's Silicon Valley for IT, Kista Science Park in Stockholm for wireless communication, and the Stuttgart area in southern Germany for machine tool manufacturing. Recently, competence clusters of this nature have also sprung up in low-cost countries thanks to smart industrial settlement policies. Taiwanese companies such as TSMC and UMC are now technology leaders in the semiconductor industry, while Bangalore in India has become a top-caliber cluster for low-cost, high-quality software development.

The importance of such knowledge clusters is particularly well illustrated by the automotive industry. Pioneering innovations in this sector in the past decade have been generated mainly in Central Europe and Japan (Figure 9.16).

Based on the location factors described above, high-cost countries are often not only the most successful research locations today, but also the most efficient. If we compare the number of published patents with the resources allocated to R&D, it is apparent that some high-cost countries can easily compensate for their high factor costs in research (Figure 9.17). The

Most automotive innovations in the last decade were developed in Europe and Japan

Fig. 9.16: Examples of recent innovations in the automotive industry

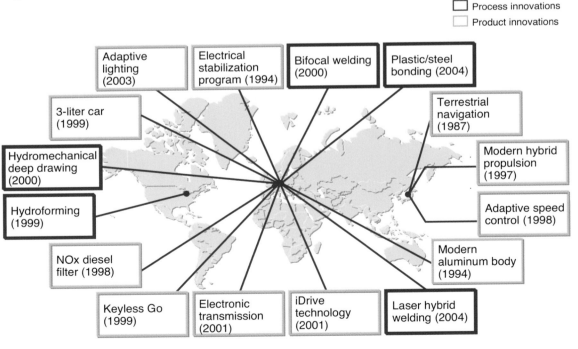

☐ Process innovations
☐ Product innovations

Adaptive lighting (2003)

Electrical stabilization program (1994)

Bifocal welding (2000)

Plastic/steel bonding (2004)

3-liter car (1999)

Terrestrial navigation (1987)

Hydromechanical deep drawing (2000)

Modern hybrid propulsion (1997)

Hydroforming (1999)

Adaptive speed control (1998)

NOx diesel filter (1998)

Modern aluminum body (1994)

Keyless Go (1999)

Electronic transmission (2001)

iDrive technology (2001)

Laser hybrid welding (2004)

Source: Press clippings

globalization of R&D will therefore not automatically lead to a network of low-cost locations. It will favor those locations throughout the world that have the best know-how base and appropriate proximity to the market.

> Since other location factors for R&D will even out, the importance of knowledge clusters will grow further

The high- versus low-cost country factor differentials governing where R&D is best located will even out to a great extent in the medium term. New markets will emerge in today's low-cost countries and education standards will converge. Excessive salary differences for mobile, top-caliber personnel will no longer be tenable. With these R&D location factors evening out, the importance of knowledge clusters to a country's or region's leadership in a specific industry will grow even further.

High-cost countries use their research resources more efficiently

Fig. 9.17: Comparison of R&D efficiency of selected countries
(Average 1990 - 99)

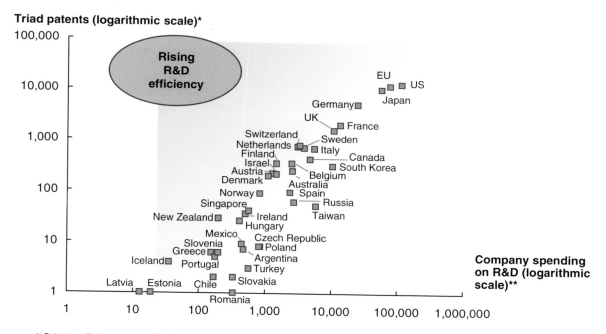

* Patent applications to the EPO, USPTO, and JPO. Figures for 1999 are based on estimates
** GERD (Gross Domestic Expenditure on R&D) in USD millions (1995), based on purchasing power parities (average for the period 1990 - 99)
Source: OECD Patent and R&D Database, September 2004

Innovation-Driven Growth – Interview with Alexander von Witzleben, Former CEO of Jenoptik, about the R&D Strategy of the Photonics Division

Source: Franz Haniel & Cie. GmbH

Mr. von Witzleben, Jenoptik is an international player, yet your R&D activities are focused on Jena, Germany. Why is that?

Jena has always been associated with optical electronics. Some of our employees are third-generation: the interest in optical products and passion for precision are handed down from one generation to the next. Building a know-how base like this at another location would take many years, not to mention our employees' extraordinarily high level of identification with our products.

Top-flight research is particularly important to your industry. This involves having a dense cluster of innovative companies and research institutions around you. Which locations can provide this?

Jena offers all of that. In the immediate vicinity we have Zeiss, Schott, and various small and mid-sized companies. We also work closely with eight local institutes, including three renowned Max Planck Institutes. Jena is the epitome of a knowledge cluster. That is why it will continue to grow. And this has clear advantages for us. It means we don't have to stand in line behind large companies when it comes to collaborating with local institutes.

It is not just your R&D activities – much of the production of Jenoptik Photonics Technologies is based in Germany. Why isn't your production network more global?

So far we have always found everything we need in Germany. Abroad we have difficulty finding employees who are anywhere near as well qualified as our staff in Jena. Different cultures and languages do not make it any easier. And the cost advantages of producing abroad are not as great as

Alexander von Witzleben was the CEO of JENOPTIK AG (now Jenoptik AG) from 2003 to 2007. He studied Economics and Business Administration at the University of Passau, and worked for KPMG Deutsche Treuhandgesellschaft in Munich for three years. He then joined Jenoptik in 1993, initially to run the corporate finance department; he was appointed to the board in 1997. Jenoptik, an optical technology company based in Germany, grew out of Jenoptik Carl Zeiss Jena GmbH. In fiscal 2006, it generated EUR 486 million in revenues, operating in over 20 countries with around 3,200 employees worldwide. In July 2007, Alexander von Witzleben became the fourth member of the Management Board of Franz Haniel & Cie. GmbH, an internationally active group of companies with 2006 revenues of EUR 27.7 billion and more than 55,000 employees in around 40 countries.

you would imagine. We moved the deburring step to Eastern Europe, for example, but the wage cost advantages were less than we expected – only lower by a factor of around two. This had a lot to do with the high qualifications we demand, of course. Counteracting effects such as logistics costs, exchange rate risks, and higher long-term interest rates also cancelled out some of the savings. So now we just purchase components from Eastern Europe, that's all. That's where we really notice the factor cost advantages.

Another benefit of manufacturing in Germany is that our production location also serves as our laboratory. The same people who develop our products also assemble the final instruments, because they are very complex and are only produced in small quantities. We only manufacture very few products in series. The direct feedback between production and R&D is also an essential ingredient for the further development of these products. The cost of separating these functions at different locations would far exceed any possible savings.

Despite the benefits of being located in Germany, do you see any threat from competitors in low-cost locations?

No. Our products are currently far too complex to copy. Our niche is also not very attractive for low-cost locations – the unit volumes are simply too small. I once said jokingly, but I meant it: we have a market share of 100 percent in Thuringia because we only sell one product there. Our niche products quickly make us the market leader even with low volumes. Competitors from low-cost countries have an advantage with high-volume products. Our markets are much too small for them to put us under pressure. And this will remain the case because we're going to continue producing leading-edge technology. In other words, our growth is innovation-driven, it doesn't come from trying to squeeze our way into established markets where other companies are already crowding each other out. We may have comparatively few cash cows, but we make up for this with high growth in our extremely innovative markets.

Further reading

Abele, E., P. Radtke, and A. Zielke: *Die smarte Revolution in der Automobilindustrie.* Frankfurt am Main: Redline Wirtschaft, 2004.

Boutellier, R., O. Gassmann, and M. von Zedtwitz: *Managing Global Innovation. Uncovering the Secrets of Future Competitiveness.* 2nd Edition. New York: Springer, 2006.

Dhawan, R., and P. Gao, R. Mangaleswaran, R. Pathak, and A. Zielke: *Towards a New Global Order for Automotive Suppliers - Findings from the China and India Supplier Survey.* Düsseldorf, McKinsey & Company, 2006.

Farrell, D. (Ed.): *Offshoring: Understanding the Emerging Global Labor Market (McKinsey Global Institute Anthology Series).* Boston: Harvard Business School Press, 2007.

JÜRGEN KLUGE, HARALD PROFF

10 Macroeconomic Implications: Accelerating Growth

For companies, a global production footprint is a great opportunity to grow into new markets and strengthen their competitiveness. On a macroeconomic level, the globalization of production also has vast repercussions. Hasty judgments on the socioeconomic impact are all too common. But it is vital to first understand the underlying drivers. The evolution of an economy from industrialization through to eventual deindustrialization appears to be an inevitable development. There are good arguments for moving in step with the tide and perhaps even shaping your path through it. Evidence is mounting to suggest that countries working proactively to trigger a phase-shift can actually accelerate their progress, reaping multiple benefits throughout the transition.

While the effects are naturally country-specific, three economic groupings can be distinguished (with overlaps due to the heterogeneity of the countries), each responding differently within a changing production landscape:

- **Industrializing/emerging economies** that are pouring huge investments into infrastructure and industrial expansion, unleashing strong expansion of their secondary sector of industry (including manufacturing, energy and construction) and rising GDP

- **Highly industrialized economies** that have already successfully navigated a long period of industrialization with rising per capita incomes, and are now on the cusp of deindustrialization

- **Post-industrial economies** that have enjoyed high per capita incomes for many years, accompanied by a surge in knowledge-based service activities and the decline of their manufacturing and process industries as a share of GDP.

Figure 10.1 maps these groups along the typical industrialization development curve,[1] contrasting degree of industrialization (size of a country's industrial value added divided by its GDP) with per capita income. The share of manufacturing is closely correlated, and the key driver of the industrial development as a whole (whereas construction and other industrial activities represent a more constant share of the GDP). The first category described above is clearly recognizable: Some Asian countries including India

---

[1] *The World Bank (2003).*

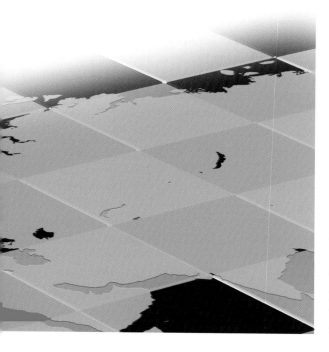

and large parts of Latin America and sub-Saharan Africa are still industrializing (or pre-industrial), with their main focus on building and expanding existing markets. Their growing industrialization raises per capita income hand in hand with thriving economic development. Interestingly, however, this development appears to cut off when the degree of industrialization reaches 45 to 50 percent of GDP. South Korea has already reached this plateau; North America, Europe and Japan are already well advanced on the path to deindustrialization. The relative importance of industry as a share of their GDP is shrinking, while their service sectors are expanding.

This evolutionary trend can be confirmed by comparing industrialization and income in any country you wish, whether Ghana, China, South Korea, Germany or the United States (only few exceptions ap-

Growing industrialization raises per capita income until industry accounts for ~50% of GDP; thereafter, services play a more important role

Fig. 10.1: Degree of industrialization and per capita income

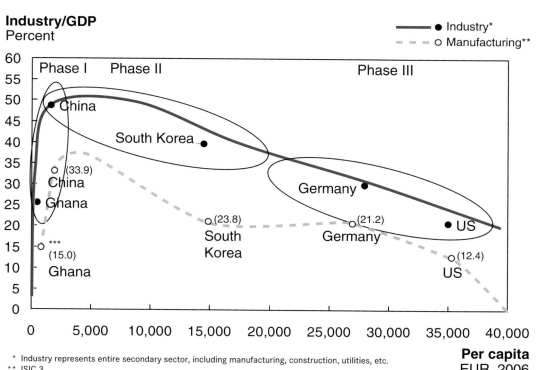

Industry/GDP
Percent

● Industry*
○ Manufacturing**

* Industry represents entire secondary sector, including manufacturing, construction, utilities, etc.
** ISIC 3
Source: EIU 2007, World Bank 2002, Global Insight (2006)

Per capita
EUR, 2006

ply, e.g., for countries that are exploiting and exporting their wealth in natural resources). This empirical evidence confirms Fourastié's hypothesis – formulated in the 1930s – that economic structures would increasingly evolve towards the tertiary sector, and that deindustrialization would set in when the share of the secondary sector had crystallized at around 50 percent of GDP.

This process can be explained by the following model of the phases of industrialization.

In Phase I, the increasing vertical integration between suppliers and manufacturers leads to rising and more complex production volumes along the entire value chain, which results in growing per capita income. Industrial clusters form, such as regional concentrations of companies engaged in steelmaking, shipbuilding, or heavy engineering. One well-known example is the cluster that has sprung up in South Korea around the local steel industry. Since beginning the production of massive steel in the early 1970s, South Korea has attracted so much shipbuilding activity that it is now one of the largest shipbuilding centers in the world, as well as home to other heavy-engineering companies, which are also major consumers of steel. Once these material interrelationships within a country reach critical mass, per capita income continues to rise only if businesses extend their networks internationally and make use of specialist suppliers, particularly in the service sector.

Phase II allows them to overcome the diseconomies of vertical scale and scope arising from excessive vertical integration as a result of spiraling management requirements – as well as their dearth of skills for handling activities outside their core business. Provided the companies in each such cluster keep their focus, they will spawn adjacent but independent companies in the service sector, from advertising agencies to caterers. Countries such as the US and the UK are perhaps the most advanced in this area, but similar developments have occurred in France and Sweden, where industry has also generated large outsourced service segments. Such service outsourc-

ing contributes to reducing the level of an economy's industrialization mainly because it counts towards a different segment: the service sector. In line with the above development, manufacturing jobs come under competitive pressure from other rising economies that are behind on the curve. Highly skilled staff whose wages are still low take on a gradually increasing share of production.

As companies outsource ("offshore") ever more services, and manufacturing activities are increasingly transferred abroad, they need to make greater use of intangible assets to achieve per capita income growth in this phase. Instead of transferring intermediate products, companies increasingly utilize their networks to transfer knowledge, ushering in Phase III.

Clusters in post-industrial economies usually consist of companies that have optimized their value creation/value chains, jointly leveraging their research and the results of independent research institutes. Companies still tend to seek physical proximity because speed plays a major role in translating a knowledge advantage into a product or service with a competitive edge. This is the case even when proximity is no longer necessary to ensure efficient input-output relationships, as knowledge diffusion will never be instantaneous. In Silicon Valley, for instance, computer and computer software firms no longer use their physical proximity for reciprocal deliveries – they have long since moved their physical production to plants in East Asia or Latin America. Instead, they use their concentrated knowledge to generate a continuous innovation pipeline.

The movement from Phase I to Phase III appears to be inexorable, though how long it takes (as we shall see) is very variable. So what is the impact of global production on each of the three country groupings? What are the breakpoints between the phases? Can an economy speed up its transition from one phase to the next? And finally, is this a win-lose game, or – as we suggest – a win-win game, where all the players gain because the wealth generated increases globally? And with a growing pie, everyone gets a bigger slice.

10.1 Impact on Industrializing Economies

Industrializing economies are the net beneficiaries of the value added by foreign companies that are optimizing their global production footprints. Obviously companies in high-cost locations also wish to take advantage of the low wage levels in developing and emerging economies, but they have additional motives with attendant ripple effects. For one thing, they are increasingly on the look-out for simpler production processes because their complex automated factories back home are reaching the limits of their efficiency. In addition to process simplifications, the transfer of production operations out of high-cost locations to emerging economies often results in product design modifications since features optimized for high-cost markets do not always make sense for customers in low-cost countries. But the shift is not just about simplification. Companies are also keen to transfer technology-intensive steps in their production processes as fast as lower-cost regions can handle them. This reinforces the technology base in the industrializing economy, while strengthening its education system via the feedback loop of the transplant's demand for qualified staff on a regular basis. The very best companies are able to create islands of high productivity in areas where wage levels are still low.

Of course there are problems – two of the foremost being the duality of economic structures and degradation of the environment in the process of development. The economies of Asia run the risk of importing the difficulties Latin America has encountered: the enormous discrepancy between poverty-stricken rural areas and wealthy enclaves, and the immediate proximity of dramatic income disparities caused by migration into urban centers. Yet the globalization of production has positive knock-on effects for local business growth, too. The presence of multinational corporations and their subsidiaries fosters start-ups by giving talented individuals the opportunity to gain experience and accumulate capital. It also leads to the upgrading of many social and legal standards, as well as wage levels. On balance, industrializing economies are clearly beneficiaries of globalization.[2]

10.2 Impact on Highly Industrialized Economies

The effects of global production present highly industrialized economies with unprecedented challenges. Until recently, these countries were themselves in the position of the first group we have just discussed, if further along the scale: sites that companies based in high-cost countries (either highly industrialized or post-industrial) often chose for their high-tech production steps. Now these regions find themselves locked in competition for such transplants with economies they themselves also consider low-cost (the industrializing economies). While production in highly industrialized economies profited heavily in the past from the wide range of production opportunities and related sourcing options, it is now undergoing redefinition in two main ways. First, companies are moving production steps that are too expensive to industrializing economies to reduce the cost gap. Second, they are investing in new technology to reduce the know-how gap vis-à-vis competitors in post-industrialized economies.

As a result, direct investments from highly industrialized countries have increased both in post-industrialized economies (for technology reasons, and to expand market share), and in industrializing economies (to meet cost targets).[3] South Korean companies are relocating their wage-intensive manufacturing steps to their neighboring countries, e.g., China. They are also investing in European and US companies that have the know-how they currently lack, especially in industrial machinery.

However, the transplants and outsourcing from post-industrialized economies to highly industrialized economies are still in equilibrium with relocation to industrializing economies. Despite turmoil at the microeconomic level, the level of industrialization at the macroeconomic level remains relatively constant. For highly industrialized economies, the changes in

---

[2] UNCTAD (2006).

[3] Ibid.

global production imply a structural shift to higher-skilled work and rising levels of production process technology.

Accomplishing this shift requires a readiness to abandon legacy industries that are losing competitiveness in their current constellation. This is painful and can lead to considerable friction, but the costs of economic adjustment can be disastrous if this shift is delayed too long. At the end of the 1990s, obsolete industry structures were propped up for too long in Southeast Asia as a result of misguided economic policy driven by excessively close-knit government, banking, union and corporate networks. This was one of the main reasons for the dramatic financial and economic crisis in the region. In eventually extricating themselves, however, governments, central banks, and other bodies in Thailand, Malaysia, the Philippines, and South Korea, all in Phase II of their industrial development, gathered a great deal of experience on how to implement the necessary changes via improved governance in the industrial and financial sectors.[4] The end result shows – as in so many other cases – that highly industrialized economies benefit greatly provided they are prepared to adapt to the globalization trends and define their specific role.

10.3 Impact on Post-Industrial Economies

New information and communication technologies have unleashed a dramatic change in value creation activities, both sharply reducing interdependencies between the various production units and increasing the manageability of complex value creation structures. This has accelerated the deindustrialization associated with the third phase of economic development. It is worth noting, however, that industrial activity in these economies has lost none of its macroeconomic importance. Although the secondary sector now accounts for a smaller relative share of GDP, inflation-adjusted industrial value added in Germany, the United States, and France has actually increased over the past 20 years – from USD 1,444 billion to USD 2,549 billion in the US, for

example.[5] Without industrial production, many services are neither useful nor conceivable. That said, there is a visible divide between these countries and highly industrialized economies. The value chains that are still largely integrated in highly industrialized economies are clearly more disaggregated in post-industrial economies.[6]

New value architectures develop when:

1. The individual value-creation activities of the previously integrated value chain disaggregate into separate units with marketable products or services

2. A business unit (or company) concentrates on a few or only one of these value-creation activities, which can result in advantages from standardization and a know-how edge

3. The activities remaining in this business unit form a new activity-centered value architecture, with a new division of labor within and between business units.

These three requirements for the development of new value architectures are usually subsumed into two phases. First is the break-up phase (deconstruction), where business units disintegrate their product-centered value chains to focus on one or a subset of the related activities. This is followed by reconstruction, rebuilding a new, activity-centered value architecture in conjunction with the value-creation activities that remain in the business unit.

One of the prime tasks of the new value architecture in the reconstruction phase is orchestrating complex value chains. This includes the management of complex production footprints. The sporting goods manufacturer Adidas provides a good illustration. Today, product development and marketing are the only

---

[4] Barton (2003 a).

[5] Global Insight.

[6] Rall (2006).

functions it performs in house. For production, for example, it manages a network of independent manufacturers.

Figure 10.2 summarizes the impact of the globalization of production on the three categories of economy. While industrializing economies experience a lengthening of local value chains (from a focus on labor-intensive assembly to ever more complex machining and metalworking activities), highly industrialized countries see a reduction in in-house production at all stages. In post-industrial economies, these value chains are differentiated and no longer integrated.

In summary, post-industrialized economies can benefit hugely from the new knowledge economy as long as they are able to keep up their innovation, and have

(and continue to build) a matching HR pool. They have to remain attractive for investments and competitive for the production of goods and services that can be exported in order to finance the imports of lower-value goods.

Alongside the three economic groups we have distinguished, there are of course also the developing countries with extremely low per capita incomes and very little significant industrialization so far. We have not included these in our analysis as they are unlikely candidates as sites for global production. This is not to say they lack attractive investment perspectives for multinational corporations, but these opportunities are primarily for companies wishing to tap new markets. Once they are able – with or without foreign assistance – to enter the cycle we have de-

While industrializing economies experience a lengthening of local value chains, highly industrialized countries see a reduction in in-house production at all stages

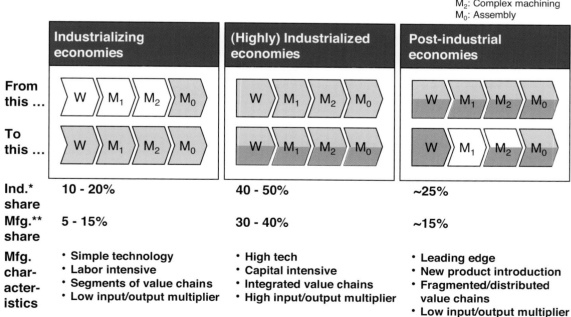

Fig. 10.2: Effects of global production on production of the 3 types of economy

Imports   Own value added   Outsourcing to (local) suppliers

W: Metalworking, shaping
M_1: Simple machining
M_2: Complex machining
M_0: Assembly

	Industrializing economies	(Highly) Industrialized economies	Post-industrial economies
From this ...	W M_1 M_2 M_0	W M_1 M_2 M_0	W M_1 M_2 M_0
To this ...	W M_1 M_2 M_0	W M_1 M_2 M_0	W M_1 M_2 M_0
Ind.* share	10 - 20%	40 - 50%	~25%
Mfg.** share	5 - 15%	30 - 40%	~15%
Mfg. characteristics	• Simple technology • Labor intensive • Segments of value chains • Low input/output multiplier	• High tech • Capital intensive • Integrated value chains • High input/output multiplier	• Leading edge • New product introduction • Fragmented/distributed value chains • Low input/output multiplier

* Includes entire secondary sector (manufacturing, construction, utilities)
** ISIC 3 (rev. 2)
Source: McKinsey

scribed, they will gain the opportunity of wealth creation and can gradually emerge from poverty.

10.4 Breakpoints Between the Phases

In the industrializing process, a company's key economic objectives are gaining economies of scale and scope, building brand names, and establishing reliable supply lines for standardized intermediate products (commodities). The ability to innovate is of little importance in this phase. When an economy reaches the threshold of maximum industrialization and the transition to deindustrialization, innovation capabilities take on greater significance. It is no coincidence that countries such as Indonesia, Brazil, or Thailand have recently stepped up their efforts to attract and generate high-tech industries and cutting-edge knowledge. In the deindustrialization phase,

the importance of entering and building markets gives way to creating new product markets and reinventing value architectures. Economic activity revolves around acquiring and transferring skills and generating high-tech products. It is in this phase that innovation capacity makes such a huge difference.

At the transition points, it follows that a change occurs in the drivers of increasing per capita income. In the industrializing phase (I), rising income is correlated with physical input-output multipliers that stimulate growth in multiple industries. In the deindustrializing phase (III), per capita income growth is driven by technology transfer, particularly via (virtual) networks. The importance of the knowledge transferred via networks continually increases relative to input-output factors until the latter barely generate any more growth, characterizing the changeover to a post-industrialization era.

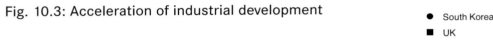

While the transition to a post-industrialized economy in the UK took ~200 years, and ~130 years in Germany, South Korea took ~60 years

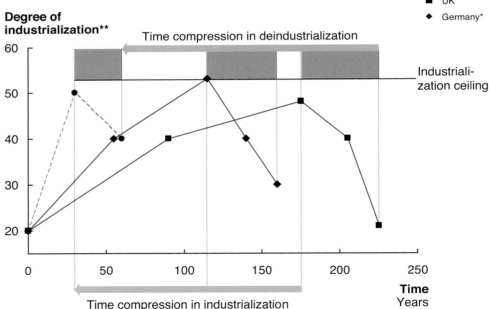

Fig. 10.3: Acceleration of industrial development

* After 1945 only West Germany; even higher degree of industrialization in East Germany due to centrally planned economy until 1989
** Share of the secondary sector (manufacturing, construction and energy) relative to the country's GDP

Source: Global insight, Mitchell (2003).

The transition manifests itself in the shifts in production between the three types of economies. In post-industrial economies, the level of vertical integration in individual companies falls hand in hand with a rising share of global sourcing, which reduces the input-output multipliers in industry clusters while the share of technology in production rises, reflected in higher R&D spending.

The globalization of production has a particularly important impact on these breakpoints: it is clearly accelerating the transition. Figure 10.3 shows the industrialization and deindustrialization process over time in the UK, Germany, and South Korea. While the transition to a post-industrial economy in the UK took about 200 years, and around 130 years in Germany, South Korea has undergone the same evolution in half the time (60 years). A large part of the reason is that Korea made faster use of the opportunities for global production than the UK and Germany did many decades before. The explosion in information and communication technology, reduced international freight costs, and lower entry barriers to markets worldwide all fast-forwarded the use of transplants and offshore outsourcing options.

Fig 10.3 shows that it is not only the industrialization phase that is shortened: the deindustrialization phase is also accelerating significantly. And a question still remains to be answered: Is there a Phase IV, and if so, what will it be?

10.5 Accelerating Industrial Development

Global production ultimately leads to an increase in living standards in all three economic groupings. Accelerating the transition process between the phases and shortening the three phases of industrialization will achieve a higher per capita income for a greater share of the population faster. The aspiration could even be conceptualized as practically "leapfrogging" over the second phase to minimize environmental burdens and the irreversible consumption of resources incurred from industrialization, as well as speeding up the rise in average incomes. This has been observed in a few regions, but never yet in an entire economy.

Given the basic prerequisites for evolving through the industrialization process shown in Figure 10.1, it is unlikely that leapfrogging will be easy. The key factor underlying per capita income growth throughout the three phases is education. Increasing the education level across the board is much easier in city states (e.g., Singapore) or limited boom areas (such as Bangalore in India) than throughout an entire country of larger geographical size. The United Nations' education indicator shows how slowly the world is converging on the eradication of illiteracy.[7]

However, the process can be nudged forward in all phases by targeted investment in education. Only higher education can achieve a broad, sustainable increase in an economy's per capita income long term. Flexible governance structures are also vital for industrial development to accelerate (rather than hinder) any discontinuities necessitated by the development process shown in Fig. 10.1. A further key driver is to stimulate innovative production-related start-ups. Appropriate nurturing, select collocation and knowledge accumulation boosts the odds of these germinating future clusters within the global economy, with an exponential impact on the region's or country's economic evolution.

* * *

There is no easy shortcut to a post-industrial economy, just as there is no easy fast-forward to an optimized value-creation network. But companies and economies can learn from the lessons of the past as well as their peers, and leverage the benefits of global production to jump-start further growth. Deciding what industry steps to locate where has long-term impact on economies, ecologies, and individual prosperity. The investment is too punishing, the time factor too crucial to make a wrong call. We very much hope that the systematic strategies outlined in this book help guide both companies and economies at every level of development as they make these complex decisions.

---

[7] UNDP (2005).

Appendix

The ProNet Initiative and Methodology

A.1.1 The ProNet Initiative

The ProNet Initiative was a collaborative project of McKinsey & Company and the Institute of Production Management, Technology, and Machine Tools (PTW) at Darmstadt University of Technology, Germany. This project included a large-scale survey, the results of which are covered in this book. The survey was conducted with over 100 representatives of 54 companies, primarily automotive suppliers, companies in the electronics industry, and machine tool manufacturers (Figure A.1).

The face-to-face setting used for the interviews with these decision-makers allowed us to better understand their company-specific situation and derive the root causes for success and failure in the development of their global production structures. The interviews unveiled the soft hurdles in establishing global production networks, such as expatriate assignments or know-how transfer. We also obtained more detailed input from around half of these companies, allowing more exact analysis of facts that the interviewees did not always have at their fingertips during the discussions.

Multiple data sources on location parameters were analyzed, and form the backbone of Chapter 2. The team also developed the ProNet model, a quantitative optimization tool for production network reconfiguration, which was used to generate the case studies with companies. A further vital input was our discussions with business partners from around the world, whether production managers in Malaysia, the head of industrial assembly at a German service provider, or a quality manager in Mexico. All yielded valuable information and insights, as did exchanges with numerous McKinsey colleagues in North America, Germany, Scandinavia, Italy, and Japan as well as in China, India, Mexico, and Eastern Europe. Their willingness to delve into myriad details and provide specific insights has enriched our work tremendously.

A.1.2 Leaders Versus Followers: The ProNet Methodology

The survey results were used to compare "leaders" – companies with extensive experience from pioneering efforts and highly successful international operations – and a group of less ambitious peers whom we term "followers." These are companies that either have only limited experience with sites abroad, or faced operational struggles in ramping up and managing production sites abroad. The "leaders" represent the upper third in the ranking, while "followers" constitute the lower third, based on three input factor categories:

Output KPIs collected during the interview (weighted at 25 percent) on the company's actual global production performance. These KPIs include cost reduction from relocation versus the respective industry benchmark, quality level of new plants, achievement of the company's own relocation targets, and ramp-up time to target capacity.

Financial performance of the company/BU (weighted at 25 percent). The evaluation of financial performance included 2003 profits, revenue growth of the previous three years, and total return on capital employed.

An outside-in assessment of the company's experience and capabilities in global production (weighted at 50 percent). This assessment was based on rankings in five subcategories. These were current share of total production in low-cost countries, adequacy of the global operations footprint as a whole (production, R&D, sourcing, support functions), success rate in establishing production sites abroad, a clearly recognizable global production strategy, and implementation rigor.

The survey results are based on statistical analysis with a confidence level greater than 99 percent in nearly every case (and always at least 95 percent).

Interestingly, the clusters of "leaders" and "followers" showed no significant bias with regard to industry, geographical region, or company size. "Leaders" and "followers" can be found in all industries, geographies, and among medium-size enterprises as well as large corporations.

Most of the interviews were conducted in Europe, dispersed relatively evenly across the focus industries of the ProNet Initiative

Fig. A.1: Interview distribution by country and industry
Percent

- 54 companies*
- 27 additional interviews and input from deep-dive expert questionnaires

Participants by region

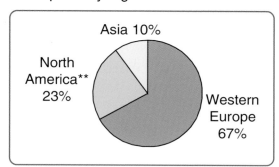

Participation by industry

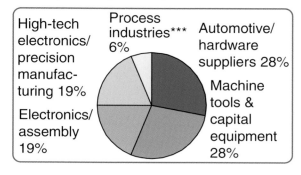

* Sometimes multiple interviewees
** Including Mexico
*** Example: fine chemicals

Source: McKinsey/PTW (ProNet analysis)

List of Figures

Chapter 3

Chapter 4

Chapter 5

Chapter 6

Chapter 7

Chapter 8

Chapter 9

List of Tables

Bibliography

Abele, E. and J. Kluge (eds.) (2005). *How to Go Global – Designing and Implementing Global Production Networks.* Projektbericht "ProNet," Düsseldorf, McKinsey & Company, Inc.

Abele, E., P. Radtke, and A. Zielke (2004). *Die smarte Revolution in der Automobilindustrie.* Frankfurt am Main, Redline Wirtschaft.

Abele, E., J. Elzenheimer, and A. Rüstig (2003). "Anlaufmanagement in der Serienproduktion," in *Zeitschrift für wirtschaftlichen Fabrikbetrieb* (ZWF), 98 (2003) 3, p. 172 - 176.

ACMA (2005). *Vision 2015 for the Indian Automotive Components Industry: An ACMA-McKinsey Report.* New Delhi, Automotive Component Manufacturers Association of India.

Aksen, D. (1998). *Teach Yourself GAMS.* Istanbul, Bogazici University Press.

Alicke, K. (2005). *Planung und Betrieb von Logistiknetzwerken.* 2nd Edition, Berlin et al, Springer.

Almeida, C., M. Kotek, M. Pereira, and P. Albuquerque (2004). *Alternative Fuel in Brazil. Flex-Fuel Vehicles.* Chapel Hill, White Paper, UNC Kenan-Flagler Business School, University of North Carolina.

Almeida, P. and B. Kogut (1999). "Localization of Knowledge and the Mobility of Engineers in Regional Networks," in *Managementscience*, 45 (1999) 7, p. 905 - 917.

Ancona, D. and H. Bresman. *X-teams (2007). How to Build Teams That Lead, Innovate and Succeed.* Boston, Harvard Business School Press.

ANFAVEA (2005). *Anuário Estatístico da Indústria Automobilística Brasileira - 2005.* Sao Paulo, Associação Nacional dos Fabricantes de Veículos Automotores.

Arntzen, B. C., G. G. Brown, T. P. Harrison, and L. L. Trafton (1995). Global Supply Chain Management at Digital Equipment Corporation. *Interfaces*, Vol. 25 (I), p. 69 - 93.

Assaf, M., C. Bonincontro, and St. Johnsen (2005). *Global Sourcing & Purchasing Post 9/11: New Logistics Compliance Requirements and Best Practices.* Fort Lauderdale, J. Ross Publishing.

Baldwin, R. and P. Martin (1999). *Two Waves of Globalisation: Superficial Similarities, Fundamental Differences.* NBER Working Paper 6904.

Barton, D., R. Newell, and G. Wilson (2003a). *Dangerous Markets. Managing in Financial Crises.* New York, John Wiley.

Barton, D., T. Hsieh, and J. Sinha (2003b). *Becoming a Global Champion: The journey to global leaderships is fraught with peril – and opportunity.* New York, McKinsey & Company, Inc.

BCG (2004). *Capturing Global Advantage.* Boston, White Paper, The Boston Consulting Group, Inc.

Bhattacharya, A. and S. Nandagaonkar (2006). "Hidden competitive advantage," in *The Economic Times*, September 26.

Bhutta, K. S. (2004). "International Facility Location Decisions: A Review of the Modeling Literature," in *International Journal of Integrated Supply Management,* Vol. 1 (I), p. 33 - 50.

Billio, M. (2002). Simulation-based Methods for Financial Times Series. Working Paper, March 2002. Venice, University di Venezia.

Bleicher, K. (1991). *Organisation, Strategien, Strukturen, Kulturen.* 2nd Edition, Wiesbaden, Gabler.

Bodnar, G. M. and R. C. Marston (2003). "'Exchange Rate Exposure' A Simple Model," in Choi, J. J. and M. Power (eds.), *International Finance Review,* Vol. 3 (Global Risk Management: Financial, Operational, and Insurance Strategies), San Francisco, Elsevier Science, p. 107 - 117.

Boudreau, J. W. and P. M. Ramstad (2007). *Beyond HR: The New Science of Human Capital.* Boston, Harvard Business School Press.

Boutellier, R., O. Gassmann, and M. von Zedtwitz (2006). *Managing Global Innovation. Uncovering the Secrets of Future Competitiveness.* 2nd Edition, New York, Springer.

Boyabatli, O. and L. B. Toktay (2004). *Operational Hedging: A Review with Discussion.* Working Paper, INSEAD, January 2004.

Brealey, R. and P. Myers (2000). *Principles of Corporate Finance.* Boston, McGraw-Hill/Irwin.

Breen, B. (2004). Living in Dell Time. *Fast Company*, 88 (November 2004), p. 86.

Breitman, R. L. and J. M. Lucas (1987). PLANETS: A Modeling System for Business Planning. *Interfaces*, Vol. 17 (I), p. 94 - 106.

Breuer, P. and C. Eltze (2005). "Die letzten Handelsschranken für Textilien sind gefallen," in *Handelsblatt* (Online Edition), March 30, 2005.

Bryan, L. L. and C. Joyce (2007). *Mobilizing Minds: Creating Wealth From Talent in the 21st Century Organization.* New York, McGraw-Hill.

Campbell, R. M., J. Hexter, and K. Yin (2004). "Getting sourcing right in China," in *The McKinsey Quarterly*, Special Edition: China Today.

Chandler, A. (1969). *Strategy and Structure: Chapters in the History of the American Industrial Enterprise.* Boston, The MIT Press.

China Today (2004). *The McKinsey Quarterly,* Special Edition: China Today, New York, McKinsey & Company, Inc.

CII (2004). *Made in India: The Next Big Manufacturing Export Story.* CII-McKinsey Report, Mumbai and New Delhi, Confederation of Indian Industry.

Coenenberg, A. G. (1997). *Jahresabschluß und Jahresabschluß-analyse.* 16th Edition, Landsberg/Lech, Verlag moderne industrie.

Coenenberg, A. G. (2003). *Kostenrechnung und Kostenanalyse.* 5th Edition, Stuttgart, Schäffer-Poeschel.

Cohen, M. A. and A. Huchzermeier (1998). "Global Supply Chain Management: A Survey of Research and Applications." in Tayur, S. et al (eds.), *Quantitative Models for the Supply Chain Management.* Boston, Kluwer Academic Press.

Cook, T. A. (2006). *Global Sourcing Logistics: How to Manage Risk and Gain Competitive Advantage in a Worldwide Marketplace.* New York, AMACOM/American Management Association.

Copeland, T. E. and Y. Joshi (1996). Why Derivatives Don't Reduce FX Risk. *The McKinsey Quarterly,* No. 1, p. 66 - 79.

Cross, R. and A. Parker (2004). *The Hidden Power of Social Networks: Understanding How Work Really Gets Done in Organizations.* Boston, Harvard Business School Press.

Czichy, R. (2002). *Leichtbau in Luft- und Raumfahrt.* Skriptum zum Vortrag im Rahmen des Innovationsforums am Leichtbauzentrum PantaRhei, June 21/22, 2002, Technische Universität Cottbus.

Das, G. (2002). *India Unbound: The Social and Economic Revolution from Independence to the Global Information Age.* New York, Anchor Books.

Davis, I. (2005). "The Biggest Contract," in *The Economist,* May 28, 2005, p. 87 - 89.

Davis, I. (2006) "Plot Your Course for the New World," in *The Financial Times,* January 13.

Deutsches Zollamt (2005). *Gewerblicher Rechtsschutz.* Jahresbericht 2004, Berlin.

Dhawan, R., P. Gao, R. Mangaleswaran, R. Pathak, and A. Zielke (2006). *Towards a New Global Order for Automotive Suppliers – Findings from the China and India Supplier Survey.* Düsseldorf, McKinsey & Company, Inc.

Dimitri, N., G. Piga, and G. Spagnolo (2006). *Handbook of Procurement.* Cambridge, Cambridge University Press.

Domschke, W. and A. Drexl (1998). *Einführung in Operations Research.* 4th Edition, Berlin et al, Springer.

Dorf, R. C. and A. Kusiak (eds.) (1994). *Handbook of Design, Manufacturing and Automation.* New York, John Wiley.

Dority, J. (1994). *HP Saves $3 Million per Month on Bulk Shippers for Deskjet.* http://www.lansmont.com/newsletters/html/94-10-p l.htm.

Drew, J., B. McCallum, and S. Roggenhofer (2004). *Journey to Lean.* Basingstoke, Palgrave MacMillan.

Drewry (2003). *The Annual Container Market Review and Forecast 2003/2004.* London, Drewry Shipping Consultants, Ltd.

Drewry (2004). *The Annual Container Market Review and Forecast 2004/2005.* London, Drewry Shipping Consultants, Ltd., 2004.

Drexl, A., B. Fleischmann, H.-O. Günther, H. Stadler, and H. Tempelmeier (1994). "Konzeptionelle Grundlagen kapazitätsorientierter PPS-Systeme," in *Zeitschrift für betriebswirtschaftliche Forschung (ZBF),* Vol. 46, p. 1022 - 1045.

Droege & Company/PTW (2004). The Emergence of China as an International Competitor to German Machinery Manufacturers. Singapore/Darmstadt/Frankfurt am Main.

Drucker, P. F. (2001). *The Essential Drucker.* Oxford, Butterworth-Heinemann.

Dubbel, H., W. Beitz, and K.-H. Küttner (eds.) (1994). *Handbook of Mechanical Engineering.* London, Springer.

Economic Times (2005). Productivity-related Salaries. *The Economic Times,* March 18, 2005, New Delhi, p. 1 and p. 12.

EIU (2003). *Country Commerce India.* New York, The Economist Intelligence Unit Limited.

EIU (2004). *Country Reports.* London, The Economist Intelligence Unit Limited.

EIU (2007). *Country Reports.* London, The Economist Intelligence Unit Limited.

Ernst & Young (2004). *Automobilstandort Deutschland in Gefahr?* Automobilbranche auf dem Weg nach Osteuropa und China. Eschborn, White Paper, Ernst & Young.

Eversheim, W. and G. Schuh (eds.) (1996). *Hütte: Taschenbuch für den Betriebsingenieur (Betriebshütte).* 7th Edition, Akademischer Verein Hütte e.V., Berlin et al, Springer.

Eversheim, W. and G. Schuh (1999). Gestaltung von Produktionssystemen. *Produktion und Management,* Vol. 3, Berlin, Springer.

Farrell, D. (ed.) (2007). *Offshoring: Understanding the Emerging Global Labor Market (McKinsey Global Institute Anthology Series).* Boston, Harvard Business School Press.

Fawcett, P. E., L. Birou, and B. Cofield Taylor (1993). "Supporting Global Operations through Logistics and Purchasing," in *International Journal of Physical Distribution & Logistics Management,* Vol. 23, Nr. 4, p. 3 - 11.

Fortune (2004). *Asian Edition,* December 27.

Franck, I. and D. M. Brownstone (1986). *The Silk Road: A History.* New York, Facts on File Publications.

Fritz, A. H. and G. Schulze (eds.) (2004). *Fertigungstechnik [Manufacturing Technology].* 6th Edition, Association of German Engineers (VDI). Berlin, Springer.

Gassmann, O. and M. von Zedtwitz (1998). "Organization of Industrial R&D on a Global Scale," in *R&D Management,* 28 (1998), p. 147 - 161.

Gupta, S. D. (ed.) (1997). *Dynamics of Globalization and Development.* Boston, Kluwer Academic Publishers.

Gutenberg, E. (1965). *Grundlagen der Betriebswirtschaftslehre, Vol. 1: Die Produktion.* Berlin et al, Springer.

Hack, G. D. (1999). *Site Selection for Growing Companies.* Westpoint, Connecticut, Quorum Books.

Hammond, J. and H. Lee (2004). *China arid the Global Supply Chain.* Online Presentation, July 2004, http://mba.tuck.dartmouth.edu/digital/Programs/CorporateEvents/SupplyChainThoughtLeaders/session7Slides.pdf.

Harding, C.-F. (1988). "Quantifying Abstract Factors in Facility-Location Decisions," in *Industrial Development*, May/June 1988, p. 24 - 27.

Hartung, J. (1999). *Statistik.* 10th Edition, Munich, Verlag Vahlen.

Hau, H. (1999). "Comment on 'Corporate Risk Management for Multinational Corporations: Financial and Operational Hedging Policies," in *European Finance* Vol. 2, p. 247 - 249.

Haug, P. (1992). "An International Location and Production Transfer Model for High Technology Multinational Enterprises," in *International Journal of Production Research*, Vol. 30, No. 3, p. 559 - 572.

Henzler, H. and W. Rall (1985). "Aufbruch in den Weltmarkt (Part 3)," in *ManagerMagazin*, 11/1985, p. 166 - 173.

Hexter, J. and A. P. Narayanan (2006). "The challenges in Chinese procurement," in *The McKinsey Quarterly*, Special Edition: Serving the new Chinese consumer.

Hexter, J. and J. Woetzel (2007). *Operation China: From Strategy to Execution.* Boston, Harvard Business School Press.

Huchzermeier, A. and M. A. Cohen (1996). "Valuing Operational Flexibility under Exchange Rate Risk," in *Operations Research* (Special Issue on New Directions in Operations Management), 44, January-February, 1996, p. 100 - 113.

Huselid, M. A., B. E. Becker, and R. W. Beatty (2005). *The Workforce Scorecard: Managing Human Capital To Execute Strategy.* Boston, Harvard Business School Press.

IATA (2003) *Air Cargo Annual.* International Air Transport Association, Aviation Information & Research Department, Geneva/Montreal.

ifo (2005). *Basar-Ökonomie Deutschland – Exportweltmeister oder Schlusslicht?* ifo Schnelldienst, 6/2005, ifo Institut für Wirtschaftsforschung e.V., Munich.

ILO (2004). *Yearbook of Labour Statistics.* Geneva, International Labour Organization (also see: http://laborsta.ilo.org/).

ILOG (2005). *ILOG Optimization Suite.* Gentilly, White Paper, ILOG S.A.

IMD (2003). *World Competitiveness Yearbook 2003.* Lausanne, IMD.

International Textile Manufacturers Federation, Global Trade Information Service (China Chamber of Commerce).

Jacob, F. (2006). *Quantitative Optimierung dynamischer Produktionsnetzwerke.* (Darmstädter Forschungsberichte für Konstruktion und Fertigung), Herzogenrath, Shaker.

Japan Automobile Manufacturers Association (2004). *Common Challenges, Common Future. Japanese Automakers in an Enlarged Europe.* Bonn.

Jiang, B. (2003). "What Pulled Sony out of China," in *Supply Chain Management Review*, No. 1, p. 22 - 27.

Kalpakjian, S. and S. Schmid (2005). *Manufacturing, Engineering & Technology.* 5th Edition, Upper Saddle River, NJ, Pearson/Prentice Hall.

Karakaya, F. and C. Canel (1998). "Underlying Dimensions of Business Location Decisions," in *Industrial Management & Data Systems*, Vol. 7, p. 321 - 329.

Kaufmann, L., D. Panhans, B. Poovey, and B. Sobotka (2005). *China Champions.* Wiesbaden, Gabler.

Kirka, Ö. and M. M. Köksalan (1995). "An Integrated Production and Financial Planning Model and Application," in *IEE Transactions*, Vol. 28, p. 677 - 686.

Knapp, B. W. (2006). *A Project Managers Guide to Contracting and Procurement.* The Project Management Excellence Center, Inc.

Küpper, H.-U. and S. Helber (1995). *Ablauforganisation in Produktion und Logistik* [Process Organization in Production and Logistics]. 2nd Edition. Stuttgart, Schäfer-Poeschel Verlag.

Laseter, T. M. (1998). Balanced Sourcing. *Cooperation and Competition in Supplier Relationships.* San Francisco, Jossey-Bass.

Leavitt, H. J. (2004). *Top Down: Why Hierarchies Are Here to Stay and How to Manage Them More Efficiently.* Boston, Harvard Business School Press.

Lee, D. (2005). "Former Stock Capital Has Unraveled: China's Business Strategies Give Companies the Edge in the Production Battle Against U.S. Competitors," in *Los Angeles Times*, April 24, 2005.

Lewis, W. W. (2004). *The Power of Productivity: Wealth, Poverty, and the Threat to World Stability.* Chicago, The University of Chicago Press.

Liker, J. (2003). *The Toyota Way: 14 Management Principles from the World's Greatest Manufacturer.* Boston, McGraw-Hill.

Lindemann, U. (2005). *Produktentwicklung und Konstruktion.* Lecture Script, Technische Universität Munich.

Long, D. (2003). *International Logistics: Global Supply Chain Management.* Berlin et al, Springer.

Lustig, I. J. and J.-F. Puget (2001). "Program Does Not Equal Program: Constraint Programming and Its Relationship to Mathematical Programming," in *Interfaces*, Vol. 31 (6), p. 29 - 53.

MacCormack, A. D., L. J. Newman III, and D. B. Rosenfeld (1994). "The new Dynamics of Global Manufacturing Site Location," in *Sloan Management Review*, Summer Issue, p. 69 - 80.

Maddison, A. (2001). *The World Economy: A Millennial Perspective*. Paris, OECD.

Madura, J. (2003). *International Financial Management*. 7th Edition, Mason, Thomson South-Western.

McKinsey (2002). *McKWissen (Cluster)*, 1/2002, Hamburg, McKinsey & Company, Inc.

Meijboom, B. and B. Vos (1997). "International Manufacturing and Location Decisions: Balancing Configuration and Coordination Aspects," in *International Journal of Operations & Production Management*, Vol. 17 (8), p. 790 - 805.

Mentzer, J. T. et al (eds.) (2006). *Handbook of Global Supply Chain Management*. Thousand Oaks, California, Sage Publications.

Meyer, T. (2006). *Globale Produktionsnetzwerke – Ein Modell zur kostenoptimierten Standortwahl*. (Darmstädter Forschungsberichte für Konstruktion und Fertigung), Herzogenrath, Shaker.

MGI (2003). *New Horizons: Multinational Company Investment in Developing Economies*. San Francisco, White Paper, McKinsey Global Institute, October 2003, McKinsey & Company, Inc.

Min, H. and W. P. Galle (1991). "International Purchasing Strategies of Multinational U.S. Firms," in *International Journal of Purchasing and Materials Management*, No. 3 (Summer 1991), Vol. 27, p. 9 - 18.

Mintzberg, H., B. Ahlstrand, and J. Lampel (1999). *Strategy Safari*. Vienna, Ueberreuther.

Mitchell, B. R. (2003). International Historical Statistics: Europe. 5th Edition. New York, Palgrave Macmillan.

Murphy, J. V. and R. W. Goodman (1998). *Building a Tax-Effective Supply Chain. Global Logistics and Supply Chain Strategies*, November 1998, Digital Edition (see: www.supplychainbrain.com).

Nassimbeni, G. and M. Sartor (2006). *Sourcing in China: Strategies, Methods and Experiences*. New York, Palgrave Macmillan.

Nelson, D. R., P. E. Moody, and J. Stegner (2001). *The Purchasing Machine: How the Top Ten Companies Use Best Practices to Manage Their Supply Chains*. New York, Free Press.

Nixdorf, A. (2003). Am Anfang war die Not. *McK Wissen (Operations)*, 5/2003, McKinsey & Company, Inc., Hamburg, p. 8 - 15.

OECD (2005). *Country Risk Classifications of the Participants to the Arrangement on Officially Supported Export Credits*. Paris, Organisation for Economic Co-operation and Development.

Owen, S. H. and M. S. Daskin (1998). "Strategy Facility Location: A Review," in *European Journal of Operational Research*, Vol. 111, p. 423 - 447.

Paquet, M., A. Martel, and B. Montreuil (2003). *A Manufacturing Network Design Model Based on Processor and Worker Capabilities*. Technical Report, Quebec, CENTOR Research Center, Université Laval.

Paquette, L. (2003). *The Sourcing Solution: A Step-By-Step Guide to Creating a Successful Purchasing Program*. New York, AMACOM/American Management Association.

Perlmutter, H. (1996). "The Tortuous Evolution of the Multinational Corporation," in *Columbia Journal of World Business*, Vol. 4, 1/2, 1969, p. 9 - 18.

Perridon, L. and M. Steiner (1999). *Finanzwirtschaft der Unternehmung*. 10th Edition. Munich, Verlag Vahlen.

Porter, M. E. (1990). The Competitive Advantage of Nations. *Harvard Business Review 3 - 4*, 1990, p. 73 - 93.

Porter, M. E. (1998). *The Competitive Advantage of Nations*. New York, Free Press.

Produktion (2004). "Neue Märkte: Wenn der beste Mann zu Hause fehlt," in *Produktion*, No. 22.

ProNet Survey (2005). Cf. Abele (2005).

Raghunatha, V. (2005). "Economic Life vs. Depreciation Rate," in *The Economic Times*, March 19, 2005, New Delhi, p. 8.

Rall, W. and B. König (eds.) (2006). *Branchen von morgen. Wie sich die wichtigsten Industrien neu erfinden*, Landsberg/Lech, Redline.

Rammer, C., B. Peters, T. Schmidt, B. Aschhoff, T. Doherr, and H. Niggemann (2005). *Innovationen in Deutschland. Ergebnisse der Innovationserhebung 2003 in der deutschen Wirtschaft*, Baden-Baden, ZEW Wirtschaftsanalysen, Schriftenreihe des Zentrums für Europäische Wirtschaftsforschung (ZEW), Vol. 78.

Reich, R. (1996). *Die neue Weltwirtschaft*. Frankfurt am Main, Fischer.

Ricardo, D. (1817). *Principles of Political Economy and Taxation*. New York, Prometheus Books.

Ritter, R. C. and R. A. Sternfels (2004). "When offshoring does not make sense," in *The McKinsey Quarterly*, No. 4, p. 124 - 127.

Rodrigue, J.-P. et al (2005). *Transport Geography on the Web*. Hofstra University, Department of Economics and Geography, http://people.hofstra.edu/geotrans.

Rowthorn. R. and R. Ramaswamy (1999). *Growth, Trade, and Deindustrialization*. International Monetary Fund, IMF Staff Papers, Vol. 46, No. 1, March 1999.

Sachverständigenrat (2004). *Erfolge im Ausland – Herausforderungen im Inland*. Wiesbaden, Sachverständigenrat zur Begutachtung der gesamtwirtschaftlichen Entwicklung, Jahresgutachten 2004/2005, November 2004.

Samuelson, P. A. (2004). "Where Ricardo and Mill Rebut and Confirm Arguments of Mainstream Economists Supporting Globalisation," in *Journal of Economic Perspectives*, 3 (Summer 2004) Vol. 18, p. 135 - 146.

Schaaf, J. and M. Weber (2005). Offshoring-Report 2005 – Ready for Take-off. *Deutsche Bank Research*, Vol. 52, Frankfurt am Main, Deutsche Bank.

Schulz, W. (2004). "Taiwans neue High-Tech-Offensive," in *VDI Nachrichten*, April 30, 2004, p. 30 - 31.

Silver, E. A., D. F. Pyke, and R. Peterson (1998). *Inventory Management and Production Planning and Scheduling*. 3rd Edition, New York, John Wiley.

Simchi-Levi D., P. Kaminsky, and E. Simchi-Levi (2002). *Designing and Managing the Supply Chain*. 2nd Edition, Boston, McGraw-Hill/Irwin.

Sinn, H.-W. (2005). Basar-Ökonomie Deutschland – Exportweltmeister oder Schlusslicht? *ifo Schnelldienst*, Sonderausgabe, No. 6, Vol 58., April 2005.

Stahl, G. (1998). *Internationaler Einsatz von Führungskräften*. Munich, Oldenbourg.

Statistisches Bundesamt [German Federal Office of Statistics] (2005). *Statistisches Jahrbuch für die Bundesrepublik Deutschland und für das Ausland: Ausgabe 2004 [Yearbook]*. Vol. 1 & 2, Wiesbaden, Statistisches Bundesamt.

Tapscott, D. and A. D. Williams (2007). *Wikinomics: How Mass Collaboration Changes Everything*. New York, Portfolio (an imprint of Penguin Books USA).

Tempelmeier, H. (2003). *Material-Logistik – Modelle und Algorithmen für die Produktionsplanung und -steuerung und das Supply Chain Management*. 5th Edition, Berlin et al, Springer.

The World Bank (2003). *Global Economic Prospects 2003: Investing to Unlock Global Opportunities*. Washington, The World Bank.

Thommen, J.-P. and A.-K. Achleitner (1998). *Allgemeine Betriebswirtschaftslehre*. 2nd Edition, Wiesbaden, Gabler.

Thonemann, U. (2005). *Operations Management*. Munich, Pearson.

Thonemann, U., K. Behrenbeck, R. Diederichs, J. Großpietsch, J. Küpper, and M. Leopoldseder (2003). *Supply Chain Champions – was sie tun und wie Sie einer werden*. Wiesbaden, Gabler [available in English from the McKinsey Operations Practice].

Tong, H.-M. and C. K. Walter (1980). An Empirical Study of Plant Location Decisions of Foreign Manufacturing Investors in the United States. *Columbia Journal of World Business*, No. 1, p. 66 - 73.

UBS (2003). *Prices and Earnings. A Comparision of Purchasing around the Globe. 2003 Edition*. Zurich, UBS AG, Wealth Management Research.

United Nations Development Programme (UNDP) (2005). United Nations Development Report 2005 International cooperation at a crossroads. New York et al, Oxford University Press.

United Nations Conference on Trade and Development (UNCTAD) (2006). *World Investment Report 2006. FDI from Developing and Transition Economies: Implications for Development*. United Nations, New York/Geneva.

Vahs, D. (2005). *Organisation*. Stuttgart, Schäffer-Poeschel.

Vanderbeck, E. J. (2005). *Principles of Cost Accounting*. 13th International Student Edition, Mason, Ohio, Thomson/South-Western.

Veloso, F. (2001). *Local Content Requirements and Industrial Development: Economic Analysis and Cost Modeling of the Automotive Supply Chain*. Dissertation, Boston, Massachusetts Institute of Technology (MIT).

Vidal, C. and M. Goetschalckx (1997). "Strategic Production-Distribution Models: A Critical Review with Emphasis on Global Supply Chain Models," in *European Journal of Operational Research*, No. 1/1998, p. 1 - 18.

Vietor, R. H. K. (2007). *How Countries Compete: Strategy, Structure, and Government in the Global Economy*. Boston, Harvard Business School Press.

Vos, B. and H. Akkermans (1996). Capturing the Dynamics of Facility Location. *International Journal of Operations & Production Management*, Vol. 16 (11), S. 57 - 70.

WEF (2005). *The World Competitiveness Report 2004 - 2005*. K. Schwab (ed.), World Economic Forum.

Welch, J. and S. Welch (2005). *Winning*. New York, HarperBusiness Publishers.

Welge, M. W. and D. Holtbrügge (2003). *Internationales Management – Theorien, Funktionen, Fallstudien*. 3rd Edition, Stuttgart, Schäffer-Poeschel.

Weygandt, J. J., D. E. Kieso, and P. D. Kimmel (2005). *Financial Accounting with Annual Report*. 5th Edition, New York, John Wiley.

Zedtwitz, M. (2003). "Initial Directors of International R&D Laboratories," in *R&D Management*, 33 (2003) 4, p. 377 - 393.

Zheng, L. and F. Possen-Dölken (eds.) (2002). *Strategic Production Networks*. Berlin et al, Springer.

Keyword Index

About the Editors and Authors

EDITORS

PROF. DR.-ING. EBERHARD ABELE is the Director of the Institute of Production Management, Technology, and Machine Tools (PTW) at Darmstadt University of Technology. Previous to that, he worked with manufacturing companies for more than a decade in positions such as head of the production technology department and technical plant manager. Before obtaining his professorship, he gained international experience in Spain, France, and China. His institute of 50 engineers and support staff does research and development in the fields of production management, the design and testing of machine tools, and the development of new production process technologies. PTW features the first real-scale learning factory at a university, the Center for industrial Productivity (CiP).

DR. ULRICH NÄHER is a Principal in the Tokyo office of McKinsey & Company. He leads the European Product Development Practice, and is also a member of McKinsey's High Tech and Automotive & Assembly Sectors. His clients include automotive OEMs and suppliers of electronics and semiconductors, whom he advises on R&D processes, operations management as well as on their portfolio, technology, and growth strategies. Ulrich Näher studied Physics and Mathematics in Munich and earned a doctorate from the Max Planck Institute in Stuttgart with a dissertation on Particle Physics.

DR.-ING. TOBIAS MEYER is an Engagement Manager based in the Singapore office of McKinsey & Company. He serves clients primarily in the Travel and Logistics Sector and manufacturing companies. His functional focus is on operations and strategy, covering topics such as site selection in production and logistics, strategic partnerships, network planning, distribution management, in-house logistics, and supply chain management in general. Tobias Meyer studied Industrial Engineering and Management in Darmstadt and at the University of Illinois at Urbana-Champaign and earned a doctorate in Mechanical Engineering from Darmstadt University of Technology.

DR.-ING. GERNOT STRUBE is a Director in the Munich office of McKinsey & Company. He is a member of the global leadership group of the McKinsey Manufacturing Practice, which he co-founded, and of the Operations Practice in Europe. He serves clients in the manufacturing industry, primarily in high tech/electronics and aerospace. His work centers on operations strategy and performance improvement programs, such as manufacturing strategy, performance transformation, and lean systems implementation as well as turnaround programs. Gernot Strube studied Aeronautical Engineering at the TU Munich where he earned his doctorate.

 **RICHARD K. ("RICH") SYKES** is a Director in McKinsey's Chicago office. He is a leader of McKinsey's North American Operational Effectiveness and Supply Chain Management Practices, and has worked on a variety of studies in the industrial, automotive, transportation, and consumer sectors. Richard Sykes holds an MBA in Finance and Marketing from the University of Chicago. Prior to business school, he worked at Eli Lilly and also earned a BS in General Engineering from the University of Illinois at Urbana-Champaign. Rich Sykes also serves on the Board of Directors of Shelter, Inc., a not-for-profit agency providing emergency care for children in need.

AUTHORS

 **MARINA DERVISOPOULOS** is an industrial engineer (*Dipl.-Wirtschaftsingenieurin*) and was a research assistant at the Institute of Production Management, Technology, and Machine Tools (PTW) and member of the Institute's "Production & Management" group. Her research interests include the optimization of global production networks, the commercial evaluation of production processes, and quality management.

 **JAMES HEXTER** is a Principal in the Beijing office of McKinsey & Company. He is co-leader of the Asian Operations Practice with a focus on sourcing, procurement, and manufacturing in China. His major projects have been in the areas of global sourcing, domestic purchasing improvement, and manufacturing and operations transformation programs across a range of industries.

 **RAIMUND DIEDERICHS** is a Director in McKinsey & Company's office in Beijing, where he transferred in 2005 after 23 years of McKinsey work in Europe. For 10 years Raimund has been the European leader and worldwide co-leader of McKinsey's Supply Chain Management Practice, part of McKinsey's Operations Practice. In 2005, he became the leader of McKinsey's SCM Practice in Asia.

FRANK JACOB (alumnus) was a Senior Associate in the Frankfurt office of McKinsey & Company. He worked primarily on topics such as purchasing, supply management, change management, and supply chain management, with an industry focus on the Travel & Logistics and Automotive & Assembly sectors. Some of his more recent project work included purchasing optimization for an automotive OEM and infrastructure maintenance management.

DR. MADS D. LAURITZEN is a Principal in McKinsey's Seoul office. He joined the Firm in 1999. A core member of McKinsey's Automotive & Assembly Sector (A&A) and Operations Practices, he spent four years in the Tokyo office, where he co-led A&A work and served Japanese automotive clients. After returning to Europe for a few years, he transferred to McKinsey's office in Seoul and is today the leader of McKinsey's A&A Sector and Operations Practice in South Korea. He is also the leader of McKinsey's service lines for global footprint optimization in manufacturing and R&D. He holds a PGD in Applied Mathematics and an MSc in Statistics & Decision Theory, both from the Technical University of Denmark.

PROF. DR. JÜRGEN KLUGE is a Director based in McKinsey's Düsseldorf office. He advises clients on issues of strategy, technology management, and innovation, primarily in the automotive, machinery, electrical and electronics, and office equipment industries, but also in the public sector. He has served as the global leader of the firm's Automotive & Assembly Sector, co-leader of its technology management center, and as Office Manager of the German office. He now has worldwide responsibility for McKinsey's proprietary knowledge and for recruiting. Before joining McKinsey, Jürgen Kluge studied Physics in Cologne and Essen. He holds a doctorate in Experimental Physics (laser) and spent some time during his doctoral work developing high-frequency electronics for a U.S. laser company. Since 2004, he has been an honorary professor at the Institute of Production Management, Technology, and Machine Tools at Darmstadt University of Technology.

MARKUS LEOPOLDSEDER is the Practice Manager of McKinsey's European Supply Chain Management Practice and member of the leadership group of the European Operations Practice. He has gathered extensive experience in more than 150 supply chain management projects across all industries – with a focus on the consumer goods and high tech industries. Markus Leopoldseder holds a degree in Electrical and Electronics Engineering from the Technology University of Vienna and, prior to joining McKinsey, worked in various functions in marketing, project management, and consulting for IBM.

TOBIAS LIEBECK is a Senior Engineer and a Member of the Executive Board at the Institute of Production Management, Technology, and Machine Tools (PTW), Darmstadt University of Technology. Along with his main focus on global production networks, his research has centered on production system engineering, innovation and technology management, and manufacturing strategy.

DR. SEBASTIAN SIMON (alumnus) worked with McKinsey & Company from 2003 to 2006. His main areas of expertise as an associate included product development, procurement, and global production networks within the automotive and high tech sectors. In 2006, Sebastian Simon joined Bosch Rexroth and is currently managing the production planning and scheduling department at a global lead plant within the hydraulics division.

DR. HARALD PROFF is an Associate Principal in McKinsey's European Operations Practice and based in the Stuttgart office. In recent years, he has focused on the redesign of manufacturing and the global reconfiguration of production sites. His clients range from start-ups and medium-size businesses to major multinational corporations and private equity firms. Harald Proff holds a joint master's degree in Mechanical Engineering and Business Administration and a doctorate in Economics from Darmstadt University of Technology, where he now also teaches International Economics.

MICHAEL STOLLE is a Senior Associate in the Silicon Valley office of McKinsey, which he initially joined in Munich. He has worked for clients in the automotive, assembly, and process industries. His functional focus is on operations, with a specific emphasis on purchasing and supply chain management and organization, for which he helped design a PSM self-diagnostic for clients. Michael Stolle is an industrial engineer (*Dipl.-Wirtschaftsingenieur*).

Acknowledgments

A book is always a collective effort. The editors and authors thank the participants in the original ProNet survey, whose experiences and insights laid the foundation for *Global Production* and inspired us to create an English version. We would especially like to thank the following staff members of McKinsey (some now alumni) and our independent service providers, whose commitment and enthusiasm made this edition possible:

Birgit Ansorge, Vera Arenz, Josef Arweck, Dagmar Böss, Roman Büschgens, Gilian Crowther, Neil Danby, Renate Doyle, Terry Gilman, Donna Gregory, Constanze Hoyer, Yuko Inagawa, Sandra Jeitler, Ira Kontosis, André Korn, Volker Kraus, Annette Lehnigk, Ginette Light, Rainer Mörike, Daniel Münch, Stefanie Pötzsch, Kerstin Polchow, Anja Ranscht, Hella Reese, Nicolas Reinecke, Walter Rehm, Regine Rusch, Jutta Scherer, Rudolf Schnitzer, Ulrich Scholz, Daniel Steiners, John Stuart, Asa Tomash, Barbara Wechs, Peter Weigang, Mareike Wölfl, and Pia Verbocket.

A very special thank you goes to Dr. Martina Bihn from Springer-Verlag GmbH, Katja Röser from LE-TeX Jelonek, Schmidt & Vöckler GbR as well as Stefan Sossna and Katja Leben from PTP-Berlin GmbH for their assistance, support, and patience.